CHEMICALS FROM PLANTS

Perspectives on Plant Secondary Products

CHEMICALS FROM PLANTS

Perspectives on Plant Secondary Products

Editors

NJ Walton

Institute of Food Research, Norwich Laboratory, UK

DE Brown

John Inns Centre, Norwich, UK

ICP Imperial College Press

World Scientific

Published by

Imperial College Press
203 Electrical Engineering Building
Imperial College
London SW7 2BT

Distributed by

World Scientific Publishing Co. Pte. Ltd.
P O Box 128, Farrer Road, Singapore 912805
USA office: Suite 1B, 1060 Main Street, River Edge, NJ 07661
UK office: 57 Shelton Street, Covent Garden, London WC2H 9HE

British Library Cataloguing-in-Publication Data
A catalogue record for this book is available from the British Library.

Humulus lupulus. Illustration taken from *Medicinal Plants* by Bentley and Trimen (1880), John Innes Foundation Historical Collections. Reproduced with kind permission of the Trustees of the foundation.

CHEMICAL FROM PLANTS
PERSPECTIVES ON PLANT SECONDARY PRODUCTS

ISBN 981-02-2773-6

Printed in Singapore.

CONTENTS

Preface vii

List of Contributiors ix

Chapter 1
Classes and Functions of Secondary Products From Plants 1
Jeffrey B. Harborne

Chapter 2
Characterisation and Control of Plant Secondary Metabolism 27
Nicholas J. Walton, Michael J.C. Rhodes, Anthony J. Michael and Adrian J. Parr

Chapter 3
Modern Methods of Secondary Product Isolation and Analysis 91
Teris A. van Beek

Chapter 4
Structure Elucidation of Plant Secondary Products 187
Georges Massiot, Catherine Lavaud and Jean-Marc Nuzillard

Chapter 5
Plant Drug Discovery and Development 215
Monique S.J. Simmonds and Renée J. Grayer

Chapter 6
Disease Prevention and Plant Dietary Substances 251
Gary Williamson, Michael J.C. Rhodes and Adrian J. Parr

Chapter 7
Biotransformations 277
Maurice C.R. Franssen and Nicholas J. Walton

Chapter 8
Production of Biologically-Active Proteins in Plants 327
George P. Lomonossoff

Chapter 9
Biotechnology and Plant Secondary Products: The Future 355
Vincenzo De Luca

Index 371

PREFACE

This book presents a series of review essays about plant secondary products. It makes no apology for covering topics of particular interest to the editors and to the individual authors, but in so doing it is hoped that the coverage may be more personal than is sometimes found in review articles with a principal aim of ensuring a detailed and exhaustive literature survey. Some effort has been made to collect and present material not widely reviewed elsewhere and to make it accessible to the non-specialist reader, or to the specialist in a related discipline. Inevitably, there are interesting areas which are not covered - for example, drugs of abuse, the development of phytochemically-based pest and disease control, and economic aspects of secondary-product biotechnology, to name but three. Likewise, end-products of primary metabolism, including carbohydrates and lipids, are not included. There is, however, a chapter devoted to the production of biologically-active proteins in plants and a moment of reflection will reveal that these may reasonably be considered "honorary" secondary products.

The perspective of this book is principally biochemical and phytochemical, with some emphasis on the extraction, characterisation and structure determination of secondary products, their pathways of biosynthesis, their enzymology and their applications. Plant secondary products have always been of interest in the context of pharmacognosy, but within the last fifteen years there has been a substantial growth in interest in secondary products from plant physiologists and molecular biologists interested in plant defence and in signalling processes. There is also increasing attention paid to these compounds as dietary constituents and particularly as actual or potential antioxidants. Secondary products are therefore at last widely recognised as a legitimate topic of mainstream interest, rather than as an arcane speciality. One of the purposes of this book is to provide an essentially biochemical context to these developments, directing the reader to specialised information which can be found elsewhere. The present time sees an unprecedented development of interdisciplinary science related to plants (and other organisms), when the effective study of major questions increasingly requires the complementary input of specialists in several disciplines.

We are very grateful to all who have contributed to this volume, particularly the chapter authors for their attention to detail and not inconsiderable patience and the Scientific Editor of World Scientific, Ms Sook-Cheng Lim. We are indebted to the late Harold Woolhouse for his encouragement and suggestions at an early stage and for his lasting influence affectionately remembered. We thank present and past colleagues and visitors to Norwich for many discussions related indirectly to this book. We are grateful to Shirley Newman and Angie Fox for assistance with the artwork. Finally, we thank Ingrid Walton and James Brown for their invaluable suggestions and encouragement.

Nicholas J. Walton
Diane E. Brown

Norwich, 1998.

LIST OF CONTRIBUTORS

Teris A. van Beek
Department of Organic Chemistry, Wageningen Agricultural University, Dreijenplein 8, 6703 HB Wageningen, The Netherlands.

Diane E. Brown
John Innes Centre, Norwich Research Park, Colney, Norwich, NR4 7UH, UK.

Vincenzo De Luca
Institut de recherche en biologie végétale, 4101 rue Sherbrooke est, Montréal (Québec), Canada, H1X 2B2.

Maurice C.R. Franssen
Department of Organic Chemistry, Wageningen Agricultural University, Dreijenplein 8, 6703 HB Wageningen, The Netherlands.

Renée Grayer
Royal Botanic Gardens, Kew, Richmond, Surrey TW9 3AB, UK.

Jeffrey B. Harborne, FRS
School of Plant Sciences, University of Reading, Whiteknights PO Box 221, Reading RG6 2AS, UK.

Catherine Lavaud
Laboratoire de pharmacognosie, UPRESA CNRS 6013, UFR de Pharmacie, 51 rue Cognacq Jay, 51096 Reims cedex, France.

George P. Lomonossoff
John Innes Centre, Norwich Research Park, Colney, Norwich NR4 7UH, UK.

Georges Massiot,
UPRESA CNRS 6013, UFR de Pharmacie, 51 rue Cognacq Jay, 51096 Reims cedex, France.

Anthony J. Michael
Institute of Food Research - Norwich Laboratory, Norwich Research Park, Colney, Norwich NR4 7UA, UK.

Jean-Marc Nuzillard
Laboratoire de pharmacognosie, UPRESA CNRS 6013, UFR de Pharmacie, 51 rue Cognacq Jay, 51096 Reims cedex, France.

Adrian J. Parr
Institute of Food Research - Norwich Laboratory, Norwich Research Park, Colney, Norwich, NR4 7UA, UK.

Michael J.C. Rhodes
Institute of Food Research - Norwich Laboratory, Norwich Research Park, Colney, Norwich NR4 7UA, UK.

Monique S.J. Simmonds
Royal Botanic Gardens, Kew, Richmond, Surrey TW9 3AB, UK.

Nicholas J. Walton
Institute of Food Research - Norwich Laboratory, Norwich Research Park, Colney, Norwich NR4 7UA, UK.

Gary Williamson,
Institute of Food Research - Norwich Laboratory, Norwich Research Park, Colney, Norwich NR4 7UA, UK.

Chapter 1

CLASSES AND FUNCTIONS OF SECONDARY PRODUCTS FROM PLANTS

JEFFREY B. HARBORNE

1. Introduction

1.1. Historical Outline

Interest was taken in the chemical constituents of plants in the very early days of modern science. Plants were used for very many purposes besides providing food, for example for treating medical ailments and for dyeing cloth. Plant alkaloids such as strychnine, brucine and quinine were first isolated in crystalline form during 1817-1820 in the laboratory of Pelletier and Caventou in France (1). The triterpene betulin from birch bark was first described in the scientific literature in 1788 (2). Anthocyanin pigments were first investigated by Robert Boyle in 1664, while the co-occurring flavones present in many plants used for dyeing were extensively studied by British chemists in the mid-nineteenth century (3).

Plant chemistry became an established discipline in universities around the turn of the last century and developed up to about 1950 through the regular discovery of new constituents in crystalline form and their characterisation, largely on the basis of the results of chemical degradation and biogenetic reasoning (4). A 1957 dictionary (5) comprehensively lists some 2750 compounds, excluding alkaloids, as being present in plants. If one adds the number of alkaloids known at this time, the total number of plant chemicals described in the literature then stood at about 5000.

The subject of plant chemistry was revolutionised after the Second World War through the development of a range of sophisticated chromatographic techniques for purifying plant constituents and of a variety of spectral methods for their rapid identification. Many, many new structures were described, even when the substances were only available in less then milligram amounts and in non-crystalline form. Hence, the rate of structural identification has increased exponentially each decade since 1950, so that the number of natural products from plants has now reached the 100,000 mark (see e.g. 6,7). It is this enormous profusion of chemical variation in the plant kingdom that constitutes the subject of this present volume.

1.2. The Classification of Plant Chemicals

Plant chemicals can be classified as primary or secondary constituents, depending on whether or not they have an essential role in plant metabolism and are universally present in all plants. Primary constituents include the common sugars, the protein amino acids, the purines and pyrimidines of nucleic acids, the chlorophylls and so on. Secondary constituents make up all the remaining plant chemicals from alkaloids to terpenoids and acetogenins to phenolics and represent substances which do not appear to have an essential role in metabolism and which vary in their distribution from plant to plant.

Such a classification has certain disadvantages. The very term "secondary" suggests that such molecules are less important to the plant, and yet as will be apparent in later sections of this chapter, they have a key role in protecting the plant from environmental pressures or in controlling plant growth. There is also no sharp division between primary and secondary metabolism. For example, erucic acid is a characteristic fatty acid of the Cruciferae, with a very restricted distribution elsewhere in nature so that it would be classified as a secondary constituent. And yet, in members of the Cruciferae it constitutes an important storage form of energy as a major acid of seed lipid and is clearly involved in primary metabolism as the young seedlings develop (8).

Once the structure of a new plant chemical has been elucidated, it can be classified according to its carbon skeletal type and the functional groups that may be present. A new quinonoid, for example, could be classified as a benzo-, naphtho- or anthraquinone, and then described as a pentahydroxy or a carboxylic acid derivative. Likewise, compounds with an acetylenic linkage are conveniently placed together, because of their chemical similarity (great reactivity), although they may vary from simple hydrocarbons to furans, amides and thiophenes (9). The problem with a fully comprehensive chemical nomenclature for plant products is that the resulting names are far too cumbersome for practical use. From the very earliest days of natural product chemistry, plant constituents have been named from the original plant sources, using either the generic or species name as a basis. Thus strychnine is named from the genus *Strychnos* and parthenolide from the species name of *Tanacetum parthenium.* Names for derivatives of such compounds can be coined by using such simple names as a basis. Benzoylecgonine is the benzoyl derivative of the alkaloid ecgonine, while cinnamoylcocaine co-occurs with cocaine in coca leaves. This system of naming after plant source also has its problems, particularly when a particular source yields more and more compounds of a similar nature, as its chemistry is more fully investigated. For example, the neem seed was originally thought to contain a single active triterpenoid insect antifeedant, called azadirachtin. Recent studies have shown the presence of six other closely related molecules, so the compounds have been renamed azadirachtins A-G (10).

Intellectually, the most satisfying system for classifying plant constituents is according to their biogenesis or biosynthetic origin. Thus, terpenoids can be classified as a single group of plant constituents on the basis of their common origin from the precursors, mevalonic acid and isopentenyl pyrophosphate, with its five carbon atoms. Within the terpenoids, it is thus easy to distinguish monoterpenoids derived from the union of two C_5 units from sesquiterpenoids, derived from three and diterpenoids derived from four. A biosynthetic system of classification, which will be employed here, has its own disadvantages. These include the sometimes limited information available on biosynthetic origins and also the fact that some of the more complex molecules are the products of more than one type of biogenetic precursor. For the present purposes, plant chemicals will be considered under three main biogenetic classes: terpenoids, alkaloids and related nitrogen compounds and phenolics. It should be remembered, however, that there are many miscellaneous constituents, which do not fit under these particular headings.

2. Major Classes of Plant Chemicals

2.1. The Terpenoids

The terpenoids or isoprenoids are characterised by their biosynthetic origin from isopentenyl and dimethylallyl pyrophosphates and their broadly lipophilic properties. They occur notably in leaf glandular trichomes, in bud exudates and bark resins. Chemically, they are mainly cyclic unsaturated hydrocarbons, with varying degrees of oxygenation (alcoholic, ketonic, etc.) in the substituent groups attached to the basic carbon skeleton. They are subdivided according to the number of five-carbon (C_5) units that are present, from the monoterpenoids (C_{10}) to the tetraterpenoids (C_{40}). Their biosynthetic relationships are illustrated in Figure 1.

The terpenoids are conveniently listed in order of increasing molecular weight (Table I). Within the monoterpenoids and sesquiterpenoids, it is usual to separate those structures with lactone substitution as iridoids (monoterpene lactones) and

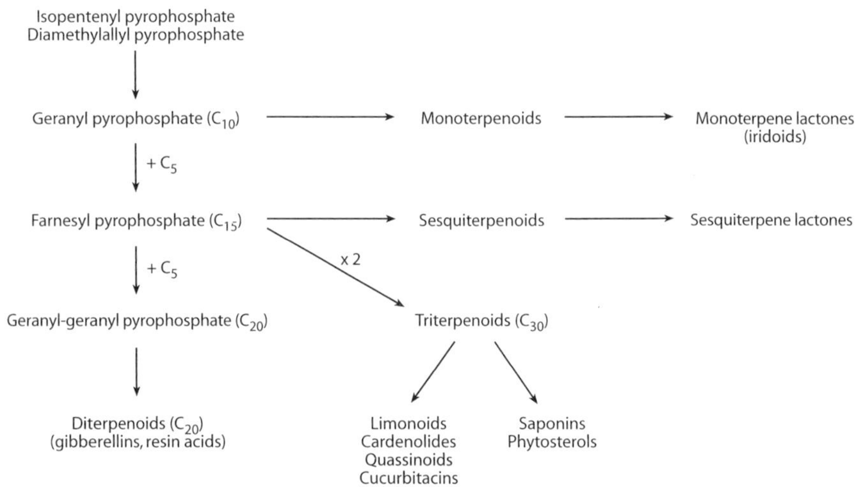

Figure 1. Biosynthetic relationships in the terpenoid series.

sesquiterpene lactones. The latter class are particularly numerous, with well over 3000 structures. Among the diterpenoids are the gibberellins, a group of natural plant growth regulators. These have been systematically numbered from 1 to 102 in order of discovery. The best known is gibberellic acid or GA_3. The largest single class of terpenoid are the triterpenoids, with many saponins being known. Within the triterpenoids are several subclasses, described as nortriterpenoids to indicate that some degradation has occurred during the latter stages of biosynthesis, with progressive loss of four to eleven carbon atoms. Thus, there are the C_{26} limonoids, the C_{24} bufadienolides, the C_{23} cardenolides (or cardiac glycosides), and the C_{20} and C_{19} quassinoids.

The number of fully characterised terpenoids is now probably in excess of 20,000. A 1982 two-volume encyclopedia of terpenoids (11) has been superceded by a completely comprehensive set of volumes, published in 1991 (12). The biochemistry and function of many of the known terpenoids have been reviewed in two symposium proceedings (13, 14).

Table I. Subclassification within the terpenoids

Terpenoid class	Description
Monoterpenoids	Volatile, essential oil constituents
Iridoids	Bitter tasting lactones, usually in glycosidic form
Sesquiterpenoids	Higher boiling essential oil constituents
Sesquiterpene lactones	Characteristic of the family Compositae
Diterpenoids	Resin acids and gibberellins
Triterpenoid saponins	Haemolytic glycosides
Steroid saponins	Haemolytic glycosides
Cardenolides and bufadienolides	Heart poisons and toxins
Phytosterols	Membrane constituents
Cucurbitacins	Bitter principles, especially of the Cucurbitaceae
Nortriterpenoids	Limonoids and quassinoids
Other triterpenoids	Lupanes, hapanes, ursanes, etc.
Carotenoids	Yellow to red pigments

2.2. Alkaloids and Other Nitrogen-Containing Metabolites

Alkaloids are the best known of the nitrogen-containing metabolites of plants. They are organic bases which have a nitrogen atom as part of their structures, usually linked into a five- or six-carbon cyclic system. They are exceedingly numerous, some 10,000 structures having been recorded in the most recent dictionary devoted to these substances (15).

Alkaloids have a restricted distribution in higher plants, being found in about 20% of angiosperm species. They are only relatively rarely found in gymnosperms and ferns and have not been recorded in mosses or liverworts. Other classes of nitrogen-containing metabolites (see Table II) are also of limited natural occurrence.

Cyanogenic glycosides do occur in about 2000 species from 100 plant families, but their distribution is sporadic. They are common in some families (e.g. Rosaceae) but quite rare in others (16). Likewise, the glucosinolates or mustard oil glycosides are mainly found in the Cruciferae and four closely related families, although there are occasional records in six other families (17).

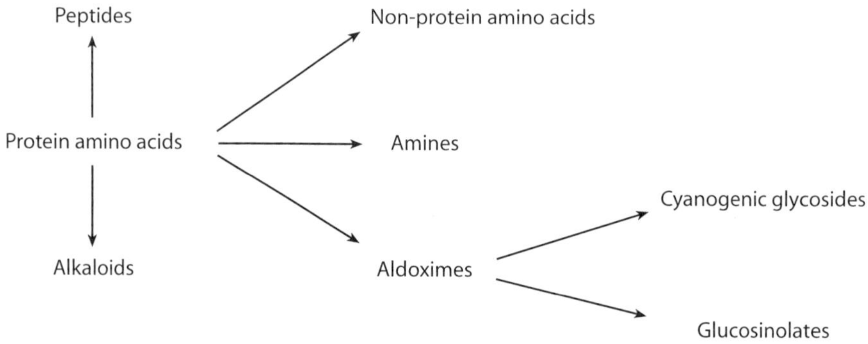

Figure 2. Biosynthetic relationships of nitrogen-containing plant metabolites.

One reason for the relatively restricted distribution in plants of secondary metabolites based on nitrogen is that the supply of this element to the plant is always limiting, even in those plants such as legumes which fix their own nitrogen. Nitrogen-containing metabolites are all derived from one or other of the protein amino acids (Figure 2) so that there is always competition for these precursors which are also required for more basic processes such as protein synthesis. When plants do produce secondary metabolites containing nitrogen, the quantities can be quite low compared to, for example, phenolic production. The low concentration of alkaloids in vegetative tissues is offset by their high biological potency. Additionally, alkaloid levels may be significantly higher in storage tissues (roots, fruits, seeds) than in the green leaves.

Although in general alkaloids are biosynthetically derived from protein amino acids, they are usually classified into subclasses according to the carbon skeleton, e.g. quinoline, isoquinoline, etc. (Table II). Nevertheless, it is clear that indole alkaloids are generally formed from the amino acid tryptophan. As will be clear from Table II, most subclasses of alkaloid have quite restricted natural occurrences. Some

Table II. Alkaloids and other nitrogen-containing plant constituents

Alkaloids*	
Amaryllidaceae	275 structures, restricted to daffodil family
Betalain	Yellow or purple pigments of the Centrospermae
Diterpenoid	Mainly in the Ranunculaceae; often poisonous
Indole	Many structures; in Apocynaceae, Loganiaceae
Isoquinoline	Largest group of alkaloids, widely distributed
Lycopodium	150 structures, restricted to club mosses
Monoterpene	Related and co-occurring with iridoids (q.v.)
Sesquiterpene	Commonly present in some orchids
Peptide	130 structures, especially in the Rhamnaceae
Pyrrolidine and Piperidine	Includes polyhydroxyalkaloids, inhibitors of sugar-hydrolysing enzymes
Pyrrolizidine	Common in Boraginaceae and in *Senecio,* Compositae
Quinoline	Common in the Rutaceae
Quinolizidine	Characteristic in the Leguminosae
Steroidal	In Apocynaceae, Buxaceae, Liliaceae and Solanaceae
Tropane	Mainly found in Solanaceae (e.g. *Atropa*)
Non-protein amino acids	Often present in seeds, especially in Leguminosae
Amines	Fishy volatiles; sporadic distribution
Cyanogenic glycosides	Wide but sporadic occurrences
Glucosinolates	Commonly present in the Cruciferae
Purines and pyrimidines	Includes caffeine of coffee and tea

* There are a variety of other alkaloids (e.g. colchicine) which do not fit into any of the main classes listed.

subclasses occur in a single family (e.g. the Amaryllidaceae alkaloids), whilst others are found in only a few families (e.g. the betalains and steroids) (1).

2.3. Phenolic Metabolites

Phenolic compounds are aromatic structures bearing one or more hydroxyl groups. Most are polyphenols, having several hydroxyl group substituents, one or more of which may be substituted by methyl or glycosyl groups. Phenolics share a common biosynthetic origin from phenylalanine (Figure 3), one of the three protein amino acids formed from sedoheptulose via the shikimate pathway.

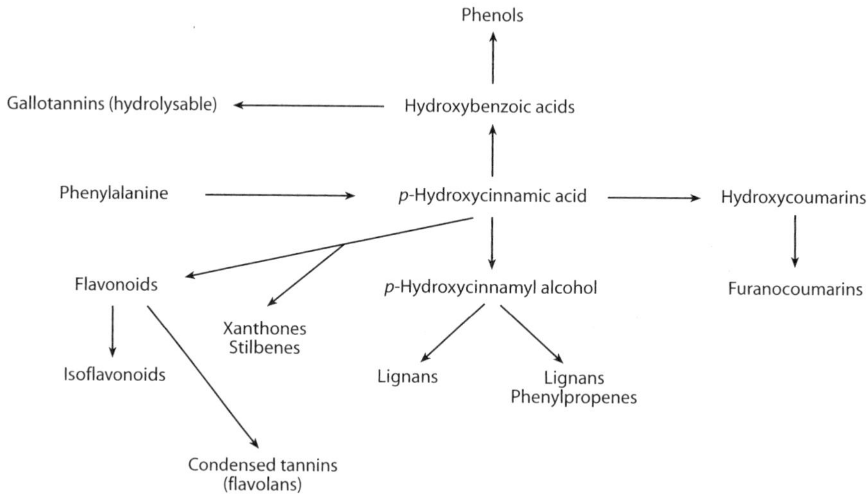

Figure 3. Biosynthetic relationships of different phenolic classes.

p-Hydroxycinnamic acid, formed from phenylalanine by deamination and *p*-hydroxylation, occupies a central role in the formation of the various classes of plant phenol (Figure 3). The biosynthesis of flavonoids follows from the condensation of *p*-hydroxycinnamic acid (as the CoA ester) with malonyl coenzyme A to give chalcones. There is a regulated stepwise pathway from chalcone to flavanone, flavone and flavonol. The flavolans or condensed tannins are end products of this pathway, formed via flavanonols, flavan 3,4-diols and flavan-3-ols.

Table III. Classification of phenolic constituents

Subclass	Description
Anthocyanins	Red to blue flower pigments
Anthochlors	Yellow flower pigments: chalcones and aurones
Benzofurans	Occur in higher plants and in lichens
Chromones	A small group, of therapeutic importance
Coumarins	Over 700 structures, widespread in plants
Minor flavonoids	Flavanones and dihydroflavonols
Flavones and flavonols	Many structures, especially in glycosidic combination
Isoflavonoids	Characteristic of the Leguminosae, in the free state
Lignans	Commonly present in wood and bark
Phenols and phenolic acids	Some acids are universal in plants
Phenolic ketones	Notable occurrences in hops and in ferns
Phenylpropanoids	Many structures, widespread
Quinonoids	Benzoquinones, naphthoquinones and anthraquinones
Stilbenoids	Includes also dihydrophenanthrenes
Tannins	Condensed and hydrolysable
Xanthones	Mainly in Gentianaceae and Guttiferae

These flavonoids are the best known group of polyphenols, sharing the same carbon skeleton of flavone or 2-phenylbenzopyrone, and comprise some 4000 structures (18). The total number of phenolics including all the non-flavonoids (see Table III) is well over 8000. The vast majority of these are of plant origin, although a few simple phenols (e.g. *m*-cresol) are found in the pheromonal secretions of elephants and some flavonoids (e.g. quercetin) are stored in butterfly wings after dietary ingestion.

The different classes of polyphenol (Table III) range in molecular weight from simple compounds such as phenol itself and salicylic acid (2-hydroxybenzoic acid) through complex anthocyanin pigments with molecular weights of 2000 or more to the polymeric condensed tannins where the molecular weight may reach 30,000.

Among the important properties of phenolics are their ability to ionise in the presence of base. Many polyphenols have catechol (adjacent *o*-dihydroxyphenyl) groups and hence have the ability to chelate divalent or trivalent metal ions. Some anthocyanins occur naturally chelated to magnesium or iron. Phenols with either *o*- or *p*-dihydroxy substitution (e.g. catechol or hydroquinone) are readily oxidized to the corresponding quinones and many *p*-quinones are natural products (19). Such oxidations can occur enzymically through the operation of the ubiquitous phenolases or peroxidases, or non-enzymically by aerial oxidation.

Since most phenolics are potentially toxic to life, they are frequently stored in plants in conjugation with sugar, sulphate or both and such water-soluble forms are located in the vacuole. Most classes of phenol occur naturally as glycosides and many different forms have been described. Over 135 different glycosides of the common flavonol quercetin have been characterised (18). Quercetin can also occur free in the leaf surface wax of some plants. Flavonoids can be found in almost any part of the plant and the same is true of other commonly occurring structures, such as phenylpropanoids and hydroxybenzoic acids. By contrast, tannins are mainly found in bark and leaf and are confined largely to these tissues in woody plants (20).

3. Functions of Secondary Constituents

There are difficulties in establishing the function of secondary metabolites in plants which are not encountered with primary products. The importance of shikimic acid as a primary metabolite is readily established by labelling experiments as a biosynthetic intermediate in the aromatic pathway. Likewise, pyruvic acid is clearly an important intermediate in carbohydrate metabolism, because it acts as a link between glycolysis and the Krebs cycle. By contrast, a secondary product may be shown to be an antifeedant and insecticidal, but it is less obvious whether it has that function in the life of the plant in which it occurs. The well known triterpenoid azadirachtin, actually a mixture of several closely related molecules (see 10), is established as a commercially useful pesticide and yet the neem tree, its source, is attacked by insects. This may be because there are always some insect species able to overcome the feeding barrier represented by azadirachtin. It also may be because this toxin is concentrated in the seed of the neem tree, so that it is only the seed that is so protected from insect attack.

Another complication with establishing a function for secondary metabolites is that some are apparently multifunctional in nature. Salicylic acid, for example, is regarded as a signal molecule often produced when a plant is stressed, e.g. by pathogen attack. It is a metabolic signal for heat production during pollination in the Arum lily. Salicylic acid accumulates in various bound forms in leaves of *Salix* spp. and has been shown to protect these trees from herbivory by the European hare. Finally, it occurs in the tree *Quercus falcata* and is released from it to produce an allelopathic effect on neighbouring plants (21).

A further complication is that some plant chemicals can serve a function in both primary and secondary metabolism. The non-protein amino acid canavanine accumulates in significant concentration in the seed of *Canavalia ensiformis* and of many related legumes. It can protect them against bruchid beetle attack, although there are some species which can overcome this defence. Additionally, canavanine acts as a storage form of nitrogen and it is metabolised during seed germination and the nitrogen present is recycled into protein amino acid (22). Such recycling of nitrogen is thought to occur with other nitrogen-containing metabolites, including alkaloids. It has been suggested to occur in the case of caffeine synthesis in the coffee leaf, where the concentration drops dramatically as the leaf matures (23).

Finally, the most disconcerting feature of secondary compounds is their frequent presence in plants as complex mixtures of structures which increases the difficulty of establishing a unique function. For example, well over eighty alkaloids have been described from the Madagascar periwinkle, *Vinca rosea,* over fifty flavonoids may be characterised from a plant of the Compositae while over 150 volatiles have been identified in the floral fragrance of an orchid (24). Some of this complexity however, it should be remembered, is simply related to the way that secondary products are biosynthesised on a template system and that intermediates of a biosynthetic pathway often accompany the main end product of that pathway. Furthermore, there is also evidence of selective advantage in plants which accumulate mixtures of several related substances over the presence of just one of those substances. It has been shown that a mixture of six furanocoumarins in the leaves of the wild parsnip, *Pastinaca sativa* is more effective against insect herbivory than one of those compounds, xanthotoxin, at the same concentration as the mixture of six furanocoumarins (25).

In the present account of secondary metabolite functions, substances with growth regulatory properties and those involved in pollination and seed dispersal will first be considered. Accounts of antifungal agents and anti-herbivore agents follow. Plant compounds with animal hormonal activity will then be considered. The final section will list some of those plant compounds involved in symbiotic associations in plants.

3.1. Plant Growth Substances

One clear function of secondary metabolites in plants is in the control of growth and development. Over the years, a variety of structures have been recognised as growth substances or plant hormones (Table IV). Characteristically, they exert their effects at very low concentrations (e.g. 1×10^{-5} M) and can be monitored by means of an *in vivo* bioassay (e.g. the oat coleoptile test for auxins). Five main groups are widely recognised based on ethylene, indole-3-acetic acid, gibberellic acid, zeatin and abscisic acid (26). More recently, two further classes have been added: the brassinosteroids based on brassinolide (27) and the polyamines based on spermine (28).

A significant feature of all classes of plant hormone, apart from ethylene, is the structural variation that is encountered when studying these substances as they occur within the plant. There is structural variation in the hormone itself; indole-3-acetic acid (IAA) occurs as the 4-chloroderivative in pea seeds and as the nitrile in cabbage. There is also considerable variation in the different conjugated forms that may be present, as a "storage" form of the hormone. Thus IAA can occur in association with arabinose, galactose, glucose or myo-inositol. The most pronounced structural variation can be found in the gibberellic acid series, where 102 different gibberellins have been characterised to date. How far each of these structures are involved in growth regulation in the plants in which they are found is still open to further experimentation (29).

In addition to the recognised classes of growth substance, there are many other secondary constituents that have individually been shown to produce an effect on plant growth or development. This is commonly in only one or two plants and the effect is often one of growth inhibition or of interacting with one of the established plant hormones. Some examples of those putative hormones are listed in Table IV. One illustration is gallic acid (3,4,5-trihydroxybenzoic acid) which has been described as a flowering inhibitor from its occurrence in leaves of *Kalanchoe blossfeldiana* (30). Another case is dihydroconiferyl alcohol, which synergises with gibberellic acid in inducing hypocotyl elongation in the lettuce plant (31).

Table IV. Plant hormones and other growth substances

Structural class	Chemical variation; effect on growth*
Established hormones	
Ethylene	Single structure; ripening hormone
Indole-3-acetic acid	Some 30 conjugates known; growth hormone
Gibberellins	102 structures; stem elongation hormone
Cytokinins	18 structures; cell division hormone
Abscisic acid	2 structures; stomatal opening hormone
Brassinosteroids	14 structures; cell elongation factor
Polyamines	4 structures; interacts with ribosomal RNA
Putative hormones (growth inhibitors)	
Phenolic acids	e.g. gallic acid
Phenylpropanoids	e.g. chlorogenic acid, dihydroconiferyl alcohol
Coumarins	e.g. coumarin, umbelliferone
Lactones	e.g. parasorbic acid
Stilbenoids	e.g. batatasin I
Sesquiterpene lactones	e.g. xanthinin
Diterpene lactones	e.g. podolactone E
Fatty acids	e.g. palmitic, oleic acids

* All established hormones have several effects on plant growth and interact with each other to control the various aspects of plant growth and development.

Secondary compounds may also have an indirect effect on plant growth, by mediating in the regulation of hormone synthesis or turnover. Phenolic compounds have been implicated in IAA turnover in some plants through an effect on the enzyme of degradation, IAA oxidase. This enzyme is stimulated *in vitro* by monophenols and inhibited by catechol-derived biphenols. How important this effect is *in vivo* is still controversial (26).

3.2. Floral Pigments and Odours

Flower colour and odour are important attractants to pollinating vectors (bats, birds, bees, butterflies, etc.) and have a key role in pollination biology (32). Thus, different pollinators have particular colour and odour preferences which are provided by the plants which they visit. Bees, for example, prefer blue and yellow flowers or those with ultraviolet absorption patterns (due to flavones) and also fragrant odours. Bats, by contrast, fly by night and pollinate white flowers with fruity or sulphurous odours. A series of methyl sulphides and other sulphur derivatives have been characterised from bat-pollinated species (33).

Secondary constituents are thus functionally important as floral pigments. A wide range of anthocyanins, other flavonoids and carotenoids have variously been characterised in floral tissues (34). A few typical examples of floral pigments involved in pollination are shown in Table V. However, it is important to stress that mixtures of related pigments are commonly found together in flowers. Anthocyanins are always accompanied by flavone co-pigments and may occasionally occur as metal chelates, as in *Commelina communis* (35). Again, in yellow flowers, carotenoids and yellow flavonols may co-occur and function together to provide their colour to plants such as *Rudbeckia hirta* (32).

A bewildering array of volatile chemicals have been encountered in the scents of angiosperm flowers (36). As many as 150 scent constituents may be produced by an orchid flower, such as *Ophrys lutea* (24). Orchids are a particularly rich source of different odour compounds and many species of this family have now been analysed (37). Among the more unusual chemical constituents encountered in plants other than orchids are the presence of *p*-cresol, 2-heptanone, indole and germacrene B in the floral odour of the cuckoo pint *Arum maculatum* (38). Mono and sesquiterpenoids are among the most common of flower volatiles but they may well be accompanied by aliphatic (e.g. pentadecane) and aromatic (e.g. eugenol) components. *Cimifuga simplex,* which is butterfly-pollinated (Table V) has 15 terpenoids, accompanied by methyl anthranilate and isoeugenol. In some forms of this plant, these two latter volatiles are replaced by benzenoids, e.g. methyl benzoate (39).

Plant pigments and scent constituents are not confined in their occurrence to floral tissues, but may be present also in leaves (e.g. in the mint plant) and in the fruits. The substances encountered in these other tissues usually differ from those found in the flowers, since their purpose is not exactly the same. In the case of fruits, distinctive

Table V. Secondary metabolites involved in plant-pollinator interactions

Compound	Plant	Role
Delphinidin 3,5-diglucoside	*Delphinium nelsonii*	Blue flower colour attracts bumble bee and humming bird
Delphinidin and cyanidin 3,5-diglucoside	*Phlox drummondii*	Violet flower colour attracts butterfly pollinator
Pelargonidin 3,5-diglucoside	*Ipomopsis aggregata*	Scarlet flower colour attracts humming bird pollinator
Patuletin and carotenoids	*Rudbeckia hirta*	Visible yellow and UV patterning increases efficiency of bee pollination
Indole and *p*-cresol	*Arum maculatum*	In scent attracts fly away from cow dung
Geraniol, geraniol and others	*Ophrys lutea*	Attracts male bee for "pseudocopulation"
Carvone oxide	*Catasetum* spp.	Attracts Euglossine male bees, which use it as a pheromone
Dimethyl disulphide and 2,4-dithiapentane	*Crescentia cujete*	Attractive to bat pollinators
Methyl anthranilate and isoeugenol	*Cimifuga simplex*	Synergise to attract butterfly pollinators
Ethyl acetate	*Zygogynum* spp.	Attracts moth pollination

colours and appropriate odours serve a role in attracting herbivores to eat them and thus to distribute the seed within the fruit. Some herbivores, of course, may damage or eat the seed as well as the fruit and hence fruit tissues may contain other types of secondary constituent to discourage the visitations of inappropriate herbivores. The secondary chemicals that are found in fruit and seed tissue have been outlined in a recent review (40).

3.3. Antiherbivore Agents

It is now widely accepted that the raison d'être of the majority of secondary metabolites in plants is to provide a defence against animal herbivory. Such herbivory is a major threat to plant survival and as a consequence of plant-animal coevolution, plants have developed various defence strategies including the ability to accumulate chemicals that are hostile to animal feeding. It is the flexible response of herbivores to detoxify these chemicals or otherwise circumvent their effects that allows the limited grazing that can be witnessed today in the natural world (32).

Plant chemicals may be apparently targetted to affect one particular type of herbivore (e.g. cardiac glycosides against mammals), although they may also be toxic to all forms of life, including the plant itself (e.g. cyanogenic glycosides). To avoid autotoxicity, plant chemicals are therefore usually located on the surface of plant tissues or sequestered in bound form within the plant in the cell vacuole. Some examples of plant chemicals that have been established as hostile to mammalian herbivory are given in Table VI. Many experiments have been carried out, for example, on the defensive chemistry of Alaskan trees to snowshoe hare grazing, and phenolics and other terpenoids have been identified as protective, as well as the triterpenoid papyriferic acid (41). Many of the compounds (e.g. sinigrin) are avoided because of their bitter or acrid taste. Many secondary constituents are directly poisonous (e.g. fluoracetate, senecionine, miserotoxin) and much work has been carried out characterising the toxic principles of those pasture plants which cause livestock fatalities (42).

Some of those plant chemicals that limit insect herbivory on plants are listed in Table VII. It is a very small sample of the many chemical structures that affect insect food choice, in addition to the nutritional factors. Sometimes, concentrations of insect repellent chemicals may increase as a result of insect feeding. Thus, the concentrations of xanthotoxin and related furanocoumarins increase by 215% in wild parsnip after moth larvae have fed on the leaves (43). Some chemicals hostile to insects (e.g. rotenone) have long been prized for their value as insecticides (44). In addition to chemicals which are immediately harmful to insects there are substances which arrest feeding through their taste properties and are known as antifeedants. Such compounds may also be toxic and this is true of the best known antifeedant, azadirachtin (10).

Table VI. Some plant chemicals hostile to mammalian herbivory

Compound and source	Animals affected
Fluoracetic acid in *Dichapetalum cymosum*	Most mammals (respiratory poison)
Miserotoxin in *Astragalus atropubescens*	Cattle
Oenanthetoxin in *Oenanthe crocata*	Cows and man
Ouabain in *Acokanthera oubaio*	Most mammals (heart poison)
Papyriferic acid in *Betula resinifera*	Snowshoe hares
Salicin in *Salix* spp.	Western hare
Senecionine in *Senecio jacobaea*	Livestock (liver poison)
Sinigrin in *Brassica* spp.	Livestock

Terpenoids, alkaloids and phenolics have all been implicated as antiherbivore agents and there is almost no class or subclass of secondary metabolite that does not have an effect on herbivore feeding. Within a given class, some molecules may be more effective than others; among monoterpenoids, oxygenated derivatives are usually more toxic than the hydrocarbons. Nevertheless, almost every secondary compound which accumulates in plant tissues is potentially hostile to herbivory. More details of the effects of plant chemicals on both insect and mammalian herbivory are available elsewhere (32, 45).

Table VII. Some plant chemicals hostile to insect and molluscan herbivory

Compound and source	Animals affected
Demissine in *Solanum demissum*	Colorado beetle
L-DOPA in *Mucuna* spp.	Bruchid beetles
Lactupicrin in *Cichorium intybus*	Locusts
Linamarin in *Lotus corniculatus*	Slugs
β-Ocimene in *Astronium graveolens*	Leaf-cutting ants
Rotenone in *Derris elliptica*	Insects
Tannins in *Quercus robur*	Aphids and moths
Xanthotoxin in *Pastinaca sativa*	Armyworms

3.4. Antifungal Agents

Higher plants contain certain low molecular weight metabolites which appear to be involved in protecting them from microbial infection. Disease resistance in plants is provided by a number of barriers, including defensive proteins, lignification, physical appressoria, etc. (32). Additionally, low molecular weight chemicals provide barriers to microbial invasion either by their presence on plant surfaces, where they inhibit spore germination or else by their occurrence in bound form within the leaf and their subsequent release in the free state during the infection process (46).

Some examples of lipophilic plant constituents present at the surface of leaf, stem or fruit are illustrated in Table VIII, with an indication of their fungitoxicity. In many cases, more than one secondary metabolite may be involved. In the case of *Lupinus albus,* for example, a range of other fungitoxic isoflavonoids in addition to luteone have been characterised on the leaf or root surface (47). Again, a mixture of oxygenated fatty acids together with epoxylinoleic acid on rice leaves have been suggested as providing resistance to rice blast, *Pyricularia oryzae* (48). Some of these compounds are also probably effective against bacterial infection, and this has been demonstrated in the case of parthenolide and its occurrence in leaf glands of feverfew, *Tanacetum parthenium* (49).

Other secondary metabolites may be present in the vacuole of upper epidermal cells of the leaf and undergo enzyme hydrolysis to yield the free antifungal agent. Some examples of these bound toxins are also shown in Table VIII (46). The occurrence of different saponins in the leaf and the root of the oat plant *Avena sativa* is noteworthy. In both cases, the original saponin loses one of its sugars to give the antifungal agent, which may then undergo further metabolism by fungal enzymes to produce an inactive sapogenin (32). In the case of 1-tuliposide A in tulip bulbs, during infection this undergoes isomerisation to 6-tuliposide A and then hydrolysis and lactonisation to tulipalin A before it is an effective fungal toxin (32). It is not clear yet how widespread bound toxins are, but a recent study of plants of Rosaceae suggested that they are commonly present in this family (50). Two examples - the flavone chrysin 7-glucoside and the simple hydroquinone glucoside arbutin - are shown in Table VIII.

Table VIII. Constitutive antifungal agents in plants

Compound and source	Fungus affected
SURFACE CONSTITUENTS	
Catechol in *Allium cepa*	*Colletotrichum circinans*
12,13-Epoxylinoleic acid in *Oryza sativa*	*Pyricularia oryzae*
Luteone in *Lupinus albus*	*Helminthosporium carbonum*
Nobiletin in *Citrus* spp.	*Deuterophoma tracheiphila*
Parthenolide in *Tanacetum parthenium*	*Botrytis cinerea*
5-Pentadecylresorcinol in *Mangifera indica*	*Alternaria alternata*
Sakuranetin in *Ribes nigrum*	*Botrytis cinerea*
Sclareol in *Nicotiana glutinosa*	*Alternaria longipes*
VACUOLAR CONSTITUENTS*	
Arbutin in *Pyrus communis*	*Nectria cinnabarina*
Avenacin AI in *Avena sativa* root	*Gaeumannomyces graminis*
Avenacoside A in *Avena sativa* leaf	*Fusarium avenaceum*
Chrysin 7-glucoside in *Malus sieboldii*	*Phomopsis perniciosa*
Sinigrin in *Brassica oleracea*	*Peronospora parasitica*
Tomatine in *Lycopersicon esculentum*	*Fusarium solani*
1-Tuliposide A in *Tulipa* cultivars	*Fusarium oxysporum*

* Bound toxins which undergo enzymic hydrolysis (usually with ß-glucosidase) to release the true antifungal agents.

One further chemical barrier to microbial invasion is induced in many plants during the infection process. Such plants synthesise antifungal compounds *de novo* in response to infection and large quantities may accumulate around the immediate site. These compounds may, however, undergo further metabolism in the case of a successful pathogen invasion and thus can usually only be recovered during the early stages of the infection process. Much work has been done on these substances, which are called phytoalexins, and some 300 structures have been characterised from plants belonging to approximately 30 families (51).

The structures are similar to constitutive antifungal agents but the compounds are universally lipophilic in nature. In some families such as the Leguminosae, the phytoalexins are nearly always isoflavonoids and are very similar to isoflavonoids which occur constitutively within the family (52). In other cases, the phytoalexins are uniquely produced as a result of the infection process. This is true of the sulphur indole-based phytoalexins of the Cruciferae (53) and of some of the dibenzofuran phytoalexins of Rosaceae (50). Two other chemical changes that occur during infection deserve brief mention. Firstly, the synthesis of some plant constituents can be increased during infection so that substances such as the coumarin scopoletin, normally present in trace amount, may increase several-fold in infected tissue. Secondly, new toxins produced by the fungal invader may be formed in the infected plant. These phytotoxins are often responsible for the damaging symptoms (e.g. wilting) which appear in the leaf as the disease takes hold of the plant (32). Another class of fungal metabolite - the mycotoxins - may be specifically produced in plant storage tissue (fruits, seeds) following infection and these mycotoxins may be poisonous to animals (birds, farm animals, Man) when nuts or seeds (e.g. cereal grain) are eaten (54).

3.5. Plant Compounds with Animal Hormonal Activity

One of the most surprising discoveries in plant science, made about 30 years ago, was the fact that insect moulting hormones are present in plant tissues. Relatively massive amounts of the sterol 20-hydroxyecdysone, which had just been characterised in trace amounts in the silkworm, were found in the leaves of the common yew *Taxus baccata* and the fronds of the fern *Polypodium vulgare* (55). Plant surveys were initiated and these steroidal hormones were found to occur regularly in ferns and gymnosperms. Subsequently, they were also found in angiosperms and a recent review indicates that these are present in 27 families of ferns, 10 families of gymnosperms and 74 families of angiosperm (56).

In addition to the presence of the actual moulting hormones of insects, such as ecdysone and 20-hydroxyecdysone, a range of ecdysones unique to plants were uncovered. About 100 of these phytoecdysones have been characterised (56). A fascinating property of many phytoecdysones is their enormous hormonal activity and some are 20 times as active as zooecdysones. As a result, they can have very damaging effects on insect development. There is thus good evidence that at least for sensitive insect species, they exert various harmful effects in insects feeding on plants containing these constituents. There is thus a reasonable argument that they are produced specifically to protect plants from insect herbivory, although much further work is needed to confirm this hypothesis.

During metamorphosis, insects also have a requirement for a second terpenoid-based hormone, juvenile hormone. Four closely related structures based on farnesol have been characterised from insects. Again, it is remarkable that one of these four compounds, juvenile hormone III, has been found in a plant in large amounts, in the sedge *Cyperus iria* (57). In addition, several related structures have been encountered in plants which disrupt metamorphosis at this critical juvenile stage. These plant substances may cause precocious metamorphosis and hence have fatal effects on insects feeding on plants containing them. Their ecological purpose, again, would seem to be one of protecting plants against insect herbivory (58).

Yet another hormonal link between plants and animals lies in the record of oestrogen-like molecules in plants. In fact, the mammalian hormones oestriol and oestrone have been described in trace amounts from several plant sources, including the seeds of the apple and the pomegranate. More effective quantities of oestrogen-mimic structures have also been encountered, notably the isoflavone formononetin in legume forage plants. This compound is responsible for continual reduction in lambing percentages in Australian sheep, because of its presence in significant amount in the Australian clover plant *Trifolium subterraneum*. Other sources of plant oestrogen include the shrub, *Pueraria mirifica*, which contains the oestrone-mimic miroestrol, and which is a potent abortifacient in humans (32).

3.6. Plant Chemicals in Symbiotic Associations

No account of phytochemistry would be complete without at least a brief mention of chemistry associated with plants living in symbiosis. This may be simply a warning to the naive phytochemist that the plant chemicals synthesised by a higher plant may be affected by microbial infection (see earlier section) or parasitic attachment.

The best-known symbiosis among lower plants is the lichen, which is an association of a fungus with an alga. Lichens, as a result, are characterised by a special secondary chemistry, which includes unique structural types (e.g. phenolic depsidones and depsides) as well as some characteristic fungal pigments (especially quinonoids) (59). A much looser type of symbiosis may be observed between marine algae and their animal predators, where algal chemicals are absorbed dietarily by shellfish and other sea creatures and utilised to protect them from larger predators. One interesting example of such a toxin passed through the food chain is brevitoxin A, a polyether with a molecular weight of 867, which is synthesised by the red alga *Gymnodium breve.* It may appear in shellfish and other fish eaten by Man and cause sickness and even death (60).

Higher plant symbioses include the familiar legume-*Rhizobium* interaction, where a bacterium initiates nitrogen fixation in the roots of clover and other legumes. The secondary chemistry of nitrogen-fixing plants is not noticeably different, however, from that of the same plants lacking *Rhizobium* infection. Theoretically, higher plants might exist in symbiosis with fungi, without necessarily showing disease symptoms. Such a hypothetical symbiosis was proposed for the composite shrub *Baccharis megapotamica,* when it was found to contain several complex trichothecins, macrocyclic molecules only otherwise known to be synthesised by soil fungi belonging to the genera *Fusarium* and *Mycotheceum*. This idea became firmly established in the early literature but more careful study of the plant in the glasshouse showed that it was not infected with any micro-organisms and moreover the trichothecins were specifically produced not in the root, leaf or stem of the plant but in the seed (61).

The most distinctive symbioses between one higher plant and another are those in which a *Striga* or an *Orobanche* species becomes a root parasite on a suitable host plant (often a crop plant) and takes all its nourishment from that host. In such cases, there is some interesting chemistry in the signal molecules released from the host root to trigger attraction and attachment of the parasite but there is not usually any exchange in secondary chemicals, the parasite producing its own distinctive chemical constituents (32). However, such exchange may occur in the case of semiparasites, which attach to a host, but are also capable of photosynthesis. The uptake of host alkaloids is now well established for different species of the semiparasitic *Pedicularis.* Piperidine alkaloids may be "borrowed" from a spruce host, pyrrolizidine alkaloids from a *Senecio* host and quinolizidine alkaloids from a lupin host (62).

4. Conclusion

This chapter has provided a brief introduction to the chemical constituents of the plant kingdom. In spite of the enormous structural diversity that has been revealed to date, there is still much to be discovered. The biosynthesis of the major classes is briefly indicated and some account is given of their functions within the plant. More detailed information on individual plant chemicals mentioned in this chapter is available in the Dictionary of Natural Products (6). Representative plant substances with biological activity are listed in the Phytochemical Dictionary (63). New plant products are principally reported in two journals: *Phytochemistry* and the *Journal of Natural Products*. Later chapters in this book will consider the separation and structural elucidation of plant chemicals and they will also describe their utilisation by Man for medicinal and other purposes.

5. References

1. G. Cordell, *Introduction to Alkaloids: a Biogenetic Approach* (Wiley-Interscience, New York, 1981).
2. E.W.H. Hayek, V. Jordis, W. Moche and F. Santer, *Phytochemistry* **28** (1989) 691.
3. J.B. Harborne, *Comparative Biochemistry of the Flavonoids* (Academic Press, London, 1967).
4. R. Robinson, *Structural Relationships of Natural Products* (Clarendon Press, Oxford, 1955).
5. W. Karrer, *Konstitution und Vorkommen der Organischen Pflanzenstoffe* (Birkhauser-Verlag, Basle, 1958).
6. J. Buckingham, *Dictionary of Natural Products*, 7 volumes (Chapman and Hall, London, 1994).
7. J.B. Harborne and H. Baxter, *Phytochemical Dictionary* (Taylor and Francis, London, 1993).
8. C. Hitchcock and B.W. Nichols, *Plant Lipid Biochemistry* (Academic Press, London, 1980).
9. F. Bohlmann, T. Burkhardt and C. Zdero, *Naturally Occurring Acetylenes* (Academic Press, London, 1973).
10. H. Rembold, *Economic and Medicinal Plant Research* **3** (1989) 57.
11. J.S. Glasby, *Encyclopedia of the Terpenoids,* 2 volumes (Wiley, Chichester, 1982).
12. R.A. Hill and J.D. Connolly, *Dictionary of Terpenoids* (Chapman and Hall, London, 1991).
13. W.D. Nes, G. Fuller and L.S. Tsai, *Isopentenoids in Plants: Biochemistry and Function* (Marcel Dekker, New York, 1984).

14. J.B. Harborne and F.A.T. Barberan, *Ecological Chemistry and Biochemistry of Plant Terpenoids* (Clarendon Press, Oxford, 1991).
15. I.W. Southon and J. Buckingham, *Dictionary of Alkaloids,* 2 volumes (Chapman and Hall, London, 1989).
16. R. Hegnauer, *Plant Systematics and Evolution,* Supplement 1 (1977) 191.
17. P.O. Larsen, in *The Biochemistry of Plants,* Volume 7, ed. E.E. Conn (Academic Press, New York, 1981), p. 501.
18. J.B. Harborne, *The Flavonoids: Advances in Research since 1986* (Chapman and Hall, London, 1994).
19. R.H. Thomson, *Naturally Occurring Quinones, III. Recent Advances* (Chapman and Hall, London, 1986).
20. E. Haslam, *Plant Polyphenols: Vegetable Tannins Revisited* (University Press, Cambridge, 1989).
21. W.S. Pierpoint, *Advances in Botanical Research* **20** (1994) 164.
22. G.A. Rosenthal, *Plant Non-Protein Amino and Imino Acids* (Academic Press, New York, 1982).
23. P.M. Frischknecht, J. Ulmer-Dufek and T.W. Baumann, *Phytochemistry* **25** (1986) 613.
24. A.K. Borg-Karlson, G. Bergström and I. Groth, *Chemica Scripta* **25** (1985) 283.
25. M. Berenbaum, J.K. Nitao and A.R. Zangerl, *J. Chem. Ecol.* **17** (1991) 207.
26. J.R. Bearder, in *Hormonal Regulation of Development* ed. J. Macmillan (Springer-Verlag, Berlin, 1980), p. 9.
27. G. Adam and V. Marquardt, *Phytochemistry* **25** (1986) 1787.
28. R.D. Slocum and H.E. Flores, *Biochemistry and Physiology of Polyamines* (CRC Press, Boca Raton, 1991).
29. A. Crozier, *The Biochemistry and Physiology of Gibberellins,* 2 vols (Praeger, New York, 1983).
30. R.J. Pryce, *Phytochemistry* **11** (1972) 1911.
31. K. Shibata, T. Kubota and S. Kamisaka, *Plant Cell Physiol.* **15** (1974) 191.
32. J.B. Harborne, *Introduction to Ecological Biochemistry*, 4th edn (Academic Press, London, 1993).
33. J.T. Knudsen and L. Tollsten, *Bot. J. Linn. Soc.* **119** (1995) 45.
34. T.W. Goodwin, *Plant Pigments* (Academic Press, London, 1988).
35. T. Kondo, K. Yoshida, A. Nakagawa, T. Kawai, H. Tamura and T. Goto, *Nature* **358** (1992) 515.
36. J.T. Knudsen, L. Tollsten and L.G. Bergström, *Phytochemistry* **33** (1993) 253.
37. R. Kaiser, *The Scent of Orchids: Olfactory and Chemical Investigations* (Elsevier, Amsterdam, 1993).
38. G.C. Kite, *Biochem. Syst. Ecol.* **23** (1995) 343.
39. I. Groth, G. Bergström and O. Pellmyr, *Biochem. Syst. Ecol.* **15** (1987) 441.
40. J.B. Harborne, *Annu. Proc. Phytochem. Soc. Eur.* (1996) in press.
41. P.B. Reichardt, J.P. Bryant, T.P. Clausen and G.D. Wieland, *Oecologia* **65** (1984) 58.

42. L.F. James, R.F. Keeler, E.M. Bailey, P.R. Cheeke and M.P. Hegarty, *Poisonous Plants* (Iowa State Univ. Press, Ames, 1992).
43. A.R. Zangerl, *Ecology* **71** (1990) 1926.
44. P.A. Hedin, *Plant Resistance to Insects* (American Chemical Society, Washington, 1983).
45. J.B. Harborne, in *Plant Defences against Mammalian Herbivory* (R.T. Palo and C.T. Robbins, eds) (CRC Press, Boca Raton, 1991) pp. 45-59.
46. R.J. Grayer and J.B. Harborne, *Phytochemistry* **37** (1994) 19.
47. S. Tahara, J.L. Ingham, S. Nakahara, J. Mizutani and J.B. Harborne, *Phytochemistry* **23** (1984) 1889.
48. T. Namai, T. Kato, Y. Yamaguchi and T. Hirukawa, *Biosc. Biotechn. Biochem.* **57** (1993) 611.
49. J.B. Blakeman and P. Atkinson, *Physiol. Pl. Path.* **15** (1979) 183.
50. T. Kokubun and J.B. Harborne, *Phytochemistry* **40** (1995) 1649.
51. J.A. Bailey and J. Mansfield, *Phytoalexins* (Blackie, Glasgow, 1982).
52. J.L. Ingham, *Advances in Legume Systematics* **1** (1981) 599.
53. D. Gross, *Z. fur Pflanzenkrankheitch u. Pflanzenschutz* **100** (1993) 433.
54. J.E. Smith and M.O. Moss, *Mycotoxins* (John Wiley, Chichester, 1985).
55. K. Nakanishi, *Bio-Science* **18** (1968) 791.
56. F. Camps, in *Ecological Chemistry and Biochemistry of Plant Terpenoids,* J.B. Harborne and F.A.T. Barberan (eds) (Clarendon Press, Oxford, 1991) p. 331.
57. Y.C. Toong, D.A. Schooley and F.C. Baker, *Nature* **333** (1988) 170.
58. W.S. Bowers, *Bio-Science* **18** (1968) 791.
59. C.F. Culberson, *Chemical and Botanical Guide to Lichen Products* (Univ. of North Carolina Press, Chapel Hill, 1969).
60. Y. Shimizu, *J. Amer. Chem. Soc.* **108** (1986) 514.
61. J.O. Kuti, B.B. Jarvis, N.M. Rejali and G.A. Bean, *J. Chem. Ecol.* **16** (1990) 340.
62. M.J. Schneider and F.R. Stermitz, *Phytochemistry* **29** (1990) 1811.
63. J.B. Harborne and H. Baxter, *The Phytochemical Dictionary* (Taylor and Francis, London, 1993).

Chapter 2

CHARACTERISATION AND CONTROL OF PLANT SECONDARY METABOLISM

NICHOLAS J. WALTON, MICHAEL J.C. RHODES, ANTHONY J. MICHAEL AND ADRIAN J. PARR

1. Introduction

The aim of this chapter is to outline current understanding of plant secondary metabolism and its regulation. An exhaustive coverage of the literature is not intended; the objective instead is to illustrate concepts, historical background and some future directions, with suitable examples drawn from a range of secondary metabolic pathways. As the title implies, the perspective is principally biochemical (and with some emphasis on secondary products which find applications outside the plant) rather than physiological or plant-developmental, but attention is given to both transcriptional and post-transcriptional regulation and, very briefly, to signal transduction pathways. In recent years, a substantial number of review papers have appeared which deal with secondary products in the context of plant defence; a different treatment is intended here, but sufficient key references are cited to enable the reader primarily interested in plant defence to find additional information without difficulty. Inevitably, the treatment is to a degree selective and leans slightly towards the personal interests of the authors.

Secondary products are compounds which, whilst not essential to the survival of the individual cell, nevertheless fulfil an important - even essential - role in the life and survival of the organism as a whole (see Chapter 1). Arguments in support of important functions for plant secondary products are underlined by the organisational complexity and specificity which characterises many aspects of their biosynthesis. Furthermore - with the very important exception of substances produced in response to stress or pathogen attack - much of secondary metabolism is reproducible from one individual of a species to another and, to a remarkable degree, produces the same principal compounds in broadly the same relative proportions. For example, an *Atropa* plant extracted for alkaloids can confidently be predicted to contain predominantly atropine, rather than its immediate precursor, tropine, or a more distantly-related alkaloid, nicotine - just as the passenger boarding a train for Norwich at London can be reassured that his train will not terminate prematurely at a "precursor station" (Ipswich), or even adopt an "alternative pathway" (to Cambridge). On the other hand, the evolutionary circumstances leading to the development of atropine formation in *Atropa*, rather than to the formation of other alkaloids with comparable activity (for example, nicotine) remain substantially unknown. A similar question has been discussed by Stafford (1) regarding two families of plant pigments with apparently similar functions, the anthocyanins and the betalains. With regard to defence (and similarly but perhaps less obviously protection against ionising radiation), a general hypothesis argues that plants evolved a broad defensive strategy based upon a diversity of antimicrobial and antifeedant substances from which, as a result of chance mutation, more potent substances conferring additional selective advantage from time to time arose (see for example (2, 3)). It is conceivable that in some cases enzymes evolved or became adapted to catalyse reactions of benefit which could already occur to some extent (though in a non-stereospecific manner) non-enzymically. Mutation will then have led to the development of modified enzymes able to catalyse reactions with new or different substrates, or to catalyse alternative reactions with existing substrates.

Characterisation of metabolic pathways is a multi-disciplinary activity. It requires the identification of metabolic intermediates and the demonstration of a plausible reaction sequence, followed by the isolation and characterisation of the individual enzymes responsible and the determination of their tissue and subcellular localisation. It requires control to be studied at the metabolic level, to investigate the relationships between biosynthesis and further metabolism or degradation and to identify reactions and metabolite movements which may contribute to rate-limitation. Thirdly, it involves the study of transcriptional and post-transcriptional control - in particular, of the coordinated expression of genes in response to environmental or ontogenic factors (which may be more- or less-defined), and of the signal-transduction pathways which lie behind the transcriptional responses observed.

2. Elucidation of Secondary Metabolic Pathways

2.1. Structural Considerations and Investigations with Labelled Precursors

The most obvious pointer to a biosynthetic origin is chemical structure. Similarity of a molecule to other known, structurally-similar substances may be a useful guide to biosynthetic origin, though it is not infallible. The alkaloid, coniine, from *Conium maculatum* (hemlock) is structurally very similar to piperidine alkaloids derived from lysine, but is a tetraketide (4, 5). Conversely, divergence or elaboration of a structure from its biosynthetic origins can mask its derivation: the quinoline alkaloid, quinine, is not by inspection obviously of monoterpenoid indole-alkaloid origin.

Historically, chemical syntheses under mild conditions of temperature and pH have provided biomimetic models. A classical example is that of Robinson (6), who was able to synthesise the alkaloid, tropinone, (Figure 1) from methylamine, acetone and succinic dialdehyde mixed together and allowed to stand at room temperature. (Modifications of this reaction are still useful for the preparative synthesis of tropinone analogues.) It has since been established that none of these substances is a precursor of tropinone *in vivo* and that the nitrogen atom in this molecule derives instead from arginine or ornithine, via the diamine, putrescine (1,4-diaminobutane) (reviewed in (7)). Such syntheses are invariably illustrative, but are not an infallible guide to the metabolic route.

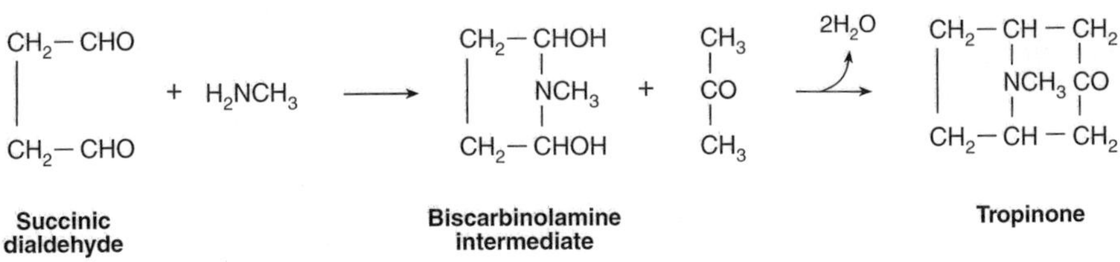

Figure 1. The classical Robinson synthesis of tropinone.

Non-isotopic feeding experiments often enabled tentative conclusions to be drawn regarding biogenic origins. Early work relating to alkaloids was reviewed by W.O. James in volume 1 of "*The Alkaloids*" in 1949 (8). Thus the tropane alkaloids, hyoscyamine and scopolamine, were suggested (correctly) to be derived from ornithine and arginine and nicotine (incorrectly) from proline. Biosynthetic studies were revolutionised in the 1950's and 1960's with the advent of radioisotopic labelling and in the following years radioactive tracer studies with the principal primary metabolites, such as amino acids, acetate and mevalonate, were applied to representatives of the major classes of plant secondary products (9-12). These experiments were not trivial to perform. Often, they were undertaken by wick-feeding to whole plants over a period of days, with subsequent isolation of the reaction products of interest. Frequently, incorporations were low. Particularly challenging was the study of the biosynthesis of compounds made only in specific differentiated tissues over a short period of time, for example the α-acids of the female flowers of hops (13, 14). Considerable skill and ingenuity were required in the degradation of radioactively-labelled products in order to ascertain the sites of incorporation of the label. An illustrative example is the biosynthesis of the lysine- and nicotinic-acid-derived piperidine alkaloid, anabasine, described in (15). This problem of chemical degradation could later be avoided by the use instead of precursors labelled in specific atoms with heavy isotopes, particularly ^{13}C and, later, ^{15}N. The products were then analysed by NMR or MS, which was able to give specific information on patterns of incorporation without degradation being required. Early examples of this approach, analysing spin-spin coupling of contiguous ^{13}C atoms, are given in (16). Two more recent example relate to the incorporation of acetate units during the biosyntheses of cocaine and the structurally-related Solanaceae tropane alkaloids (17, 18). Recently, the use of specifically-labelled intermediates has led to a revision of the final stages of hyoscyamine biosynthesis. The mechanism of formation of the tropoyl unit of this alkaloid and of its ester linkage with tropine have long been uncertain. It has now been established (19) using studies with transformed root cultures that rearrangement of the phenyllactoyl moiety to a tropoyl unit occurs after esterification with tropine and not before - i.e. by rearrangement of littorine (Figure 2). The enzymology of this process still remains unresolved.

Plant cell and organ cultures have made an important contribution to studies of this type since in principle a precursor can be supplied in the culture medium, at a precise concentration and under sterile conditions, over a relatively long period of time. Success obviously depends upon active expression of the biosynthetic pathway in question. When the pathway is expressed generally in the plant body, it may be expressed also in plant cell cultures; but unfortunately cell cultures as a rule do not readily express those pathways which are specific to highly-differentiated cell types (20). In contrast, organ cultures, such as "hairy root" cultures established by transformation with *Agrobacterium rhizogenes*, appear to express the pathways normally present in the corresponding organs of the intact plant (21-24); the obvious limitation here is that it is not possible at present to generate organ cultures of every type one might require (for example floral organ cultures).

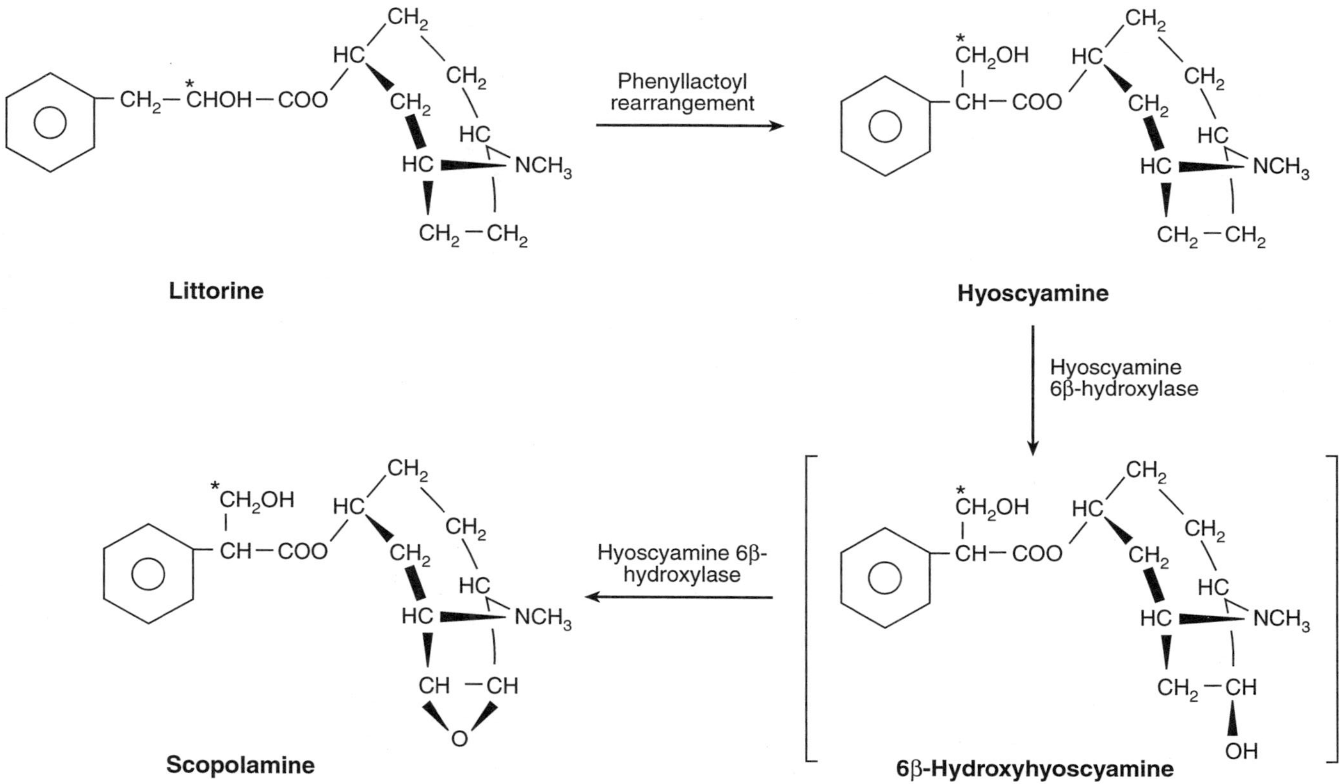

Figure 2. Relationships between the tropane alkaloids, littorine, hyoscyamine and scopolamine.

Because new and complex secondary products are constantly being discovered and characterised, the biosynthetic map is never complete. A biosynthetic investigation may be stimulated or awakened as a result of interest in the pharmaceutical activity of a compound. For example, the biosynthetic origins of the diterpenoid anti-cancer drug from yew, Taxol®, continue to be investigated (25). More rarely, whole biosynthetic pathways may be called into question. The discovery of a new bacterial pathway to isoprenoids (26) has caused the mevalonate pathway to be reexamined in plants. It has now been established that at least some isoprenoids may be biosynthesised by a non-mevalonate pathway (Figure 3) involving a condensation of a triose phosphate with pyruvate, as demonstrated by NMR studies of the labelling pattern of marrubiin, a terpenoid of *Marrubium vulgare*, synthesised from [1-^{13}C]-glucose supplied as precursor (27). Full details of the intermediates of this new pathway and its enzymology have still to be elucidated. It is possible that this pathway is widespread and explains the poor incorporation of mevalonate into some isoprenoids which has frequently been obseved in the past. Non-mevalonate and mevalonate pathways may coexist in the same plant; in *Marrubium*, it was established that steroids showed a labelling pattern consistent with the established mevalonate pathway (27).

An obvious problem with labelled-precursor experiments was - and remains - that incorporation of a labelled compound does not prove beyond doubt that the compound is a metabolic intermediate *in vivo*. For example, labelled quinic acid is incorporated into the aromatic amino acids phenylalanine, tyrosine and tryptophan, but is not on the direct shikimate pathway (28, 29). In another instance, δ-*N*-methylornithine is incorporated into the tropane alkaloid, hyoscyamine (7, 30), but is almost certainly an intermediate in a minor metabolic pathway.

A recent development is the use of *in vivo* NMR to study secondary metabolism. Ford and coworkers have studied the metabolism of nicotine and tropane alkaloids in transformed root cultures (31, 32). The principal advantage of this approach is that it is non-destructive; its principal disadvantage at present is relative lack of sensitivity.

2.2. Enzyme Isolation and Characterisation

2.2.1. Substrates

Enzyme assays of course require substrates and provision of substrates is a major issue. Because of the specialised nature of secondary metabolism many substrates

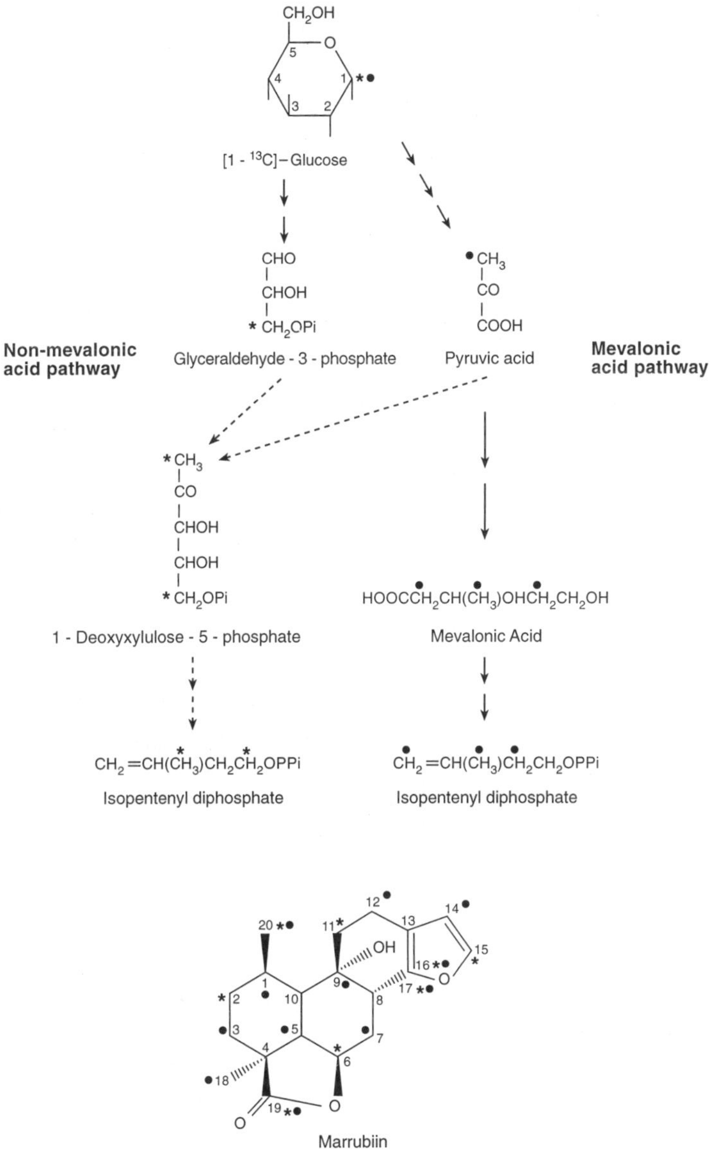

Figure 3. Schematic outlines of mevalonate and non-mevalonate pathways of isoprenoid biosynthesis. Carbon atoms of marrubiin labelled from [1-^{13}C-glucose], consistent with operation of the non-mevalonate pathway, are shown as *; the pattern expected from the mevalonate pathway is shown as •. See text for details.

must be extracted or synthesised. Sometimes, an intermediate may be isolated from another species where it accumulates in larger quantities. An important example is the secoiridoid glucoside, secologanin (Figure 4), which accumulates to substantial levels (*ca.* 1.5% of fresh weight) in the leaves of *Lonicera* species (Caprifoliaceae) but is metabolised further in many species, particularly members of the Apocynaceae such as *Catharanthus* and *Rauvolfia*, to give rise to a large and diverse range of alkaloids (33-35). Both enzymic and biomimetic reactions of secologanin were intensively studied, with a view to understanding the chemistry and biochemistry which occurs following its condensation with aromatic amines (particularly tryptamine) and subsequent deglucosylation; such experiments could provide substrates and analogues, either directly *in situ* or after isolation, with which to investigate later enzymic reactions (36, 37). A recent example of versatile biomimetic synthesis is the preparation of a range of piperidine alkaloids and related compounds from 2-cyano-6-oxazolopiperidine (38).

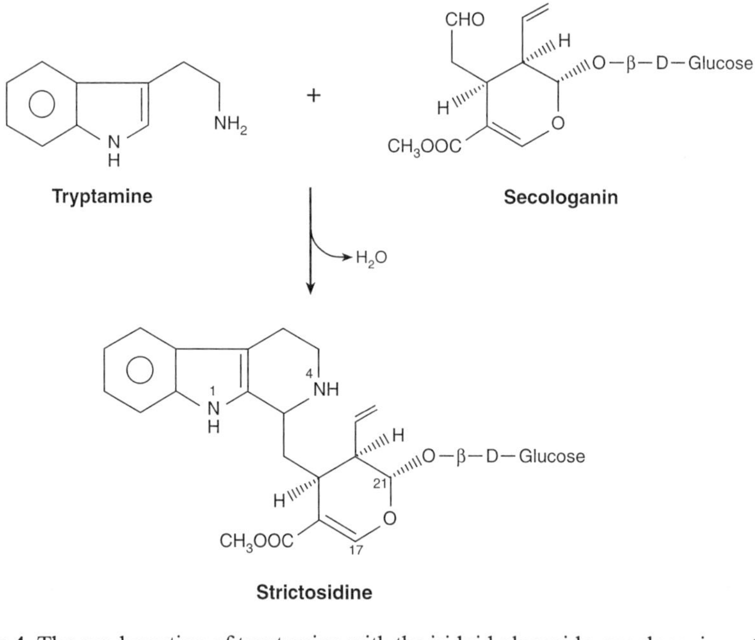

Figure 4. The condensation of tryptamine with the iridoid glucoside, secologanin, catalysed by strictosidine synthase. See section 2.4.1. for discussion of the hydrolysis of strictosidine.

Plant cell and organ culture facilitates enzymological investigations by allowing large quantities of biosynthetically-active tissue to be provided at will in the laboratory. Such tissue can be used for studies with labelled precursors (see above, Section 2.1.) or for enzyme extraction (see below, this section); in addition, cell-free extracts or purified enzymes (including those produced by heterologous gene expression - see below) may be employed in their own right to generate substrates. These are essentially examples of biotransformations (see Chapter 7). For example, strictosidine synthase, the initial enzyme of monoterpenoid indole alkaloid production (Figure 4), was immobilised and used to generate gram quantities of strictosidine from secologanin and tryptamine (39); as indicated above, this strictosidine could then be used as a substrate supplied to cell-free extracts in order to begin to elucidate the biosynthetic steps which commence with the hydrolysis of strictosidine by strictosidine β-D-glucosidase. Another typical example was the synthesis of labelled naringenin for use in the assay and purification of flavanone 3β-hydroxylase, an early enzyme of the flavonoid pathway, from flowers of *Petunia hybrida*; an enzyme preparation from parsley cell cultures was used to provide the chalcone synthase and chalcone isomerase necessary to generate naringenin from *p*-coumaroyl-CoA and malonyl-CoA (40). Cell cultures and their extracts were extensively used in the enzymic characterisation of flavonoid biosynthetic pathways; see, for example (11).

In other cases, microbial enzymes can be used, as in the application of a partially-purified preparation of *p*-coumarate: CoA ligase from *Pseudomonas* for the preparation of *p*-hydroxycinnamoyl CoA thioesters for studies of phenylpropanoid metabolism (41, 42).

2.2.2. Enzyme purification

There are four principal reasons to purify enzymes. The first is the need to relate catalytic activities to individual proteins. How many enzymes are responsible for an observed reaction? Is the reaction enzymic? Is a single reaction catalysed by more than one isoenzyme? Can a single enzyme catalyse more than one reaction in a metabolic sequence, or operate with a range of substrates? The second reason is that purification is the starting point for the detailed understanding of the catalytic process. Does the enzyme require cofactors? What is the reaction mechanism? How efficient is the enzyme as a catalyst? How stable is it? Thirdly, antibodies to a purified enzyme provide a classical route to the screening of a gene expression library and the isolation of a gene. Finally, the purified enzyme may be of use in the provision of substrates or in biotransformation reactions.

Many plants containing interesting and valuable secondary products would not be first-choice subjects for the biochemist on account of their high content of fibre and tannins. The former leads to extraction difficulties; the latter are apt to cause severe enzyme inactivation (43). Inactivation problems may be avoidable to some degree by the use of cell- or organ-cultured tissue which may have a lower tannin content than normal plant tissue, but are usually combated by the incorporation in extraction media of thiol antioxidants, metal-chelators and protective agents such as polyvinylpolypyrrolidone and by working as expeditiously as possible. The choice and initial treatment of the tissue are very important. Ideally, this will be enriched in the enzyme of interest. For example, the levels of monoterpene cyclase in conifers such as *Abies grandis* were tenfold higher in wounded trees relative to control trees (44). Where the activity is localised in a particular cell type, it may be possible to achieve selective extraction by mechanical means, as in the release of the secretory cell clusters of glandular trichomes, the leaf epidermal structures responsible for essential oil synthesis and storage in *Mentha* and related genera (45).

The initial step in many purification procedures continues to be a conventional concentration and fractionation step with ammonium sulphate or, much less commonly, polyethylene glycol. Most subsequent procedures are chromatographic - non-denaturing electrophoretic techniques are still comparatively uncommon - and proprietary columns packed with near-monodisperse chromatographic media covering the major chromatographic separation techniques - anion-exchange, hydrophobic-interaction, chromatofocusing and gel filtration - can in favourable circumstances allow high flow rates and effective resolution and provide the basis of many enzyme-purification strategies. Comprehensive accounts of the purification, properties and assay of many enzymes of the major plant secondary product pathways are compiled in a single volume (volume 9) of the series "*Methods in Plant Biochemistry*". Alonso and Croteau, for example, have summarised the application of many of these techniques to prenyltransferases and cyclases of isoprenoid biosynthesis (45). Anion-exchange chromatography is particularly suitable for these enzymes on account of their low pI values, which means that they remain ionised and therefore able to bind to the matrix at low pH values at which most proteins cannot bind; on the other hand, hydrophobic-interaction chromatography is frequently problematic for these enzymes because the binding is too tight, leading to poor recoveries. Many effective enzyme purifications have depended upon various forms of affinity chromatography, employing a matrix-bound reaction substrate, product or analogue as ligand. Dye-ligand matrices are attractive because they are available commercially and can therefore readily be screened without designing and synthesising a specific affinity matrix with no guarantee that it will prove successful (see for example (46), describing the purification of casbene synthase), although the basis of their action is not well understood. In other cases, substrate-based affinity matrices have proved crucial where other strategies have failed. Putrescine-*N*-methyltransferase, the initial enzyme of nicotine and tropane alkaloid biosynthesis from putrescine, was purified using an affinity matrix of carboxyl-bound ornithine, effectively putrescine in immobilised form (47). Immunoaffinity chromatography depends upon the use of a specific antibody bound covalently to a matrix. It has the limitation that the pure protein has often already to be available in order to raise the antibody, unless the antibody has been raised to an immunologically closely-related protein or a conserved epitope.

Particular difficulties are often encountered for membrane-bound or multi-component enzymes. Cytochrome-P450 monooxygenases, responsible as a class for a diversity of hydroxylation reactions, pose particular problems since purification generally involves separation of the oxygenase component from the reductase and from the lipid components required for activity and hence the enzyme activity of the oxygenase may need to be followed indirectly during purification (48, 49).

Short of crystallisation of the protein, the usual criteria adopted for purification to apparent homogeneity are electrophoretic: the detection of a single band on sodium dodecylsulphate polycrylamide gel electrophoresis (SDS-PAGE) and, much to be preferred, of a single spot on 2-D gel electrophoresis, using the most sensitive detection technique available (usually silver-staining). Even with these criteria, it is possible for closely-similar proteins to co-purify. It is of course important to recognise that enzyme activity determined in an assay may require the presence of more than one polypeptide.

As much ingenuity may be applied to the detection and assay of enzymes as to their purification. Choice of assay methods is closely allied to the availability of substrates (see above) and the need for sensitivity. This need arises from the low activities to be measured, from enzyme or substrate instability, from the scarcity of substrates and - not uncommonly - from uncertainty about the precise identities of the products formed. Radiometric methods offer sensitivity within the limitations of the specific radioactivity of the substrate and may be preferred if they are a direct development from tracer-feeding experiments. Separation of the radioactively-labelled enzyme reaction product(s) may in principle be achieved by radio-HPLC, radio-TLC or other methods described in Chapter 3 of this volume, as in the study of enzymes of gibberellin biosynthesis (50). A particularly simple assay of condensation reactions, such as that catalysed by strictosidine synthase (see Figure 4), may be possible if the substrate can be radioactively-labelled in such a way as to release water as the labelled product (51, 52). The requirement for a labelled substrate may, however, be a considerable limitation to the general applicability of an assay method and for routine use non-radiometric chromatographic methods, particularly using diode-array HPLC (Chapter 3), are often sought. There is also an advantage if a continuous spectrophotometric or fluorometric assay can be developed, since this can facilitate the rapid screening of fractions during enzyme purification. Microtitre-plate readers can permit the rapid spectrophotometric or fluorometric assay of large numbers of samples in small reaction volumes.

2.3. Secondary Metabolic Pathway Gene Isolation and Characterisation

2.3.1. Genetics

Classical genetic approaches based on the analysis or complementation of mutants have not been applied universally to plant secondary metabolism because in many cases such mutants are difficult to detect. Mutations in secondary product pathways are very likely to be non-lethal and may not be easy to select by simple screening procedures. In addition, the long generation times of plants in relation to microbes make mutant-based approaches relatively long-term. There are, however, many important exceptions to this rule.

The first exception rests upon the fact that plant genes can in principle be used to complement mutations in other organisms which are more easily screened. This approach is powerful when applied to primary metabolism, but of limited application in secondary metabolism, where the pathways and genes may not be present in other organisms. An elegant example is the isolation of genes for the α-subunits of anthranilate synthase, the initial enzyme both of the specific pathway leading to acridone alkaloids in *Ruta graveolens* and more generally of the primary metabolic pathway to the essential amino acid, tryptophan (53). A cDNA library was prepared from mRNA isolated from young plants and initially screened using a probe based on the C-terminal sequence of yeast anthranilate synthase α-subunit. Clones were then selected for their ability to complement *Escherichia coli* deletion mutants defective in anthranilate synthase and therefore unable to synthesise tryptophan. Using this approach, young *Ruta* plants were found to express two genes for anthranilate synthase alpha subunits; ASα-2 was constitutively expressed in *Ruta* cell cultures, whereas the transcription of ASα-1 was found to correspond to elicitor-induced acridone alkaloid accumulation.

The second exception rests simply upon the fact that mutations in some secondary product pathways do result in phenotypes which are easily screened (though conversely the presence of the phenotype does not necessarily indicate a mutation in biosynthesis). This is the case with the pathway to the flavonoid pigments, including anthocyanins. It is also true of the pathways to plant growth substances and regulators: for example, gibberellins (54, 55), auxins (56); abscisic acid (57), and the brassinosteroids (58). In maize, a now-classic experiment demonstrated biochemical complementation of mutants in anthocyanin biosynthesis simply by pressing together tissues cut from developing seeds; a putative colourless intermediate accumulating behind a metabolic lesion in one of the tissues was able to diffuse to the boundary with the second tissue, which was blocked in an earlier reaction, and was there able to be metabolised into visible coloured pigment (59). By ordering pairwise combinations of the mutants, it was possible to construct a hypothetical metabolic sequence although the intermediates remained unidentified. Subsequent elucidation of the flavonoid pigment pathway in various laboratories and several plants was achieved by a combination of both genetic and biochemical approaches (60-64). The genetic approach was facilitated by the realisation that many mutations in anthocyanin biosynthesis were the result of gene disruption caused by the insertion of mobile transposable elements and that once these elements had been characterised and sequenced their site of insertion into the genome would lead to the identification of the mutated gene. For example, the use of the transposable element *Activator* to clone the *bronze* locus in maize was described by Federoff and coworkers in 1984 (65). Subsequently, transposon tagging was very successfully applied to the isolation of anthocyanin biosynthetic pathway - and regulatory - genes associated with flower colour in *Petunia* and *Antirrhinum* (see e.g. 66, 67).

Because of its short generation time and small genome, *Arabidopsis thaliana* is rapidly becoming a favourite organism for genetic approaches to secondary metabolism, including studies of flavonoid biosynthesis and glucosinolates (68); full advantage can then be taken of the mapping of the *Arabidopsis* genome (69). The use of *Arabidopsis* is, of course, limited to the study of those pathways which are encoded in the *Arabidopsis* genome.

2.3.2. Molecular genetics

The classical biochemical and molecular genetic approach to the isolation of a pathway gene, for example tryptophan decarboxylase from *Catharanthus* cited in (52), remains the purification of an enzyme from its native plant source (or from plant cell or organ culture), followed by its use to raise antibodies for the screening of the protein expression products of a gene library. Increasingly, however, genes are isolated by strategies which do not depend upon initial protein purification, either commencing with an enriched mRNA population, by amplification from DNA using the Polymerase Chain Reaction (PCR) and synthetic oligonucleotide primers generated from known gene sequences already available in the sequence databases, or by a combination of these approaches. Amino acid sequence data may also be used to design PCR primers, taking account of the degeneracy of the genetic code and of codon usage frequencies. The isolation of genes without prior isolation of the protein may well leave the identity of the gene and its function unresolved - a biochemical problem which may prove facile, but which can be very challenging if the substrate is unknown, unavailable and chemically complex, or if the gene encodes a regulatory, transport or binding protein. Meijer and colleagues isolated 16 different cDNA clones encoding cytochrome P450 monooxygenases from *Catharanthus roseus* by PCR amplification, using degenerate primers based on the highly-conserved cytochrome P450 haem-binding region and a second, non-specific primer complementary to the mRNA polyA tail or to the phage vector sequence; the reactions catalysed by the encoded enzymes, however, remained unknown (70).

The success of gene isolation using gene libraries derived from mRNA populations depends upon the basis on which the library has been made and on the manner in which it has been screened. (It also depends critically upon the quality of the mRNA used to prepare the library.) Nucleic acid hybridisation techniques allow cDNA libraries generated from populations of extracted mRNA to be screened against other mRNAs or cDNAs used as probes. It is therefore possible in principle, by judicious choice of mRNA populations and probes, to select primarily those cDNA clones corresponding to genes expressed in an organ-specific manner (e.g. in roots, but not in leaves); or expressed selectively in tissues which differ in some obvious way (e. g. pigmented as distinct from non-pigmented); or expressed in response to a specific stimulus (e.g. light or wounding); or expressed in a cultivar- or species-dependent fashion.

A ripening-associated mRNA population was the basis of the now-classical work by Grierson and colleagues to isolate ripening-associated genes from tomato (71, 72). Individual clones were found to encode a number of genes of cell-wall, ethylene and carotenoid metabolism. One, pTOM 13, showed significant homology to the gene for the 2-oxoglutarate-dependent dioxygenase, flavanone-3-hydroxylase and was found to encode the ethylene-forming enzyme, 1-aminocyclopropane-1-carboxylic acid oxidase (ACC oxidase). An antisense construct of this clone inhibited ethylene biosynthesis in transgenic plants (73). This has rightly been cited as an example of success by molecular genetic means in the isolation of a gene encoding an enzyme intractable to isolation by classical biochemical approaches. A differential screening approach applied to high- and low-nicotine -producing cultivars of *Nicotiana tabacum* was reported by Hibi and colleagues (74). Two cDNA clones were isolated; one showed homology to 2-oxoglutarate-dependent dioxygenases; the other was demonstrated by heterologous expresssion in *Escherichia coli* to encode putrescine *N*-methyltransferase, the initial enzyme of the nicotine biosynthetic pathway leading from the diamine, putrescine. The difficulty and uncertainty inherent in differential screening approaches is illustrated, however, by an attempt to isolate clones associated with the biosynthesis of thiophenes in *Tagetes patula;* in this case, a root / leaf differential screening did not lead to the isolation of root-specific clones with the characteristics expected for involvement in the thiophene pathway (75).

Southern hybridisation using genomic DNA and a labelled DNA probe can be used to isolate homologues of genes or sequences which are already available. Southern hybridisation can give direct information on the number of gene copies or variants present in the genome, some of which may be under distinct transcriptional control, or may encode isoenzymes, or which may encode more distantly-related enzymes which nevertheless retain substantial sequence homology. Such information is also an intrinsic product of whole-genome sequencing projects, such as that of the cruciferous weed, *Arabidopsis thaliana* (69). The analysis of sequence information in terms of protein or enzyme "superfamilies" is of growing importance to an understanding of the basis of enzyme mechanisms and their evolution (see e.g. 76).

The availability of a gene affords opportunity for over-expression and for facilitated purification of the encoded protein - hence bringing the protein / gene isolation sequence full circle. This is likely to be of especial value for (many) secondary-metabolic enzymes which are difficult to purify by reason of low abundance or instability. The availability of large amounts of pure protein facilitates detailed study of catalytic mechanism and raises the possibility of use in biotransformations (see Chapter 7) or for the provision of substrates (see Section 2.2.1.). For example, strictosidine synthase (see Figure 4), overexpressed in *E. coli*, was immobilised and used to perform the synthesis of the indole glucoalkaloid, strictosidine, from tryptamine and secologanin (39). The gene encoding chalcone reductase, the first enzyme of the branch pathway leading to the soybean phytoalexin, glyceollin, was cloned into an expression vector and transformed into *E. coli*; from 45 g of *E. coli* cells, it was possible to isolate 5 mg of homogeneous chalcone reductase (77). Two enzymes of benzylisoquinoline alkaloid biosynthesis, the berberine bridge enzyme and berbamunine synthase, have been heterologously expressed in large amounts using an insect baculovirus system (78). Finally, it is possible specifically to engineer a gene to facilitate subsequent protein purification; for example, a polyhistidine sequence can be attached to the C-terminus to enable purification by a nickel affinity matrix (79).

2.4. Pathway Organisation

2.4.1. Enzymes and metabolic channelling

Enzymes are the basic elements of metabolic organisation and as catalysts facilitate reactions which would otherwise be kinetically unfavourable. They also determine the stereochemistry of reaction products. For example, condensation of tryptamine with secologanin occurs non-enzymically at near-neutral pH and room temperature to give a mixture of the glucoalkaloid stereoisomers, strictosidine and vincoside. In contrast, the very much more rapid enzymic condensation catalysed by strictosidine synthase produces a single isomer, strictosidine (see Figure 4). Furthermore, hydrolysis of strictosidine using a non-specific β-glucosidase, or non-enzymically using methanolic HCl, normally gives rise to N-4: C-17 ring-condensation products, the vallesiachotamines, or to an N-1: C-21 product, nacycline (see 80, 81), whereas the hydrolysis of strictosidine catalysed by the strictosidine-β-glucosidases partially purified from suspension cultured cells of *Tabernaemontana divaricata* gave rise to the (physiological) N-4: C-21 condensation products, cathenamine and *epi*-cathenamine and several other compounds, but only relatively small amounts of vallesiachotamine and *iso*-vallesiachotamine (82). Kinetic control of reactions by enzymes is thus an essential determinant of the range of biosynthetic products observed, as well as of their rate of production.

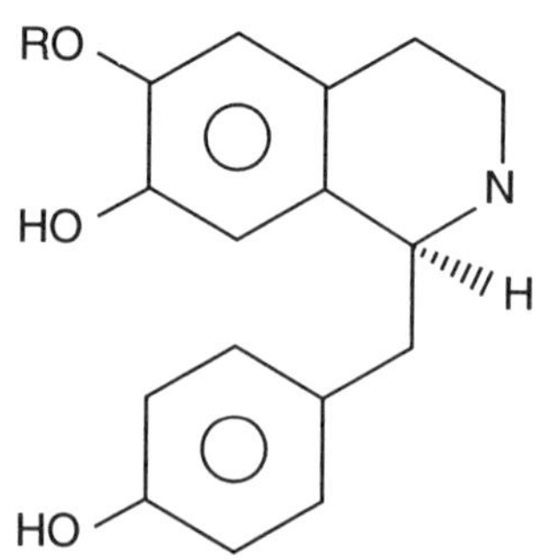

Figure 5. (*S*)-Coclaurine (R = CH_3) and (*S*)-norcoclaurine (R = H).

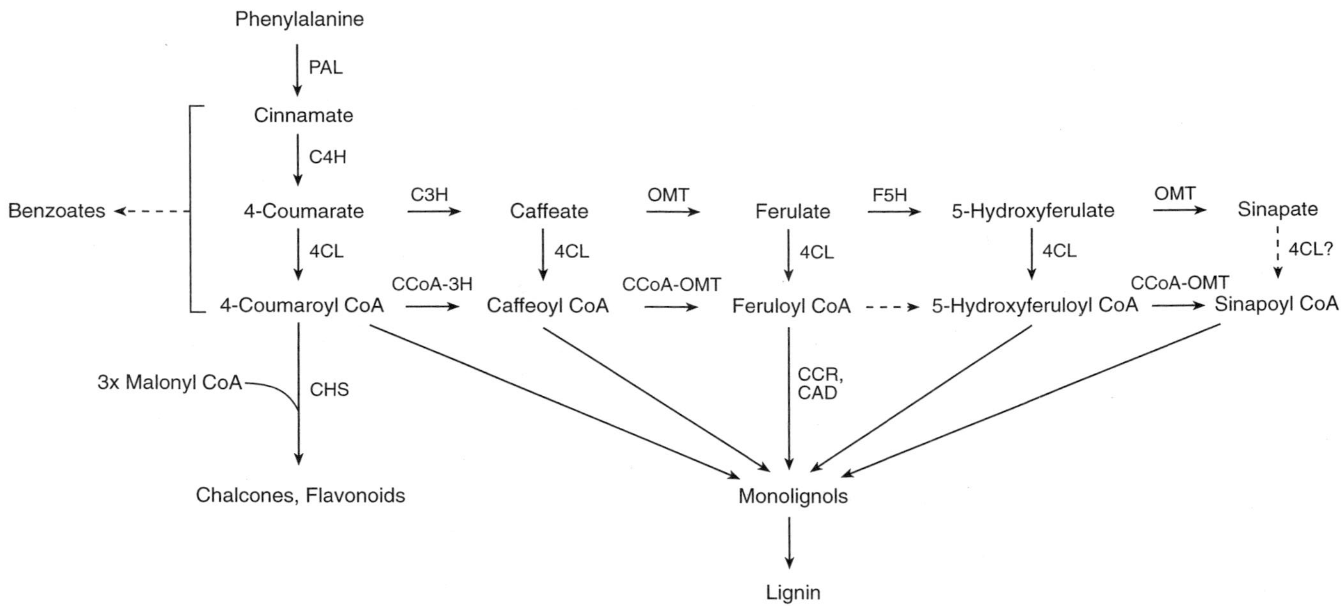

Figure 6. Possible pathways for the formation of monolignols for lignin biosynthesis. PAL, phenylalanine ammonia lyase; C4H, cinnamate 4-hydroxylase; C3H, coumarate 3-hydroxylase; 4CL, 4-coumarate: CoA ligase; OMT, *O*-methyltransferases; F5H, ferulate 5-hydroxylase; CHS, chalcone synthase; CCR, cinnamoyl-CoA reductase; CAD, cinnamyl alcohol dehydrogenase. CCR produces cinnamaldehydes which are then reduced by CAD to cinnamyl alcohols (monolignols).

Secondary metabolic enzymes vary considerably in specificity - sometimes even in the case of an individual activity compared across plant genera. For instance, *S*-adenosyl-*L*-methionine: (*S*)-coclaurine-*N*-methyltransferase of *Tinospora cordifolia*, an enzyme of benzylisoquinoline alkaloid biosynthesis, accepts as substrates only (*S*)-coclaurine and (*S*)-norcoclaurine out of 15 benzylisoquinoline alkaloids tested and shows strict stereospecificity for the (*S*)-isomers (Figure 5); yet the same enzyme in other genera accepts both (*R*)- and (*S*)- substrates (83). Other enzymes of the benzylisoquinoline pathway similarly show strict substrate specificity, leading to a clearly-defined metabolic sequence (84). In contrast, cinnamoyl-CoA reductase, the initial enzyme in the phenylpropanoid branch pathway leading to the formation of monolignols for lignin biosynthesis, shows comparable kinetic constants for four 4-hydroxycinnamoyl-CoA thioesters (4-coumaroyl-CoA, caffeoyl-CoA feruloyl-CoA and sinapoyl-CoA) (85, 86); and the similar broad substrate specificity of the CoA-thioester-forming enzyme, 4-coumarate: CoA ligase, together with the potential presence *in vivo* of "parallel" enzymes able successively to hydroxylate and methylate either the free acids or their CoA thioesters, leads to the possibility of a "metabolic grid" *in vivo* in which a molecule of 4-coumaric acid destined for monolignol formation might follow any one of several pathways (Figure 6).

Although the identity of reaction products is generally determined by the identity of the substrate supplied, this is not invariably the case. For example, at pH 8-9, an enzyme of the morphinan alkaloid biosynthetic pathway, acetyl-CoA: salutaridinol-7-*O*-acetyltransferase (Figure 7), produces thebaine by spontaneous allylic elimination from the initial reaction product, salutaridinol-7-*O*-acetate; however, at pH 7, this elimination does not occur and the reaction products are dibenzazonine alkaloids containing a nine-membered ring (87).

The shortest possible metabolic pathway consists of two reaction steps and this is also the minimum required for direct transfer of the product of one enzyme to the substrate-binding site of another. Such direct transfer constitutes "metabolic channelling" (see e.g. 88). For example, Hrazdina and Wagner (89) supplied ^{3}H-labelled phenylalanine and ^{14}C-labelled cinnamic acid to fractions from a homogenate of etiolated buckwheat seedlings extracted following induction of anthocyanin biosynthesis by irradiation. The ratio of the incorporation of the labels into 4-coumaric acid was measured. This value was generally less than 1.0, but 9.5 in the exclusion-volume fraction emerging from an agarose A-4 gel filtration column. This was adduced to be evidence of direct transfer of cinnamic acid from PAL to C4H (90, 91). Other studies have in contrast suggested only a loose association of PAL with a membrane fraction and with C4H (92). Metabolic channelling has also been

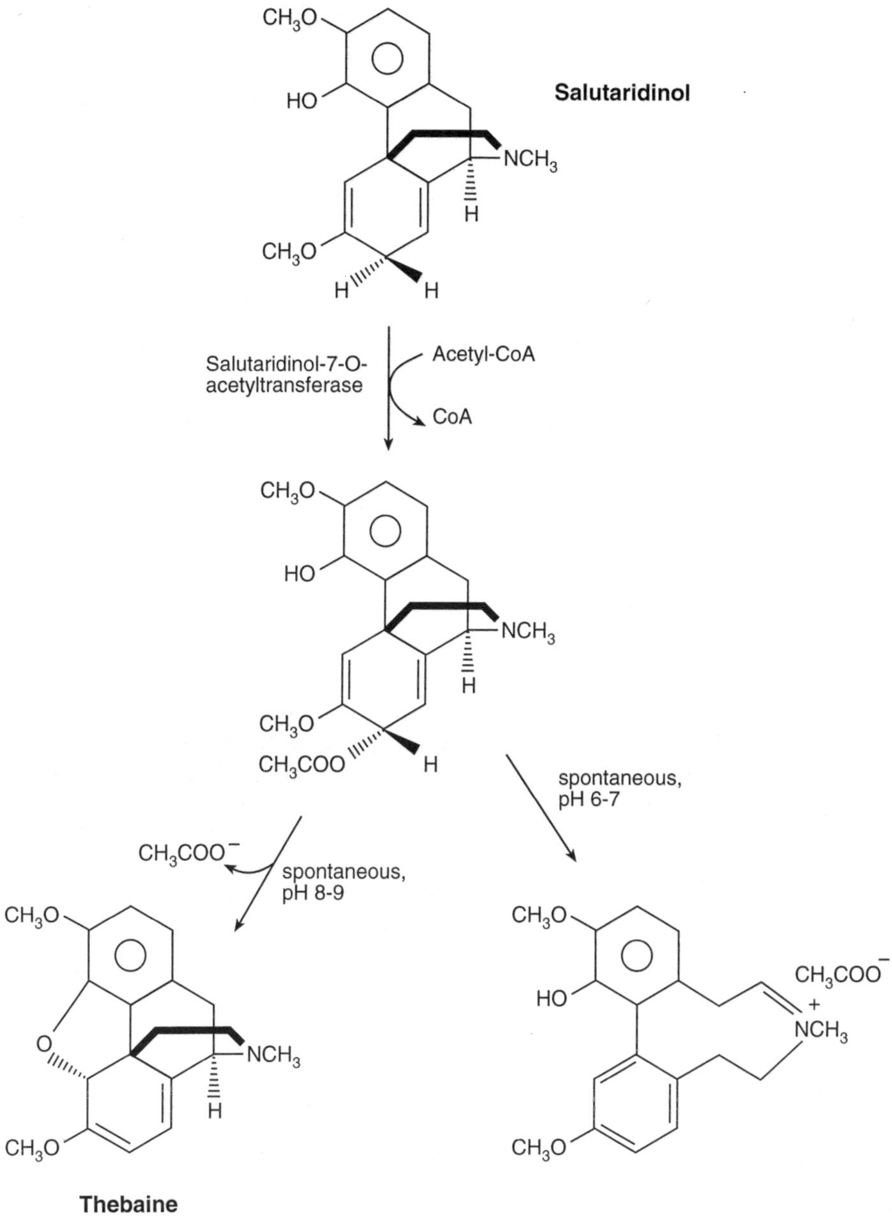

Figure 7. Reaction products of acetyl-CoA: salutaridinol-7-*O*-acetyltransferase.

implicated in the regulation of terpenoid biosynthesis (93, 94) and in the biosynthesis of ornithine- and lysine-derived alkaloids, where the evidence rests on the lack of the expected redistribution of label by metabolism through a symmetrical putrescine (1,4-diaminobutane) or cadaverine (1,5-diaminopentane) intermediate (reviewed in 7).

Metabolic channelling is inherent in multienzyme complexes or in multicatalytic polypeptides. In some cases, the same enzyme activities may exist in multienzyme-complex or multicatalytic-polypeptide forms depending upon the organism. This is true of acetyl-CoA carboxylase (see 95), the initial enzyme of the fatty-acid and polyketide biosynthetic pathways, and of enzymes of the shikimate pathway of aromatic amino acid biosynthesis (91, 96, 97). Perhaps the most highly-characterised multi-enzyme complexes in secondary metabolism are the modular polyketide biosynthetic complexes of *Streptomyces* (98, 99), where organisation at the level of the multienzyme complex is reflected in a corresponding operon structure. The enzymology of polyketide biosynthesis in plants is less intensively characterised. Chalcone synthase, the initial enzyme of the flavonoid biosynthetic pathway catalyses the stepwise condensation of three malonyl-CoA units with one molecule of 4-coumaroyl-CoA (resveratrol (stilbene) synthase catalyses a similar reaction) and is formally analogous to a β-ketoacylsynthase condensing activity in the polyketide biosynthetic pathway (100), but with the following important distinctions: it uses 4-coumaroyl-CoA instead of acetyl-Acyl Carrier Protein (acetyl-ACP) as the starter unit and malonyl-CoA instead of malonyl-ACP in chain elongation (Figure 8). Sequence similarity between chalcone synthase and the β-ketoacylsynthases of polyketide and fatty-acid biosynthesis is limited, suggesting no close evolutionary relationship. (Besides resveratrol synthase, the chalcone synthase family of enzymes (101, 102) also includes acridone synthase, which utilises *N*-methylanthraniloyl-CoA; benzophenone synthase, which uses a benzoyl-CoA substrate; benzalacetone synthase, where only a single malonyl-CoA molecule is condensed and carbon dioxide is released; and phloroisovalerophenone and phloroisobutyrophenone synthases, involved in the formation of hop α-acids, where isovaleryl-CoA and isobutyryl-CoA, respectively, are substrates in place of the aromatic CoA thioester.) In legumes such as soybean (*Glycine max*) which synthesise isoflavonoid phytoalexins such as glyceollin in response to pathogen attack, the initial polyketide chalcone synthase reaction product is reduced to 4,2',4'-trihydroxychalcone by the enzyme, chalcone reductase. The amino acid sequence of this enzyme shows no homology to the ketoacylreductases of fatty acid biosynthesis (103), but contains a "leucine zipper" region (usually associated with nucleic acid binding). There is evidence that the enzyme is physically associated with chalcone synthase and reduces the bound polyketide intermediate prior to its cyclisation on the synthase.

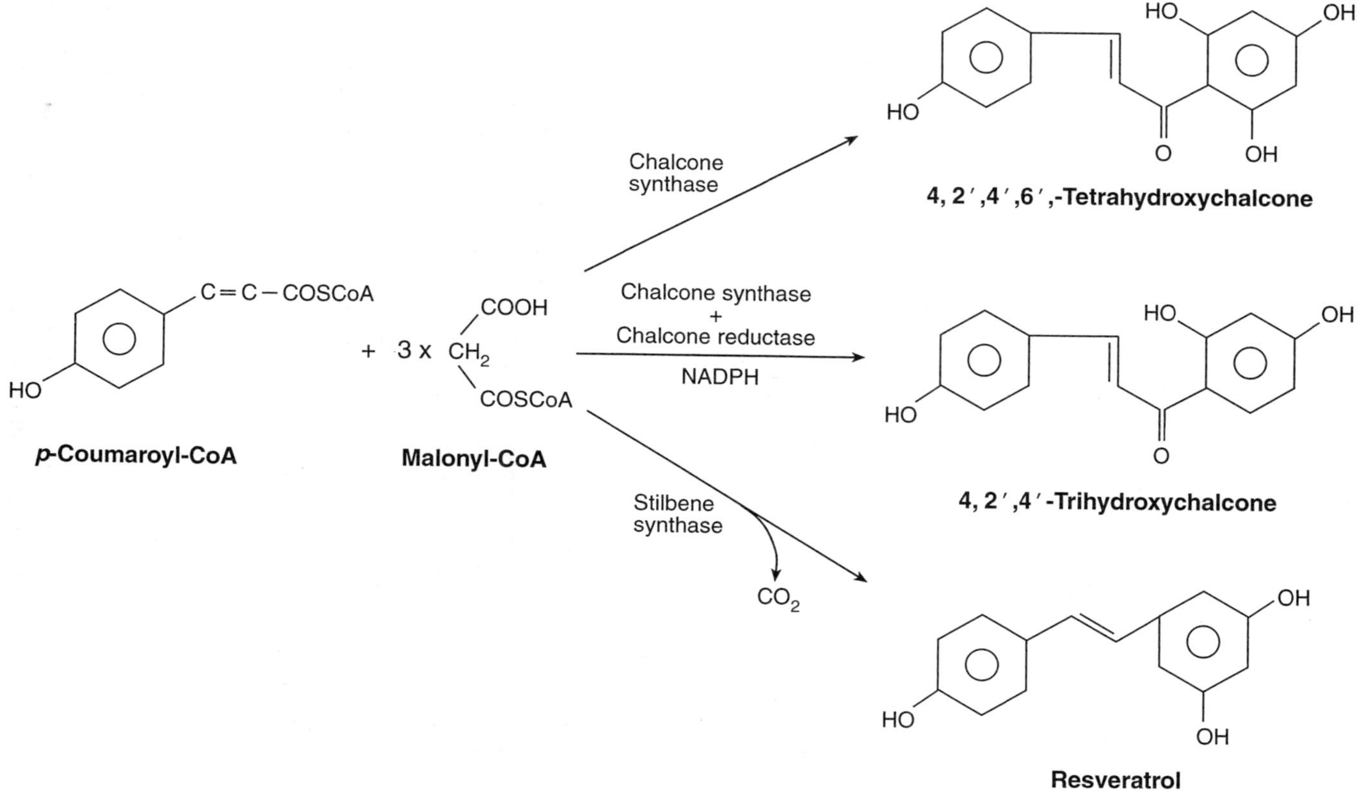

Figure 8. Reactions of chalcone synthase, chalcone reductase and stilbene synthase.

A particular case of "metabolic channelling" would appear to be the control of oxidative-coupling reactions such as those which lead to the polymerisation of monolignols in the formation of lignin (85, 86). It is assumed that this is not a random, uncontrolled process and debate continues as to the precise identity of the oxidative enzymes responsible. The initial steps in the process are likely to involve the movement of monolignol glucosides to the apoplastic space and their hydrolysis to monolignols. Important analogous oxidative coupling reactions occur in the formation of diferulate cross-links in cell walls, in the production of suberin (104) and in the formation of lignans. A significant recent development has been the isolation from *Forsythia* of a "dirigent protein" which is able to direct the stereochemistry of the oxidative condensation leading to lignan formation (105).

2.4.2. Cellular and subcellular localisation

Metabolic channelling may result from subcellular compartmentation. Such compartmentation is highly complex, particularly when compounds are formed as a result of the convergence of individually-elaborate pathways. Despite impressive analytical "*tours de force*" in specific cases, knowledge in general remains scant and the cellular and subcellular localisation of many entire pathways still remains to be determined.

Insight into the localisation of enzymes can be obtained by several approaches. These include classical biochemical subcellular fractionation techniques, using as markers enzymes known to be characteristic of individual organelles; studies of the metabolism of isolated organelles obtained by the fractionation of whole-plant tissues or of protoplasts (106); immunocytochemical localisation (107); and, to an increasing extent, deduction from the presence of leader sequences known to be responsible for targeting newly-synthesised polypeptide to the appropriate cellular compartment (see e.g. 108). Establishing the localisation of enzymes is not trivial, and because of purity considerations it is not feasible using classical biochemical techniques to establish rigorously that an enzyme is *absent* from a particular cell type or organelle. Determining reliably the concentrations of metabolites in subcellular compartments is very difficult; although methods have been developed for the measurement of subcellular metabolite levels by fractionation of freeze-stopped tissue in nonaqueous media (109, 110) or following rapid protoplast fractionation (106), these have been infrequently applied to secondary metabolism.

The application of immunocytochemical techniques has been shown nicely in a study of the tissue and subcellular localisation of three enzymes associated with lignin biosynthesis - PAL, cinnamate 4-hydroxylase and a cationic cell-wall peroxidase - in hypocotyl tissue of *Phaseolus vulgaris* (French bean) during differentiation and in response to wounding (107). Immunogold labelling was used to visualise the cellular and subcellular distribution of the enzymes. As expected of a cytochrome P450 monooxygenase, cinnamate 4-hydroxylase (C4H) was entirely membrane-bound, with some associated with the Golgi; PAL subunits were cytosolic, though some appeared to be membrane-associated. There was no direct evidence for a specific association of PAL and C4H. All the enzymes were associated with the pre-xylem and phloem tissues and with the metaxylem; in addition PAL and C4H were found in cells adjacent to the metaxylem, presumably to provide phenylpropanoid precursors for lignification. Following wounding, PAL and C4H were found especially in epidermal cells adjacent to the wound site.

Association of part of a secondary metabolic pathway with a distinct subcellular organelle or macromolecular complex was shown particularly elegantly in the case of the formation of the benzyl isoquinoline alkaloid, berberine, in *Berberis* cell cultures. Berberine is synthesised from (*S*)-reticuline in four enzymic steps, commencing with ring-closure to (*S*)-scoulerine, catalysed by the "berberine bridge" enzyme. These four enzymes are located in cell particles which band on a sucrose gradient at a density of 1.14 and which, if supplied with (*S*)-reticuline (but not the (*R*)-isomer), can be visualised by their production of yellow berberine (111). Similar subcellular particles can be demonstrated in *Coptis*, which also synthesises berberine, but interestingly, two of the four steps occur in the converse order (112), a reminder that a pathway determined in one species is not necessarily identical in another (Figure 9).

Isolated plastids, including chromoplasts and etioplasts, have been extensively used in studies of mono- and di-terpenoid biosynthesis (94, 113, 114). Plastids contain geranyl diphosphate synthase (115), geranylgeranyl diphosphate synthase (116) and terpene cyclases (113). It has been suggested that sesquiterpenes, in contrast, may be synthesised at the cytosol / endoplasmic reticulum boundary, using isopentenyl diphosphate exported from the plastid, since the mevalonate pathway appears blocked in the cytosol at the level of β-hydroxymethylglutaryl-CoA biosynthesis (117). As indicated above (27) (Section 2.1.), the discovery of a pyruvate / triose phosphate pathway of isoprenoid biosynthesis which does not involve mevalonate as an intermediate is causing previously-held assumptions concerning isoprenoid biosynthesis to be reexamined. Hydroxylation of terpenoids is generally catalysed by microsomal cytochrome P450 monooxygenases, raising the question of the mechanisms by which terpenoids are transported following initial biosynthesis; and

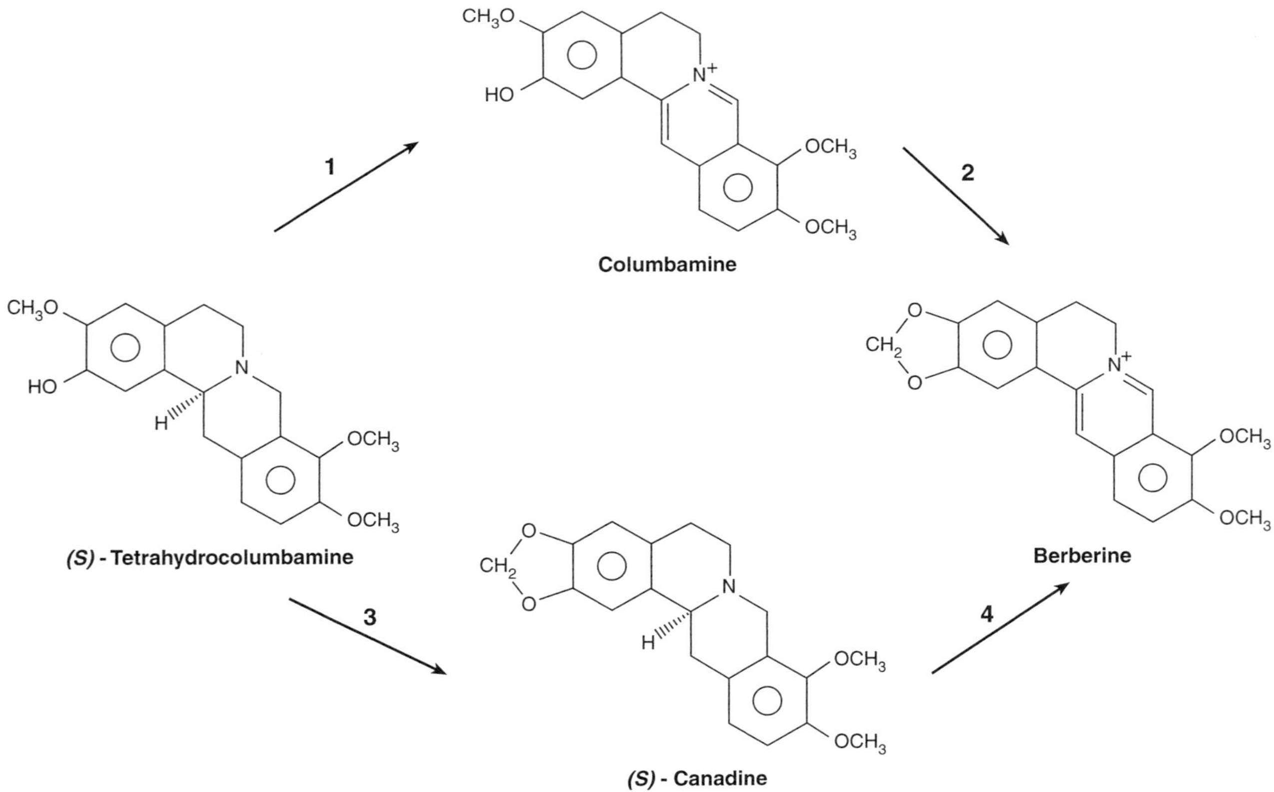

Figure 9. Alternative final routes to berberine in *Berberis* (upper route) and *Coptis* (lower route). **1**: (*S*)-tetrahydroprotoberberine oxidase; **2**: berberine synthase; **3**: (*S*)-canadine synthase; **4**: (*S*)-canadine oxidase

following hydroxylation, glycosylation is commonplace.

The formation of glycosides and other conjugates is extremely widespread amongst secondary metabolites and enables many toxic substances, such as lipophilic terpenes which can disrupt membranes, to be sequestered in a more water-soluble form in the vacuole. Plant growth substances, for example auxins (118) and phenolic compounds, including flavonoids (61), are very commonly glycosylated or otherwise conjugated. Where such compounds are stored in vacuoles, the mechanism of deposition may be variable. For example, glucosyl and malonyl transferases associated with pterocarpan biosynthesis appear tonoplast-associated (119, 120), whereas 1-sinapoyl: L-malate sinapoyltransferase, responsible for producing sinapoyl malate esters, appears to be located exclusively in the vacuole (121). There is evidence too for active transport of anthocyanins and flavonoids into vacuoles, mediated by specific carrier proteins (122). Alkaloids appear to be transported actively into vacuoles (123, 124), where they are stored as salts of commonly-occurring organic acids, or in association with phenolic compounds (125).

The potential for complexity in secondary metabolic compartmentation is amply illustrated by the pathway to the indole alkaloid, vindoline, one of the precursors of the important anti-cancer bis-indole alkaloids, vincristine and vinblastine, which are produced by the Madagascar periwinkle, *Catharanthus roseus* (summarised in 126). Four cellular compartments are involved and, in common with other indole alkaloids derived from strictosidine (see Figure 4), the convergence of two pathways, one providing tryptamine and the other the terpenoid (secoiridoid) glucoside, secologanin. Three of the early steps in the biosynthesis, catalysed by tryptophan decarboxylase, geraniol 10-hydroxylase (terpenoid pathway), and strictosidine synthase, occur in the cytosol, the endoplasmic reticulum and the vacuoles, respectively. The six final reactions leading sequentially from tabersonine to vindoline are a microsomal cytochrome P450 monooxygenase, a cytosolic *O*-methyltransferase, an unidentified hydratase, a plastid-thylakoidal *N*-methyltransferase, a cytosolic 2-oxoglutarate-linked dioxygenase (hydroxylase) and a cytosolic acetyltransferase. Furthermore, there are several enzymic steps linking 10-hydroxygeraniol with secologanin and strictosidine with tabersonine. The mechanisms responsible for ordering the transfer of metabolites from organelle to organelle and enzyme to enzyme within this lengthy biosynthetic sequence remain to be unravelled.

Finally, a very important aspect of subcellular localisation is to keep substrates and enzymes apart until they are required to meet. The best example of this is the hydrolysis of vacuolar cyanogenic glycosides (127) and glucosinolates (128) when tissues are damaged, giving rise to toxic and antifeedant nitriles and isothiocyanates. In these cases, compartmentation serves to separate substrate and enzyme until brought together by herbivore activity; this represents a "primed" defensive strategy which is an alternative to synthesis *de novo* in response to attack.

2.4.3. Isoenzymes and gene families

Isoenzymes are distinguishable populations of enzymes able to perform the same catalytic function. They - or their subunits - appear to be encoded by distinct genes and may be localised in distinct subcellular organelles. The question of whether individual isoenzymes are associated with the formation of specific classes of metabolite has been the subject of continual debate. This issue can be approached - though not necessarily definitively answered - when secondary metabolites are specifically induced in response to a stimulus, for example phytoalexins in response to pathogen-derived elicitor compounds.

The demand for synthesis of increased amounts of lignin and phenylpropanoid-derived defence compounds as a result of wounding or pathogen attack places significant additional demands on the shikimate pathway to provide phenylalanine. This is reflected in the induction of expression of genes of the shikimate pathway (129-131). In cultured tomato cells, Görlach and coworkers (132) investigated the effects of elicitor treatment on the transcription of six genes of the prechorismate section of the shikimate pathway. The levels of some mRNA transcripts, including also the level of PAL transcripts, increased 10-20-fold within 6 h of elicitor treatment. In the cases of both 3-deoxy-D-*arabino*-heptulosonate 7-phosphate synthase (DAHP synthase) and chorismate synthase (CS), where two genes encoded isoenzymes, the transcript level of only one gene (encoding DAHP synthase 2 and CS1, respectively) was increased by elicitor treatment. Similar results were obtained when whole tomato plants were analysed following infection by *Phytophthora infestans*. Genes encoding particular isoenzymes of a primary metabolic pathway can therefore be induced in concert with the induction of a secondary metabolic pathway. Comparable results have been obtained for the induction of anthranilate synthase α-subunit genes in *Arabidopsis*, where the gene encoding ASα1 is induced in concert with the elicitation of indole phytoalexins (133).

Functional differentiation between isoenzymes of primary metabolism in relation to secondary pathways is also likely to occur in other cases, for example that of acetyl-CoA carboxylase (95). Here, the plastidial isoform may be involved primarily in supplying malonyl-CoA for *de novo* fatty acid biosynthesis, whereas the cytosolic form may supply malonyl-CoA for other pathways, including flavonoid and isoflavonoid biosynthesis. In alfalfa (*Medicago sativa*) cells, elicitor treatment leading to isoflavonoid phytoalexin accumulation causes a *ca.* threefold rise in the activity of acetyl-CoA carboxylase (134). A gene encoding a cytosolic acetyl-CoA carboxylase has been isolated and cloned from alfalfa (135). A number of genes for the enzyme of mevalonic acid biosynthesis, 3-hydroxy-3-methylglutaryl-CoA reductase (HMG-CoA reductase), have been isolated and their expression studied in various tissues and organs and at different stages of plant development (113). In potato tubers, specific HMG-CoA reductase genes are induced in response to pathogen attack and wounding.

Phenylalanine ammonia lyase (PAL), the initial enzyme of the phenylpropanoid pathway, is a tetrameric enzyme; its subunits are encoded by multigene families, usually comprising between two and six members, although there are *ca.* 40 PAL genes in potato and apparently only one in Loblolly pine (136). There is now abundant evidence from studies in French bean (137-139), parsley (140) and *Arabidopsis* (136, 141) that individual members of the gene family are expressed in a distinct manner with respect to organ specificity, plant development and environmental cues, including particularly uv light, wounding and pathogens. Promoter sequences associated with environmentally-induced or organ-specific PAL gene expression have been identified (discussed further in Section 3.3.1.). These conclusions have been reached using several experimental approaches. A classical approach has been the study of the pattern of expression in transgenic plants (for example tobacco or *Arabidopsis*) of reporter-gene fusions, generated by placing the β-glucuronidase (*gus*) reporter gene under the control of promoter sequences of individual PAL genes; expression of *gus*, shown by the appearance of a blue colour in intact tissue when the β-glucuronidase substrate is provided, indicates that the promoter is active (138, 139). The effects of light, wounding, pathogen-derived elicitor compounds etc. can then be specifically investigated. Another has been the use of RNAse protection analysis, employing specific probes to identify the production of those transcripts corresponding to the respective (iso)enzymes, since the probes afford specific protection to RNAse digestion (137). A third approach has been to use isoenzyme-specific antisense constructs to down-regulate the expression of specific PAL genes, examining the resultant effects on the levels of individual isoenzymes using isoenzyme-specific antibodies (137).

The biochemical significance of individual PAL isoforms remains unclear, more particularly because (unlike enzymes of the shikimate pathway) there is no evidence that they are located in different subcellular compartments. For several years, the kinetic behaviour of PAL was thought to exhibit negative cooperativity, but this was eventually shown to result from the interaction under assay conditions of isoforms which, separated and assayed individually, exhibited normal Michaelis-Menten behaviour (142, 143). The expression in *Escherichia coli* of parsley cDNA clones encoding isoenzymes of PAL demonstrated that these isoenzymes had indistingiuishable kinetic properties (144).

Chalcone synthase is also encoded by a multigene family, at least in some plants, such as French bean (*Phaseolus vulgaris*) and *Petunia*, though in parsley (*Petroselinum crispum*) it is encoded by a single gene, there being two alleles differing in a promoter-insert. This is a particularly abundant enzyme in many tissues which respond to uv light or to pathogen-derived elicitor substances by synthesising flavonoids or isoflavonoids, this abundance apparently reflecting catalytic inefficiency (62). In spinach, two forms of the enzyme were purified which differed in molecular molecular weight but which showed similar kinetic properties. Both were cytosolic enzymes located mainly in the lower and upper epidermis of the leaf and both were light-inducible in suspension culture (145, 146). The expression of chalcone synthase genes in relation to organ specificity, plant development and environmental factors has been intensively studied, particularly in parsley, French bean, alfalfa (*Medicago sativa*) and *Petunia* (see also Section 3.3.1.). In *Petunia*, the expression of around ten members of the chalcone synthase gene family was analysed; only two members, designated *CHS-A* and *CHS-J*, were normally expressed to any appreciable extent, though the expression of two further genes (*CHS-B* and *CHS-G*) was uv-inducible in young seedlings (144). In contrast to other systems, none was inducible by pathogen-derived elicitor substances. In alfalfa, which synthesises isoflavonoid phytoalexins (principally the pterocarpan, medicarpin) in response to elicitors or to infection, the expression of at least five different chalcone synthase genes could be demonstrated (134, 148).

Another sizeable multigene family, consisting of 10-14 members, encodes tyrosine decarboxylase in opium poppy (*Papaver somniferum*) (149). This enzyme catalyses an early step in the biosynthesis of benzyl isoquinoline alkaloids, including morphine, and in undifferentiated cell suspension cultures of *P. somniferum* is induced in response to elicitors which promote the formation of antibiotic benzyl isoquinoline alkaloids, such as sanguinarine (150). The various members of the multigene family showed differential, tissue-related expression.

In a number of cases it is clear that multigene families lead to populations of isoenzymes with very similar, or identical, biochemical properties and that the essential distinction between them relates to their induction, or their tissue or organ distribution. In other words, understanding isoenzymes becomes an issue of understanding transcriptional control and its specificity. (This is also discussed in Section 3.3.1.). Beyond this it is likely, though there are few demonstrated examples, that the isoenzyme pattern may relate directly to the chemistry of pathway end-products. One example may be 4-coumarate: CoA ligase (4CL); in poplar, three isoforms of this enzyme were demonstrated which differed in their specificity for 4-hydroxycinnamoyl-CoA substrates, and the ratio and distribution of these isoforms could be correlated with the composition of extracted lignin (151).

2.4.4. Tissue and organ localisation and transport

The occurrence of many secondary products is closely associated with organ differentiation. Examples are obvious, including the glands of the female flowers (cones) of hops (*Humulus lupulus*), responsible for the production of α-acids (14), the laticifers associated with the production and translocation of long-chain isoprenoids such as rubber (152, 153), the glandular trichomes of leaves of the Lamiaceae, such as *Mentha* spp., which produce the monoterpenes and essential oils characteristic of these species (113) and, perhaps most familiar of all, the widespread and highly-visible associations of anthocyanins and other pigments, such as carotenoids, with defined parts of flowers and fruits. Conversely, the absence or lack of development of the characteristic organs may prevent secondary product biosynthesis. This is consistent with many observations with plant cell cultures, which frequently - though by no means invariably - fail to biosynthesise in substantial amounts the secondary products they were intended to produce (20). Shoot or root organ cultures, for example those established by genetic transformation with *Agrobacterium* species, are differentiated and in general produce the compounds characteristic of normal plant roots or shoots (21-24), but in accordance with this principle appear rapidly to lose the capacity for organ-associated secondary-product formation once tissue differentiation begins to be lost. Robins and coworkers (154) observed a decline in the capacity of *Datura* transformed root cultures to biosynthesise tropane alkaloids shortly after treatment of the cultures with growth hormones; the first detectable event (within 24 h) was a rapid decline in the activity of putrescine *N*-methyltransferase, the first enzyme of the pathway leading specifically from putrescine to the tropane alkaloids.

Secondary products are variable in the degree to which they are translocated and this reflects their physiological functions. Phytoalexins, produced in response to localised pathogen attack, may accumulate specifically adjacent to the site of infection, whilst flavonoids induced in response to uv light may be localised to upper epidermal cells. The co-localisation of gene transcripts, enzyme proteins and furanocoumarin and flavonoid glycoside end-products induced as a result of these stimuli has been systematically demonstrated in parsley cell suspension cultures and seedlings by Hahlbrock and coworkers, using *in situ* nucleic-acid hybridisation, immunocytochemical and microspectrophotometric techniques (155-158). These studies have shown that transcripts, proteins and end-products are all produced in the same cells.

In other cases, long-distance transport of secondary products occurs. For example, it has long been known that nicotine and the tropane alkaloids are biosynthesised in the roots of the solanaceous plants in which they are found and transported to the aerial parts, where they accumulate (8). Reciprocal grafting experiments showed that a tomato plant grafted onto a *Nicotiana* or *Atropa* rootstock accumulated, respectively, nicotine or hyoscyamine in the aerial parts; conversely, a *Nicotiana* or *Atropa* plant grafted onto a tomato rootstock did not accumulate these alkaloids. The site of biosynthesis of scopolamine (see Figure 2), the 6,7-epoxide of hyoscyamine, has since been investigated by modern techniques in root cultures of *Hyoscyamus*. Hyoscyamine 6β-hydroxylase, the 2-oxoglutarate-dependent dioxygenase responsible for both the 6β-hydroxylation and subsequent epoxidation of hyoscyamine to scopolamine, has been localised immunohistochemically to the cells of the pericycle, and RNA gel blot analysis has demonstrated that the transcript of the gene is similarly localised (159). Transcript was not found in the aerial parts of the plant, or in cell cultures. Long-distance transport of alkaloids is probably a widespread phenomenon, consistent with the role of these compounds as antifeedants and toxins. There is evidence that the pyrrolizidine alkaloids of the Asteraceae are transported as their *N*-oxides (160). In *Ginkgo biloba*, roots appear to be the site of synthesis of ginkgolides and bilobalides, with transport then occurring to the aerial parts (161).

Finally, secondary products of many classes are very commonly liberated beyond the living plant altogether (162). Examples include flower scents, frequently terpenoids or phenolic substances (see e.g. 163); the flavonoid nodulation-inducers important in the interaction of legume species with the nitrogen-fixing bacterium, *Rhizobium* (164); jasmonate esters, which may be both flower scents and gaseous signalling compounds (165); and the chemically-simple plant hormone, ethylene (166). The possibility that plants might "communicate" via gaseous signals has recently been discussed (167).

3. Regulation of Secondary Metabolism

3.1. Synthesis and Degradation

The level of any metabolite represents the resultant, over time, of synthesis and further metabolism, including catabolism. It indicates nothing of rates of synthesis or degradation; and many metabolically-active compounds, for example, CoA thioesters and the methyl donor, *S*-adenosylmethionine, are present only in low concentrations.

The catabolism of individual plant secondary products *in vivo* may be very appreciable, but in general this remains experimentally a neglected area. For example, fruit ripening can lead to a significant loss of some secondary products, just as it may be associated with the synthesis or biotransformation of others (e.g. flavours or pigments); tomatoes, for example, lose the toxic glycoalkaloid, tomatine, as they ripen (168) and in chilli pepper fruits (*Capsicum annuum*), a decline in the content of the pungent principle capsaicin has also been reported (169). The degradation of the tropane alkaloid, atropine, in *Duboisia* seedlings has been investigated by Kitamura and coworkers (170). The half-life of carboxyl-labelled atropine supplied as a tracer was about three weeks and a variety of transformation products were detected, including unidentified esters of tropic acid. There is evidence that such tropic acid esters might be recycled to re-form tropane alkaloids (171). In contrast, however, Baldwin and Ohnneiss have concluded that in *Nicotiana* the pyridine alkaloid, nicotine, is an essentially stable end-product and is not remobilised even in cases of severe nitrogen shortage (172).

Finally, degradation of a non-specific nature may occur in cell cultures supplied exogenously with secondary metabolites. Cultures of hop cells (*Humulus lupulus*) fed with alpha-acids very rapidly degraded them by peroxidative action (173). Even if alpha-acids had been produced by the cultures, they would never have been detected in the culture medium! These non-specific peroxidative reactions have however been turned to advantage in specific instances to catalyse biotransformations (this volume, Chapter 7).

3.2. Metabolic Control

The aim of Metabolic Control Analysis (MCA) is to describe mathematically the behaviour of metabolic pathways and to establish those factors, in particular enzyme activities and substrates, which are most critical in determining metabolic flux.

Alternative models have been described and recently reviewed (174-176) and their specific application to plant metabolism has been discussed (177). MCA is important both in its own right and because of its bearing on the design of experimental strategies for the directed and effective modification of secondary metabolism by genetic engineering.

In brief, a metabolic System may be described: i) in terms of the fractional changes in either flux or metabolite concentration which occur in response to fractional changes in enzyme activity (referred to respectively as Flux Control and Concentration Control Coefficients); and ii) in terms of the fractional changes of local reaction rates in response to fractional changes in substrates or other effectors, with all other variables fixed (referred to as Elasticity Coefficients).

Under steady-state conditions, at constant flux, the sum of the Flux Control Coefficients of all the enzymes in a System is unity and for each metabolite the sum of the Concentration Control Coefficients is zero. Flux Control Coefficients and Elasticity Coefficients are further related by a Connectivity Theorem which states that, for every enzyme that responds to a given metabolite, the summation of the products of the Flux Control Coefficients and Elasticity Coefficients is zero. A corresponding relationship can be drawn between Concentration Control Coefficients and Elasticity Coefficients.

ap Rees and Hill (177) have discussed in detail the theoretical and practical requirements which need to be met before MCA might confidently be applied. First, the metabolic pathways should be adequately described and all branches known. Secondly, metabolites must be evenly distributed. Thirdly, metabolites must be in a free form and "metabolic channelling" must not occur. Fourthly, enzymes must not show cooperative behaviour or catalyse multiple reactions.

Each of these conditions poses a major problem in many secondary metabolic pathways, where, as we have seen, an adequate description of the pathway may be unavailable, where compartmentation and channelling are suspected to occur and where enzymes may be active with several related substrates. The measurements themselves must be as reliable as possible (though MCA may be sufficiently robust to accommodate routine experimental error - 175). In the case of metabolic flux measurements, it is necessary to be aware of the possibility of non-equilibration of added labelled precursors with endogenous metabolites and, conversely, of the possibility of incorporation of label through equilibration rather than net flux. In the case of metabolite determinations, harvested tissue must be frozen promptly and the potential degradation of metabolites occurring when tissue is thawed prior to analysis must be avoided. If possible, and particularly if compounds are known to be unstable, tissue should be "spiked" with a small quantity of the pure metabolite in question. Non-aqueous fractionation (110) and rapid cell fractionation (106) techniques are available to begin to approach the issue of compartmentation of metabolites, but these have rarely been used in studies of secondary metabolism. Comparable considerations apply in the case of enzymes, where it is important to establish not only that loss, degradation or some other change in the properties of the enzyme does not occur during extraction (best achieved in a co-extraction experiment using pure or separately-isolated enzyme), but that assay conditions are thoroughly standardised so that, so far as is practicable, maximum catalytic activities are always determined.

A final condition demanded by MCA is that the fractional perturbations should indeed be fractional, i.e. they should not appreciably perturb the system from the steady state. This is perhaps particularly important where metabolic inhibitors are used, or where genetic manipulation is used as a tool to increase or decrease the activity of a target enzyme. The danger here is of non-specific or pleiotropic effects; large changes in the activity of one enzyme may cause unsuspected effects on the activities of others, for example as a result of altered gene expression. These and other considerations have been discussed by Stitt and colleagues (178, 179).

A principal benefit of MCA has been the growing recognition that metabolic pathways are not simply subject to flux limitation by individual and clearly-identifiable "rate-limiting" enzymes; rather, flux determination is the resultant of partial rate limitations by each of several enzymes. Nevertheless, Stitt (178) has made a distinction between "regulatability" - the potential for the activity of an enzyme to respond to regulatory mechanisns *in vivo* - and "regulatory capacity" - the potential for an alteration in the activity of an enzyme to cause a change in metabolic flux. An important objective of MCA is to identify enzymes belonging to the latter capacity (i.e those found to have the largest Flux Control Coefficients). It is important to recognise, however, that raising the activity individually of such an enzyme in a pathway, for example by genetic manipulation, may act to transfer its component of flux limitation to the remaining enzymes - in just the same way as the construction of a by-pass road may cause a new bottleneck in traffic flow to appear. This was clearly demonstrated in attempts to increase the biosynthesis of tryptophan in yeast; the activities of all five enzymes of the pathway needed to be increased by genetic manipulation before any increase in tryptophan synthesis occurred (180). This realisation - together with the phenomenon of cosuppression (Section 3.4.1.) - is a disincentive to those who would seek to increase the content of desirable plant secondary metabolites simply by overexpressing individual structural genes and, conversely, a corresponding encouragement to adopt other approaches, for example the construction of new pathways using foreign genes, or the ectopic expression of transcription factors to increase the expression of pathway genes *en bloc* (see Section 3.3.2.).

Three examples illustrate both the insights into metabolic regulation which can be gained from the generation and study of transgenic plants and, at the same time, the need for care in interpretation. First, Bate and coworkers (181) and Howles and coworkers (182), in successive experiments, studied the correlations between the activity levels of phenylalanine ammonia lyase (PAL) and the accumulation of phenylpropanoid-derived end-products in transgenic tobacco (*Nicotiana tabacum*) plants transformed with exogenous *PAL* genes. The plants investigated included both plants with severely-diminished levels of PAL activity (0.2% of wild-type - see 181), as a result of the effects of cosuppression (see Section 3.4.1.) and individuals showing, in particular tissues, PAL activities *ca.* 2-5-fold greater than wild-type levels. The major soluble end-product of phenylpropanoid metabolism in tobacco is chlorogenic acid, a conjugate of caffeic acid with quinic acid, and a linear correlation was observed over a range of PAL activities between PAL activity and chlorogenic acid content. In contrast, such a strong correlation was not observed in the case of the flavonoid, rutin. These results were interpreted as demonstrating that PAL was the major overall determinant of metabolic flux into the phenylpropanoid pathway, but that this was severely modulated by additional limiting steps in the case of individual end-products, such as rutin. Whilst this interpretation is almost certainly correct, it remains possible that the correlation between PAL activity and chlorogenic acid content masks a more complex situation in which chlorogenic acid formation is similarly correlated with the activities of other, specific enzymes and, further, that the activities of these enzymes might not be independent of PAL activity.

In the second example, Burtin and Michael (183) investigated the effects of varying expression of the *adc* gene, encoding arginine decarboxylase, on the formation of polyamines and pyrrolidine alkaloids in *Nicotiana*. In this case, it was clear that the activity of arginine decarboxylase did not make a major contribution to the control of flux to either group of compounds; upon investigating the effects of over-expression and antisense constructs of the *adc* gene, the sole demonstrable effect was an increase in the pool size of agmatine, the amine decarboxylation product of arginine: there was no effect on the levels of polyamines or of alkaloids. They concluded that other activities were (or had become) rate-limiting, essentially buffering completely the effects of over-expression of *adc*. In the terminology of Stitt

(178), arginine decarboxylase in these studies showed no regulatory capacity.

The final example is also drawn from the nicotine biosynthetic pathway. As indicated earlier (Section 2.3.2.), the gene for putrescine *N*-methyltransferase (PMT) was isolated by a subtractive hybridisation approach using high- and low-nicotine-producing strains of *Nicotiana tabacum* (74). This could be taken to indicate a key role for PMT in flux regulation of the nicotine biosynthetic pathway, consistent with the rapid loss of PMT activity during the hormonally-induced dedifferentiation of transformed root cultures (see 154). Instead, the correct deduction is that although PMT is specifically associated with nicotine production - it catalyses an essential reaction in the pathway - it need not be the sole gene upregulated in the high-nicotine-producing line. This interpretation is supported by other experiments showing that overexpression of the gene for PMT causes no increase in nicotine formation (184).

Against this background and relatively primitive attempts towards genetic manipulation, our knowledge of the regulatory properties of secondary-metabolic enzymes still remains very fragmentary. It is likely, though not yet widely demonstrated, that transport and sequestration mechanisms (see Section 2.4.2.) are important in the overall regulation of secondary-metabolic pathways. It is also clear, in those cases where studies have been carried out, that secondary pathways have a substantial influence on primary metabolism. This is in line with the theory of MCA, which *a priori* does not distinguish between primary and secondary pathways. Examples are the increased transcription of genes encoding shikimate pathway enzymes in response to uv light or elicitors, already discussed in Section 2.4.3. and corresponding effects on genes encoding glucose-6-phosphate dehydrogenase and 6-phosphogluconate dehydrogenase, enzymes of the pentose phosphate pathway, in elicitor-induced alfalfa (134); in the latter case, increased activity of the pentose phosphate pathway is thought to be required to provide NADPH required for pterocarpan phytoalexin biosynthesis. Finally, in general little is yet known of the mechanisms by which intermediates or end-products of plant secondary metabolic pathways might exert feedback regulation, or influence gene transcript levels; perhaps the best-documented instances at present occur in the phenylpropanoid pathway in *Phaseolus vulgaris* suspension cultures, where the product of the phenylalanine ammonia-lyase (PAL) reaction, cinnamic acid, has been shown to cause inactivation (though not degradation) of the PAL enzyme (185) and to repress the accumulation of PAL and chalcone synthase transcripts (186, 187), whereas *p*-coumaric acid, the product of the succeeding enzyme (cinnamic acid 4-hydroxylase) is a "feed forward" activator of the chalcone synthase gene, *chs15* (188).

3.3. Transcriptional Control

Frequently, a strong correlation is observed between the onset and level of transcription of a secondary-metabolic gene and the appearance and level of its encoded enzyme activity. On the other hand, some of our understanding of transcriptional control has come from the study of inducible secondary-product pathways, and these may not necessarily prove to be the most appropriate models for understanding the fine-tuning of pathways which are constitutively expressed. The expression of pathway genes is increasingly recognised to be modulated additionally by post-transcriptional and post-translational events.

The chain of events leading to transcription can be considered, perhaps arbitrarily, to begin with the reception of a signal, generated either externally or internally, and to end with a response at the metabolic level. In the case of constitutively-expressed pathways, the nature of the "signal(s)" is likely to be non-obvious. Since transcription is often relatively easily measured, it is a good starting-point for analysis of pathway gene expression; some aspects of signal transduction mechanisms which may lead to transcription are addressed in outline in Section 3.6.

3.3.1. Cis-elements and trans-acting factors

A central tenet of transcriptional control is the coordinated regulation of expression of groups or banks of pathway genes in response to a smaller number of regulatory genes (see 189). These regulatory or "master" genes encode transcription factors (189, 190), which are DNA-binding proteins. (Their expression is in turn under transcriptional control; and a regulatory hierarchy can therefore be envisaged.) Transcription factors bind to specific and characteristic regions (motifs) of DNA sequence associated with the coding sequences of their target genes and in general are thought to act by modulating DNA conformation, thereby activating the basal transcriptional machinery. They are referred to as *trans*-acting factors, since they are encoded by genes which are spatially remote from the target gene(s) upon which they act. The motifs of DNA to which they bind are referred to as *cis*-elements, since in contrast they are directly associated with the coding sequences of the target genes. Genes encoding transcription factors have been detected by the genetic analysis of mutants or by transposon-tagging (see above) and using sequence-homology-based approaches (Southern blotting, or PCR). In contrast to mutations in pathway genes, mutations in regulatory genes characteristically cause changes in the expression of groups or banks of genes, i.e. in the expression of entire pathways or parts of pathways.

Transcription factors can be isolated by exploiting their characteristic property of specific binding to *cis*-elements. Weisshaar and colleagues (191) were able to isolate three cDNAs encoding transcription factors which bound to Box II, a *cis*-element (one from a total of four) found in parsley to be associated with light-responsiveness of the chalcone synthase promoter. Yu and co-workers (192) isolated two transcription factors, KAP-1 and KAP-2, by a combination of ion-exchange chromatography and DNA-affinity chromatography from French bean cell-suspension cultures (*Phaseolus vulgaris*). Each of these bound specifically to the H-box, a *cis*-element occurring three times within the chalcone synthase *chs15* promoter.

As implied above, *cis*-elements are identified by being shown to be required for - or implicated in - gene expression in response to specific stimuli or effectors, such as uv light, osmotic stress, pathogen-based elicitor substances, plant growth substances etc. They are also required for organ- or tissue-specific expression. Several experimental approaches, many now routine, are available. "*In vivo* DNA footprinting" (described in 193, 194) was used to define the sequences necessary for light-activation of the chalcone synthase promoter in parsley. Gel-retardation or mobility-shift determinations exploit the observation that DNA fragments complexed with protein show altered electrophoretic mobility; they were used, for example, in the isolation of a flower-specific transcription factor of *Antirrhinum* (Myb 305) which activated transcription of phenylpropanoid-pathway genes (195). Disruption or deletion of the suspected *cis*-element will attenuate or abolish its effect; such experiments are essential for confirmation (see for example (196, 197) for disruption / deletion analysis of the PAL promoter and G-box in *Arabidopsis*) and (188, 198, 199) for other examples. Conversely, the *cis*-element can be incorporated into a reporter-gene construct - commonly incorporating *gus*, the gene encoding β-glucuronidase - and its effects on reporter-gene expression (including transiently in protoplasts) in response to environmental or chemical stimuli can be determined. Tissue- or organ-specificity of gene expression may be studied. A large number and range of experiments of this type have been reported (see e.g. 138, 139, 188, 194, 198, 199, 200). Finally, these methods can be complemented by other techniques to examine in detail the transcription of target genes and the levels of mRNA, protein and enzyme activity. For example, van Tunen and colleagues (201) used RNAse protection assays in combination with primer-extension studies and sequence data to investigate promoter usage and flower-tissue-specific expression of genes encoding chalcone-flavanone isomerase in *Petunia hybrida*. Quattrochio and co-workers (202) used a combination of nuclear transcription "run-on" assays (which can provide an estimate of transcription rates of target genes, as distinct from mRNA levels), RNA

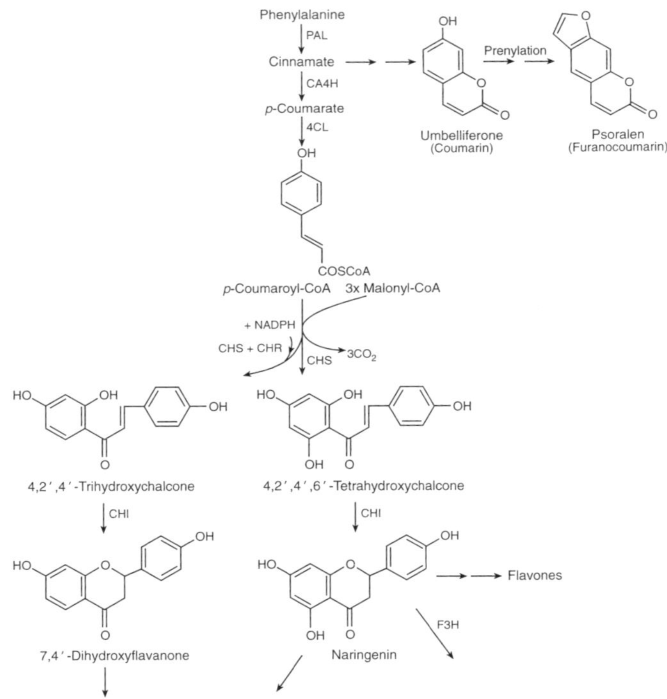

Figure 10. Some principal phenylpropanoid branch-pathways and end-products**.** Glyceollin I, an isoflavonoid pterocarpan phytoalexin which is characteristic of soybean, is a prenylated member of the class of pterocarpan phytoalexins of the Leguminosae which also includes phaseollin (French bean), medicarpin (alfalfa) and pisatin (pea). The formation of a range of flavone and flavonol glycosides in parsley is light-dependent and uv-stimulated. Furanocoumarins in parsley are phytoalexins; their formation can be induced in cell-suspension culture by fungal elicitors. PAL, phenylalanine ammonia lyase; C4H, cinnamate 4-hydroxylase; 4CL, 4-coumerate:CoA ligase; CHS, chalcone synthase; CHR, chalcone reductase; CHI, chalcone isomerase; F3H, flavanone 3-hydroxylase; DFR, dihydroflavonol reductase.

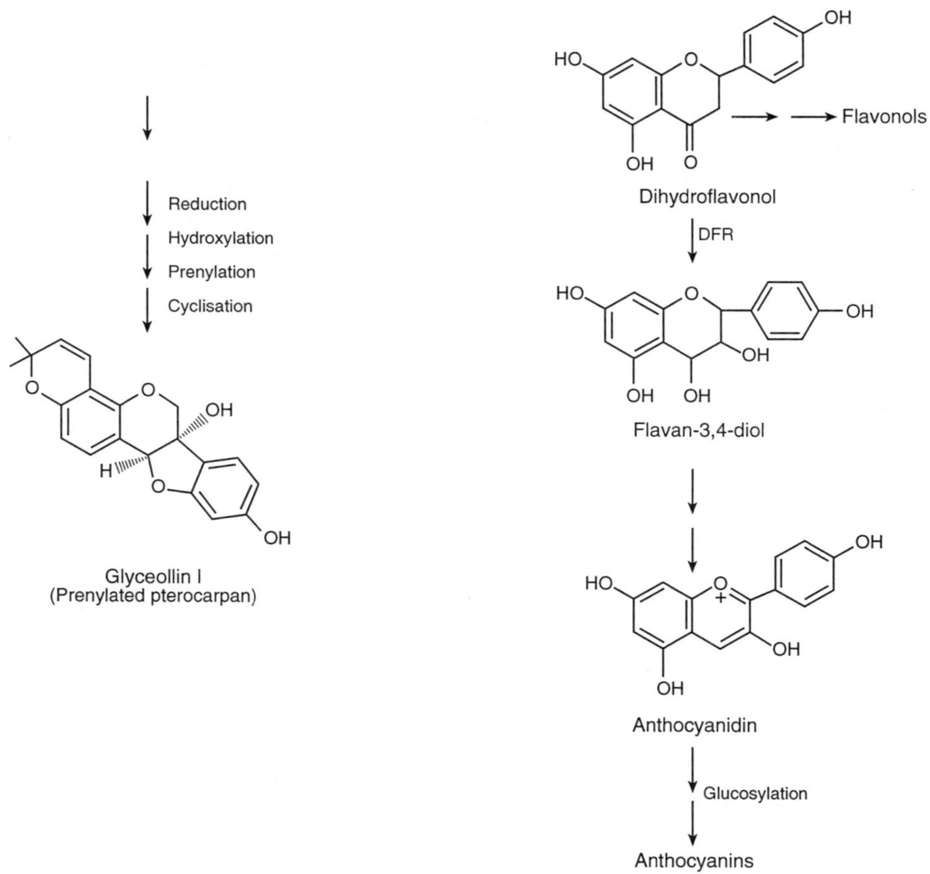

Reduction
Hydroxylation
Prenylation
Cyclisation
OH
H
OH
Glyceollin I
(Prenylated pterocarpan)
HO
OH
OH
O
Flavonols
Dihydroflavonol
DFR
HO
OH
OH
OH
OH
Flavan-3,4-diol
HO
OH
OH
OH
Anthocyanidin
Glucosylation
Anthocyanins

gel blot analyses and transient expression assays to investigate the control of anthocyanin pigmentation in *Petunia* flowers by the regulatory genes *an1*, *an2*, *an4* and *an11* and the effects of the heterologously-expressed maize regulatory genes, *Lc* and *C1*.

Using such approaches, an understanding of transcriptional control is emerging for an increasing range of secondary-metabolic pathways. The most thoroughly-studied are branches of the phenylpropanoid pathway, including those strongly implicated in plant defence: the uv-induction of flavonoid glycoside formation in parsley cell-suspension cultures and seedlings (62, 203); the pathogen-elicitor-induced biosynthesis of furanocoumarin phytoalexins, in the same system; the elicitor-induced biosynthesis of isoflavonoid phytoalexins in alfalfa (119, 134) and French bean (192, 204); and the formation of anthocyanin pigments in maize, *Petunia* and *Antirrhinum* (62, 64, 66, 67, 205, 206). Biosynthetic relationships between these classes of phenylpropanoid-derived metabolites are summarised in Figure 10. The reader is referred to the accounts cited above for comprehensive details.

Amongst the most intensively-studied *cis*-elements are the so-called G- and H-boxes, which are present in the promoters of a number of genes encoding enzymes of phenylpropanoid metabolism, including PAL and chalcone synthase (summarised in 188, 192, 195). G-boxes comprise a characteristic nucleotide sequence, CACGTG, and have been found to be important in the responses of a range of promoters to stimuli as diverse as light, abscisic acid and anaerobiosis. H-boxes have been characterised by their possession of the sequence, CCTACC, (see 188, 192) but a closely-related consensus sequence is more widely distributed amongst the promoters of genes of phenylpropanoid metabolism (summarised in 207) and is involved in responses to uv irradiation and to elicitor substances and in the expression of phenylpropanoid-pathway gene expression in flowers (195, 198, 199).

An important question is the extent to which there is a one-to-one correspondence between the presence of any individual *cis*-element and the occurrence of a particular transcriptional response, for example in relation to elicitor treatment or uv-irradiation. Analysis of the parsley system (reviewed in detail by Hahlbrock and colleagues in (203) suggested that genes encoding metabolically closely-related enzymes (for example PAL and 4CL) were likely to share common *cis*-elements, but that this relationship did not necessarily hold for genes encoding less-functionally-related enzymes, even if they were transcribed in response to a common stimulus. It was not possible, for example, to identify a *cis*-element universally conserved amongst genes transcribed in response to a fungal elicitor and such a correspondence model is clearly over-simplistic (see 192, 195). This is likely to reflect in part the additional possibility of direct physical interaction between transcription factors themselves (see 192, 208).

3.3.2. Transcription factors and their applications

The essential elements of a transcription factor are a specific-DNA-binding domain and a transcriptional-activating domain, which are often highly-conserved. Transcription factors in plants bear substantial homology to transcription factors in other organisms. Well-described families of transcription factors, from studies in other eukaryotes, include the so-called basic leucine zipper (bZIP) proteins, the Myb- and Myc-related transcription factors, the homeodomain proteins and the MADS-box proteins (see 189, 190).

The first plant transcription factors to be described were the R (Myc-related) and C1 (Myb-related) proteins of maize. Together, these act in controlling the expression of anthocyanin biosynthesis - and hence pigmentation - in maize aleurone, and probably activate transcription of all the structural genes of the pathway. From genetic analysis and sequence determination, it is known that both the R and C1 factors comprise families of closely-related and functionally-equivalent proteins, but in individual alleles there can be substantial differences in promoter region *cis*-elements, so that the expression of individual transcription-factor genes (for example, in response to light) may be markedly different (reviewed in 205).

Transcriptional control of anthocyanin biosynthesis has been intensively examined in the flowers of *Antirrhinum* and *Petunia* (66, 67, 205). Characterisation of these experimental systems has proceeded largely independently and gives opportunity for the structures and regulatory functions of analogous transcription factors to be compared between plant species. For example, both *Antirrhinum* and *Petunia* flowers contain Myc-related transcription factors similar to the maize R proteins: in *Antirrhinum*, the *Delila* gene product and in *Petunia*, the products of two genes, *An1* and *Jaf13*. There are differences between maize, *Antirrhinum* and *Petunia* in the extent to which these genes regulate anthocyanin-pathway genes; for instance, unlike the maize R factor, the *Delila* gene product of *Antirrhinum* appears to be involved in regulating only the later stages of anthocyanin biosynthesis (209). A Myb-related factor (Myb305), showing binding to an H-box-like element, has been isolated from *Antirrhinum;* this activates PAL-gene transcription in *Antirrhinum* (and in tobacco), and presumably acts with the *Delila* gene product in achieving expression of the entire pathway (195).

Both the maize *R* homologue, *Lc*, and *Delila* have similar effects when expressed in tobacco (210, 211), suggesting that selectivity towards the target pathway genes between maize and *Antirrhinum* may be explicable by differences in the target gene promoters in the two plants, rather than by differences between the transcription factors. Similar conclusions have been reached for the products of the maize *C1* and *Petunia An2* genes, as a result of experiments with reporter-gene constructs delivered ballistically to target tissues, the tissues being examined subsequently for the development of red spots denoting anthocyanin formation (202, 205).

It is possible that where functionally-homologous transcription factors are present in the same tissue, the relationship between them may provide a mechanism for sensitive transcriptional control. Martin and coworkers (212) have examined the competitive interaction between two closely-homologous myb-related transcription factors, from *Antirrhinum* flowers, that bind to the same DNA motifs. The genes encoding Myb305 and Myb340 were each expressed in yeast and their effectiveness in activating a series of anthocyanin-pathway-gene reporter constructs was compared. (Expression in yeast was chosen for these studies on account of the absence of background interferences.) Although Myb340 was found to be a stronger activator of reporter-gene expression, and the binding constants of the two proteins to a target region of a chalcone isomerase promoter were similar, less Myb340 was found to be bound to the target sequence when the two proteins were synthesised in yeast or in *E.coli*. Evidence was presented that this was due to selective phosphorylation of Myb340, possibly at a serine residue known to be replaced by an alanine in Myb305. Both transcription factors were shown by *in situ* hybridisation to be expressed in the same cells of *Antirrhinum* flowers *in vivo*. Myb340 is expressed at an earlier stage of flower development than Myb305, during a phase of rapid increase of flavonoid-pathway gene expression, and it was suggested that this "acceleration phase" could be mediated by the more powerful Myb340 activator, which might then subsequently give way to Myb305 to provide stable, sustained expression.

Mechanisms of transcriptional control probably also incorporate direct physical interactions between transcription factors. For example, Jackson *et al* (208) proposed that the observed positive and negative effects of the *Del* gene in the transcription of anthocyanin-biosynthetic genes in *Antirrhinum* flowers might be explained on the basis of complex formation between the Del protein and another transcription factor (denoted as a "mesophyll-specific factor"); whereas the formation of this complex would activate the transcription of some biosynthetic genes, it could repress the transcription of others (in this instance, *chs*, encoding chalcone synthase), either by lowering the concentration of free "mesophyll-specific factor" or by preventing its incorporation in an alternative transcription-activating complex.

Transcription factors offer very significant opportunities for the manipulation of plant metabolism (189). First, transcription factors often regulate *banks* of pathway structural genes, as distinct from individual genes. This addresses a key problem identified from Metabolic Control Analysis (Section 3.2.): the overall limitation of metabolic flux through a pathway by partial limitation occurring at each of a number of enzymic steps. In principle, complete pathways may therefore be upregulated simply by overexpression of a single transcription factor. This has been demonstrated by expressing the maize gene, *Lc*, or the *Antirrhinum* gene, *Delila*, (see above) in tobacco or tomato; a 23-fold increase in anthocyanin levels in tomato leaves was achieved by expressing *Delila* under the control of a cauliflower mosaic virus (CaMV) 35S promoter, which gives high-level, constitutive gene expression (210). Secondly, because transcription factors contain discrete DNA-binding and transcriptional-activating domains, it is possible in principle to engineer transcriptional inhibitors by the removal of activating domains, or to perform combinatorial experiments where DNA-binding and transcriptional-activation domains of differing origins can be brought together in transcription factors with novel regulatory characteristics. Even "unengineered" transcription factors, when heterologously overexpressed, can cause severe down-regulation of a pathway, as shown by the inhibition of phenolic-acid and lignin biosynthesis caused by the overexpression in tobacco of two Myb transcription factors (Myb308 and Myb305) from *Antirrhinum* (213); in this case, the inhibition was ascribed to competition of the weak *Antirrhinum* transcriptional activators with the stronger endogenous activators of tobacco. Finally - perhaps most importantly - transcription factors encoded by homeobox genes play key roles in plant morphogenesis: they comprise "master genes" which determine, for example, the timing and induction of floral meristems and the differentiation of flower organs (189, 214, 215, 216). The use of such genes to intervene at the organ-differentiation level might obviate completely the need for manipulation at a subsidiary level. For example, Larkin and coworkers (217) demonstrated that expression of the maize *Lc* gene stimulated the ectopic production of trichomes in *Arabidopsis* flowers. As discussed above (Section 2.4.4.), trichomes are the sites of synthesis and accumulation of important essential oils in members of the Lamiaceae, such as *Mentha* (mint).

The relationships between transcription factors and the signal transduction pathways with which they are associated are only gradually beginning to emerge. A notable example is provided by the discovery and analysis of two transcription factors, KAP-1 and KAP-2, which recognise the H-box of the chalcone synthase *chs15* promoter in French bean (192). These were shown to respond to elicitation of the bean cell suspension culture, achieved using glutathione, by a rapid increase in their specific activities in the nuclear fraction, although their total activities within the cell were unchanged. Furthermore, both factors responded to alkaline phosphatase treatment, which caused a change in the electrophoretic mobility of the factor-H box complex. Both these findings were interpreted to suggest a role for these factors in rapid signal transduction in response to elicitor, involving a change in phosphorylation state and intracellular distribution.

3.4. Gene Silencing and Post-transcriptional Regulation

Although in gene expression studies much emphasis has been placed on transcriptional control, this as indicated earlier affords only a partial picture of the overall regulation of expression, which in addition includes gene-silencing mechanisms and control of mRNA availability through processing and degradation. These complex interactions are of fundamental importance - not least from a biotechnological viewpoint, in contributing to the stability and reproducibility (or otherwise) of the expression of introduced genes.

3.4.1. Gene silencing

As already discussed, multiple copies of closely-related genes encoding a single enzyme activity (or closely-related isoenzymes) are a general occurrence. Their expression may be determined at a transcriptional level, as a result of differences in promoter *cis*-elements. In addition, however, interactive behaviour between multiple copies of genes and/or their transcripts may result in gene inactivation. This is referred to as homology-dependent gene silencing. In principle, this may be the result of inactivation at the transcriptional or at the post-transcriptional level - i.e. by genetic modification (for example, by promoter methylation or changes in chromatin structure, or by inactivation or modification of mRNA transcripts. Potential mechanisms and the evidence available to support them have been comprehensively reviewed by Flavell (218) and by Meyer and Saedler (219) and Meyer (220). The further terms "cosuppression" and "sense-inactivation" refer to the gene silencing observed to occur as a result of the introduction of a transgene and which affects not only the exogenous transgene, but also the endogenous gene(s). The transformation of plants with "antisense" constructs (constructs consisting of a nucleotide sequence complementary to the transcribed "sense" strand of DNA) was found to be capable of decreasing the expression of corresponding endogenous genes. As controls for these experiments, transformations were also performed with "sense" constructs.

Unexpectedly, the "sense" constructs were also found to inhibit expression of the endogenous genes. This phenomenon, first observed in the case of chalcone synthase in *Petunia hybrida* and leading in this case to white flowers lacking anthocyanins (221, 222) was subsequently observed more generally and recognised as a potential tool to block the expression of metabolic-pathway genes.

In many cases, cosuppression appears to be caused post-transcriptionally, since reduced steady-state levels of mRNA of the endogenous gene can be measured, yet transcription is often unaffected. Although the mechanism remains uncertain, it is suggested that the formation of an antisense transcript of the introduced gene may be an essential step in the process (223, 224). This would then form a duplex with sense transcript which would then be subject to degradation. This proposal would be consistent with - but not a corollary of - the observation that homologous sequences which are untranslatable (and therefore unable to lead to the production of a protein to mediate the effect) are nevertheless able to cause silencing of endogenous genes. Three hypothetical mechanisms have been summarised (224): first, a direct transcription of antisense RNA (in the opposite direction to the production of sense RNA); second, production of antisense RNA in response to some threshold level of sense transcription of the introduced gene; and third, production of antisense RNA in response to the formation of an aberrant transcript. It is not known how a threshold level of transcription or the production of an aberrant transcript might trigger the formation of an antisense transcript.

An important observation is that cosuppression phenotypes can revert to other states, which may then be inherited somatically (218). For example, *Petunia* plants cosuppressed for chalcone synthase and bearing white flowers can develop branches bearing (uniformly) purple flowers, or flowers with purple sectors. This suggests modification of the expression pattern by underlying changes at the transcriptional level rather than the post-transcriptional level.

3.4.2. mRNA stability

As indicated in the preceding section, transcript stability may be a significant factor in gene silencing and is *a priori* an important consideration in the regulation of gene expression. It is known that there is considerable variation in the stability of mRNA molecules, ranging from lifetimes of minutes to days. It is assumed that unstable mRNA molecules encode proteins of which the levels must be tightly controlled. Examples of unstable transcripts in plants include those for phytochrome and for certain auxin-induced transcripts. The methods available for estimating the half-life of transcripts (reviewed in (225) include nuclear run-on transcription studies; measurements of the kinetics of mRNA decay; the use of transcriptional inhibitors, such as actinomycin D; RNA electroporation into protoplasts; pulse-chase and approach-to-steady-state methods; the use of repressible promoters, which can be turned off and the decay of transcript measured; and *in vitro* measurements using polysome-based systems. Each has disadvantages; for example, methods based on the incorporation and/or loss of radioactive isotope suffer from the low levels of incorporation which are characteristically often achieved into non-abundant transcripts.

Details of mRNA processing in plants, and the extent to which this may or may not be a concerted process, are gradually emerging (see 220, 225). It is clear from gene-splicing experiments that both polyadenylation signals and introns may potentially affect mRNA levels and overall gene expression, probably at least in part by affecting rates of mRNA degradation. In particular cases, for example the Small Auxin UpRNA (SAUR) genes of soybean and *Arabidopsis*, sequence elements have been identified which are thought to be associated with mRNA stability (226). *Trans*-acting factors are also likely to be involved - for example cytoplasmic Poly A-Binding Proteins (PABPs), known in yeast to promote deadenylation of the poly-A tails of mRNA molecules (reviewed in 225).

Several lines of evidence suggest that the stability of mRNAs is linked to translation - i.e., an mRNA not translated is likely to be rapidly degraded. The protein synthesis inhibitor, cycloheximide, has been observed to stabilise several unstable mRNAs (227). The observed light-induced stabilisation of the transcript of the pea ferredoxin gene, *Fed-1*, is decreased by a number of mutations which block translational initiation or elongation (228). The engineering of frame-shift mutations into plant genes, notably the major Kunitz trypsin inhibitor gene (*Kti3*) of soybean (229) and the phytohaemagglutinin, PHA, in French bean (230), strongly reduced the accumulation of mRNA, even though nuclear run-on experiments indicated a similar level of transcription to that of the normal genes. Further experiments in which premature nonsense codons have been introduced into genes have given similar results (see 225).

Degradation of mRNA molecules is assumed to occur by a combination of endonuclease and exonuclease action and in rare cases it has been possible to detect degradation intermediates. The specific identities of the RNAses involved has not been elucidated, but this question is potentially amenable to a genetic approach, analysing mutant plants with altered profiles of RNAses (225).

3.5. Control of Enzyme-protein Levels

A priori, an essential factor contributing to the observed level of gene expression is enzyme turnover - the balance between the synthesis of active enzyme (which could include the activation of previously-inactive enzyme) and its inactivation or degradation. This is as complex an issue as transcriptional and post-transcriptional regulation: not only are the general processes still incompletely understood, but (as with mRNA processing and turnover) each specific enzyme is an individual case. Furthermore, (see Section 2.4.1.), there is also ample likelihood of modulation of activity or turnover associated with subtle alterations of pathway organisation, channelling, and polypeptide aggregation.

An account of the general processes by which polypeptides acquire tertiary structure and catalytic activity is beyond the scope of this chapter and this topic has recently been reviewed (231). It has particular importance in transgenic experiments, in which proteins encoded by genes introduced from other organisms are expected to achieve correct tertiary structure and catalytic activity in a non-native environment. Protein folding is mediated by accessory proteins acting as "foldases" or "molecular chaperones" (232, 233). Foldases include the peptidyl-prolyl *cis/trans* isomerases (rotamases) which catalyse isomerisation of peptide bonds around proline residues, which is a slow step in protein folding, and protein disulphide isomerase, which catalyses disulphide bond formation. Chaperonins include a variety of classes of heat shock proteins (HSPs) present in a variety of subcellular organelles and thought to be associated variously with the prevention of mis-folding of newly-translated polypeptides (234). Their role appears to be energy-requiring, involving the hydrolysis of ATP. In many cases, insight into function has come from understanding of the role of corresponding proteins in other eukaryotes and in bacteria. The best-characterised plant chaperone is the endoplasmic reticulum HSP70 protein, BiP (Binding Protein). This protein does not recognise a consensus binding site, but appears to interact more tightly with non-native than with native regions of target proteins (231, 235). In organelles evolutionarily related to bacteria (chloroplasts and mitochondria), homologues of the well-characterised prokaryotic chaperonin GroEL and of its co-chaperonin, GroES, participate in the assembly of the photosynthetic CO_2-fixing enzyme, ribulose bis-phosphate carboxylase/oxygenase (Rubisco) (231-233). In general, plant secondary metabolic enzymes do not make obvious candidates for the study of these processes and for the time being it is probable that insights will continue to be gained using more abundant proteins as models.

Generalised mechanisms of protein degradation are emerging with the discovery and biochemical and genetic characterisation of a number of endoproteinases and, particularly, of cytosolic complexes, the 20S - 26S proteasomes, which in plants as in other organisms, appear to perform a catalytic function as part of the ubiquitin-dependent cycle of protein degradation (236, 237). Ubiquitin is a 76-amino acid protein which becomes multiply attached to proteins through the sequential action of ATP-requiring, ubiquitin-activating enzymes (E1) and ubiquitin-conjugating enzymes (E2) and, in some instances, a further protein, a ubiquitin ligase (E3). A number of genes encoding plant E1 and E2 proteins have been cloned, particularly from *Arabidopsis*. Following tagging by ubiquitination, it is assumed that proteolysis of the protein may occur, with subsequent release of the attached ubiquitin molecules for recycling and reuse. A major question, of course, relates to the process by which proteins become selected for ubiquitinylation. *A priori* this might be stochastic, or it might be determined by individual features of the protein molecule, such as damage sustained during enzyme catalysis, or loss of a prosthetic group.

Few examples of enzyme degradation have yet been studied in secondary metabolism. ACC synthase, which catalyses a step thought to have a large influence upon metabolic flux in the pathway of ethylene biosynthesis, has been found to have a relatively short half-life, ranging from 20 min to 2 h, depending upon the organ and the species (166). In tomato suspension culture cells, the increase in activity following elicitor treatment is reversed by protein kinase inhibitors and enhanced by phosphatase inhibitors, suggesting that phosphorylation events contribute to protein turnover (238). As related previously, (Section 3.2.), in the case of phenylalanine ammonia-lyase (PAL), there is evidence that inactivation of the enzyme contributes to the overall level of activity observed and that enzyme phosphorylation is an important factor in turnover (239). In this instance, the product of the enzyme reaction, cinnamic acid, causes inactivation, though not degradation, of PAL, possibly via modification of the active-site dehydroalanine residue (185); cinnamic acid can also inhibit the accumulation of *PAL* gene transcripts (186). The evidence for phosphorylation in both of these cases is suggestive of a link with protein kinases and phosphatases involved in a wider variety of cellular processes and in signal transduction mechanisms (see below).

3.6. Signal Transduction Mechanisms

As indicated earlier, expression of a secondary metabolic pathway can be considered an end-result of a chain or network of events which begins with the reception of a signal. The signal may or may not be readily identifiable and, *a priori*, need not reside in a single factor. Metabolic pathways interconvert molecules; signal transduction pathways, in contrast, transmit information. In the former case, the identity and concentration of the molecules (toxins, antifeedants, uv protectants etc.) is of central importance to physiological function; in the latter case, it is information transfer which is central and the molecular interconversions perform a vector function. (This distinction helps to place in context observations that although the recognised components of plant signal transduction pathways may be analogous, or homologous, to components found in other organisms, they may be interfaced in different ways.) Secondary metabolites feature both as end-products of signal-transduction pathways, as indicated above, but also as identified components of these pathways (plant growth substances, salicylic acid, jasmonic acid, for example - see below). Furthermore, a key function of a variety of secondary metabolites is to act as diffusible signal molecules between plants and other organisms. The diversity of secondary metabolites released from plants includes flavonoid nodulation inducers (164) in the interaction of legumes with nitrogen-fixing *Rhizobium* bacteria, simple phenolic compounds such as acetosyringone in the infection of plants by the tumour-inducing soil *Agrobacteria* (240), the gaseous hormone ethylene, flower scents (see e.g. 163), hydrocarbons (241) and - very possibly - an increasingly-recognised range of volatile derivatives of known signalling compounds, such as jasmonate and salicylate esters. The biochemistry and physiology of individual plant growth substances and signalling compounds have been comprehensively reviewed and the reader is referred to specific accounts for detail in relation to auxins (56, 242), gibberellins (55, 243), cytokinins (244), ethylene (166, 245, 246), abscisic acid (57, 247), the brassinosteroids (58), jasmonates (165, 248-251) and salicylate (252-255). Signal transduction in the context of plant defence is reviewed in (256).

The general questions relating to plant signal transduction pathways include: i) what are the components of the pathway, how do they interact (including interaction with components of other pathways) and how are they organised or compartmented; and ii) what common elements can be discerned in relation to pathways in other organisms? Several strategies have been adopted towards elucidating these questions, including biochemical techniques (for example, the isolation of cell-surface elicitor-binding proteins); the generation and characterisation of mutants; and pharmacological approaches, investigating the effects of components or inhibitors of signalling pathways characterised in other organisms. In many instances, the complementary application of these approaches is proving powerful.

The complexity of signalling pathways is illustrated by the mode of action of the light-sensing protein, phytochrome (205, 257), which is responsible, *inter alia*, for switching on the production of anthocyanin pigments in response to light; this process includes the induction of expression of chalcone synthase. Three distinct pathways have been found to operate in transducing the phytochrome signal (reviewed in 258). Induction of chalcone synthase expression and anthocyanin production occurs via a cyclic GMP-dependent pathway, whereas distinct Ca^{2+}-dependent and cyclic GMP- and Ca^{2+}-dependent pathways are required for chloroplast development and the establishment of photosynthetic function. These pathways were in part elucidated using single cells of a phytochrome-deficient mutant (*aurea*) of tomato, into which components of signalling pathways in other organisms, or known inhibitors of such components, could be microinjected. Mutants of *Arabidopsis* defective in phytochrome-mediated photomorphogenesis have allowed the identification of a number of positive and negative regulators which are components of the phytochrome signal transduction pathway. Mutations in negative regulators, such as the anthocyanin-accumulating *fusca* mutant, can lead to the constitutive expression of phytochrome-mediated responses in the absence of induction by light. Several genes for such regulators have now been cloned; one of these, *cop9* (constitutive photomorphogenic 9) encodes a polypeptide with limited homology to the β subunit of heterotrimeric GTP-binding proteins (259), well-known membrane-bound components of signal-transduction pathways in other eukaryotes.

Elements of signal transduction pathways which have become familiar in other organisms, such as heterotrimeric GTP-binding proteins, protein kinases and protein phosphatases, and second messengers, including Ca^{2+} ions, cyclic nucleotides and lipid-derived compounds, are all being found to participate in signal transduction in plants (see 260 for review). As indicated above, however, their interactions in plants are not necessarily predictable from other organisms. The ethylene transduction cascade in *Arabidopsis*, for example, incorporates both a membrane receptor similar to a bacterial "two-component" regulator and a protein kinase similar to the animal MAPKKKs (Mitogen-Activated Protein Kinase Kinase Kinases) (245, 246). The special importance for signal transduction mechanisms of the particularly complex compartmentation of plant cells has recently been emphasised (258). Metabolic compartmentation ensures that low-molecular-weight components of endogenous signalling pathways such as cytokinins, auxins, gibberellins and brassinosteroids are synthesised in the correct compartments, that common precursors are appropriately channelled and that end-products of these pathways are stored or made available in the appropriate form (for example free, or glycosylated, or complexed with amino acids). Targeting of translation products ensures that protein components (for example, protein kinases and phosphatases) are directed to the correct subcellular organelles or complexes. Finally, membrane integrity is essential for both of these aspects and for the maintenance and controlled modulation of ionic and electrical potentials.

The metabolic aspects of plant signalling substances (*inter alia*) are illustrated by two well-known compounds synthesised in response to external stimuli. The first of these, jasmonate (165, 248-251), may be produced following wounding, pathogen attack, fungal elicitation, desiccation, osmotic stress, senescence and even touch (it is involved in the tendril-coiling response in the climbing plant, *Bryonia*). In a number of instances, ethylene is co-involved in the observed responses. Jasmonate is characteristically associated with the appearance of a repertoire of specifically-induced proteins. In barley, for example, jasmonate-induced gene transcripts include those encoding enzymes of phenylpropanoid-derived phytoalexin biosynthesis (PAL, chalcone synthase and caffeic acid *O*-methyltransferase). The biosynthesis of several classes of alkaloids (for example, indole alkaloids, nicotine and benzophenanthridines) occurring variously in response to wounding or (in cell cultures) to fungal cell-wall elicitors, is mediated by jasmonate (250, 261, 262). In various species, jasmonate induces the appearance of lipoxygenase isoforms, proteinase inhibitors, pathogenesis-related (PR) proteins, osmotin and proteins involved in the regulation of ribosomal protein synthesis. Following wounding, jasmonate production may occur in remote tissues and this appears to result from a translocated signal, mediated by a phloem-translocated 18-amino-acid polypeptide, systemin (263).

An outline pathway of formation of jasmonate from the unsaturated fatty acid, linolenate, is shown in Figure 11 and involves the sequential action of 13-lipoxygenase, allene oxide synthase and allene oxide cyclase. An interesting comparison has been drawn between the jasmonate pathway and the formation of prostaglandins and other arachidonate metabolites in animals, which are involved in inflammatory processes and pain responses (see 248). A significant unresolved question concerns the origin of the linolenate, which is presumed to arise by lipolysis of membrane lipids; it is not known how this might be regulated. The final β-oxidation steps to jasmonate itself appear unnecessary for biological activity, as demonstrated by experiments with the higher homologue, trihomojasmonate (250). On the other hand, jasmonate itself exists in several physiologically-active forms, including esters and complexes with amino acids, and compartmentation and interconversion of these forms may well be important in modulating the responses mediated by jasmonates. Elucidation of elements of the signalling pathway(s) which lead to the appearance of jasmonate may help to resolve these questions. Very little is known of these elements, though a wounding-associated Mitogen-Activated Protein (MAP) kinase may be involved (248, 264).

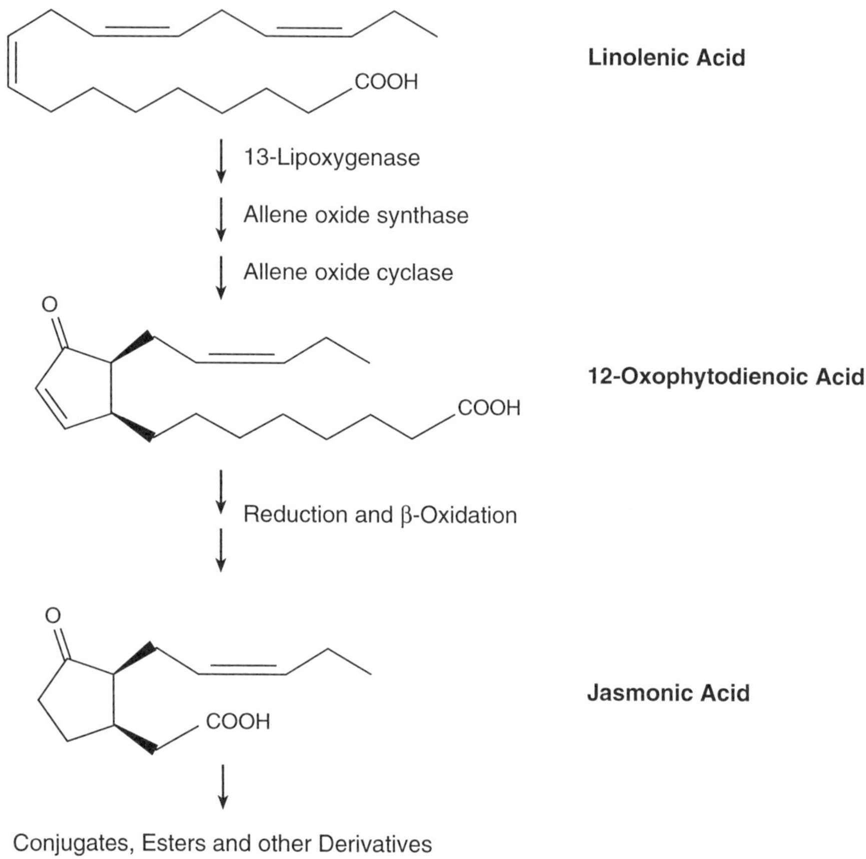

Figure 11. The formation of jasmonate.

Salicylic acid (2-hydroxybenzoic acid) was first identified as a signal molecule associated with resistance to viral infection and with the induction of a distinct set of pathogenesis-related (PR) proteins (252-255). An emerging picture of salicylic acid as the signal molecule responsible for the establishment of Systemic Acquired Resistance (resistance to a second pathogen challenge at a site remote from the initial infection - see 265) was supported by experiments in which the establishment of SAR was diminished in transgenic plants expressing a *Pseudomonas* gene, *nahG*, encoding salicylate hydroxylase, which metabolised salicylate further to catechol. The results of grafting experiments between normal plants and plants expressing *nahG* suggested, however, that salicylate might not itself be the translocated signal, even though its formation at the remote site appeared essential for resistance to secondary challenge. Subsequently, it was shown that a "salicylate-binding protein" possessed all the characteristics of catalase and it was suggested that the effects of salicylate might be mediated by H_2O_2, arising as a consequence of inactivation of catalase. An obvious question was the extent to which responses to salicylate might be related to the well-known localised "H_2O_2 burst" (266) which rapidly follows pathogen infection, and whether H_2O_2 might constitute the translocated signal. Increasingly, the "H_2O_2 burst" is recognised as a distinct phenomenon and, furthermore, it has been demonstrated that salicylate can bind to a range of haem-containing enzymes and that therefore its effects might be unattributable to the specific inhibition of catalase. It is possible that salicylate radicals arising from oxidative interaction with haem proteins represent the active species (see 252); this emphasises the importance - as in the case of jasmonate - of oxidative reactions (themselves likely to arise as a result of tissue damage) in the mediation of plant defence processes.

From a metabolic viewpoint, several significant questions remain unanswered. These include in particular the mechanism of salicylate biosynthesis and its subcellular localisation. Although it has been established that benzoate is a precursor of salicylate, the enzymology of benzoate formation from cinnamate has yet to be demonstrated. A benzoate 2-hydroxylase for the conversion of benzoate to salicylate has in contrast been characterised (255). A salicylate: UDPglucose glucosyltransferase has also been demonstrated and it is likely that the formation and hydrolysis of the glucoside, glucose ester and simple aliphatic esters of salicylate have an important bearing on the availability and physiological action of salicylate (255, 267). Downstream of salicylate, appreciable progress is now being made in the elucidation of elements of the salicylate transduction pathway by the analysis of *Arabidopsis* mutants which exhibit constitutive SAR, or alternatively in which SAR is compromised (see 252).

Though these examples demonstrate much recent progress, they also illustrate the metabolic uncertainties still associated with the biosynthesis and physiological availability of these compounds and the complexities remaining to be tackled in elucidating the signal transduction mechanisms associated with their effects. In all probability, signal transduction pathways are networked, thus providing for "cross-talk" between them, and hence the complexity of the problem increases still further. More generally, the "signals" relating to chronic or sub-acute environmental influences or nutritional status may be subtle and difficult to define. (The long-established literature in this field, for example that which relates alkaloid yields to nutritional status, might provide old problems for modern reinvestigation.) From the biotechnological perspective, progress in this area allied to studies of transcription factors is likely to bring large dividends in terms of an increased ability to control metabolic pathways in response to defined chemical or environmental switches (268).

4. Outlook

The extent of recent progress in the characterisation of secondary metabolism needs no emphasis. To a significant degree, this represents the complementary application of new approaches arising from allied disciplines, building upon a painstakingly-acquired basis of natural-product chemistry and biochemistry. During the past twenty years, emphasis has shifted further towards molecular genetics; and within the last fifteen years, the complementarity between all these disciplines has increasingly been recognised, for example with the discovery of "new" or previously-unrecognised secondary metabolites performing important functions as signalling compounds. This interdisciplinarity is increasingly essential. This is not only because new and physiologically-significant compounds are constantly being revealed, but also because old knowledge and assumptions may be reevaluated or placed in a fresh context; the recent discovery of a new pathway of isoprenoid biosynthesis following a general assumption that the mevalonate pathway was the only biosynthetic route emphasises that long-held assumptions cannot necessarily be accepted without question.

One of the major technical developments within the past twenty years has been the emergence and development of plant genetic transformation and regeneration techniques (269). The number of species amenable to transformation is constantly increasing. Plant biology has moved from the descriptive to the technological. This enables metabolic assumptions to be tested and new metabolic pathways to be introduced. Metabolic studies have gained increased prominence in this climate for the obvious reason that molecular-genetic intervention using structural genes or transcription factors requires a metabolic interpretation and, conversely, an appreciation of how secondary metabolism is regulated can inform a genetic manipulation strategy and avoid misdirected effort. Further consideration of these issues appears in Chapter 9.

A second area of technical development relates to gene expression analysis and gene cloning. The *Arabidopsis* sequencing programme will shortly be completed. Traditional cloning and sequencing technology are now joined by "DNA-chip" technology (270, 271), making possible the large-scale analyse of patterns of gene expression and the isolation and amplification of sequences in an automated or semi-automated manner. If such developments carry any danger it is perhaps that the researcher needs to be even more diligent in interpretation and in defining carefully his or her particular research problem.

The continued elucidation of transcriptional and signal-transduction mechanisms will afford quite new and powerful approaches to the directed manipulation of entire secondary metabolic pathways. These will complement anti-sense and cosuppression approaches to the down-regulation of pathways and "metabolic diversion" approaches that take advantage of novel enzymes found in other organisms, such as bacteria. They will also afford expanding opportunities for the biochemist and molecular geneticist interested in protein and nucleic-acid structure-function relationships, in the design of modified regulatory proteins and in the modulation of regulatory- and pathway-gene expression.

5. Acknowledgement

It is a pleasure to thank Cathie Martin for her comments and suggestions on this chapter.

6. References

1. H. A Stafford, *Plant Sci.* **101** (1994) 91.
2. T. Hartmann, *Entomologia Experimentalis et Applicata* **80** (1996) 177.
3. C.G. Jones and R.D. Firn, *Philos. Trans. R. Soc. London B* **333** (1991) 273.
4. E. Leete, *J. Amer. Chem. Soc.* **85** (1963) 3523.
5. E. Leete, *J. Amer. Chem. Soc.* **86** (1964) 2509.
6. R. Robinson, *J. Chem. Soc.* **111** (1917) 762.
7. R.J. Robins and N.J. Walton in *The Alkaloids*, Volume 44, ed. G.A. Cordell (Academic Press, New York, 1993), p. 115.
8. W.O. James in *The Alkaloids*, Volume 1, eds. R.H.F. Manske and H.L. Holmes (Academic Press, New York, 1950), p. 16.
9. J. Bonner and J.E. Varner, eds. *Plant Biochemistry* (Academic Press, New York, 1965), 1054 pp.
10. H. Grisebach in *Chemistry and Biochemistry of Plant Pigments*, ed. T.W. Goodwin (Academic Press, London, 1965), p. 279.

11. K. Hahlbrock and H. Grisebach in *The Flavonoids*, ed. J.B. Harborne, T.J. Mabry and H. Mabry (Chapman and Hall, London, 1975), p. 866.
12. T.W. Goodwin, ed. *Plant Pigments* (Academic Press, London, 1988), 362 pp.
13. F. Drawert and J. Beier, *Phytochemistry* **13** (1974) 2749.
14. R.J. Robins, A.J. Parr, M.J.C. Rhodes and N.J. Walton, in *Bitterness in Foods and Beverages*, ed. R.L. Rousseff (Elsevier, Amsterdam, 1990), p. 49.
15. E. Leete, *J. Amer. Chem. Soc.* **78** (1956) 3520.
16. E. Leete, N. Kowanko and R.A. Newmark, *J. Amer. Chem. Soc.* 97 (1975) 6826.
17. E. Leete, *Planta Medica* **56** (1990) 339.
18. R.J. Robins, T.W. Abraham, A.J. Parr, J. Eagles and N. J. Walton, *J. Amer. Chem. Soc.* **119** (1997) 10929.
19. R.J. Robins, N.C.J.E. Chesters, D. O'Hagan, A.J. Parr, N.J. Walton and J.G. Woolley, *J. Chem. Soc. Perkin Trans. I* **1995** 481.
20. A.W. Alfermann and M. Petersen, *Plant Cell Tiss. Org. Cult.* **43** (1995) 199.
21. A.J. Parr and J.D. Hamill, *Phytochemistry* **26** (1987) 3241.
22. J.D. Hamill and M.J.C. Rhodes, in *Plant Biotechnology*, Volume 3, ed. D. Grierson (Blackie, Glasgow, 1993), p. 178.
23. A. Spencer, J.D. Hamill and M.J.C. Rhodes, *Phytochemistry* **32** (1993) 911.
24. K.M.S.A. Salem and B.V. Charlwood, *Plant Cell Tiss. Org. Cult.* **40** (1995) 209.
25. P.H. Heinstein and C.-j. Chang, *Annu Rev. Plant Physiol. Plant Mol. Biol.* **45** (1994) 663.
26. M. Rohmer, M. Knani, P. Simonin, B. Sutter and H. Sahm, *Biochem. J.* **295** (1993) 517.
27. W. Knöss, B. Reuter and J. Zapp, *Biochem. J.* **326** (1997) 449.
28. C. Leuschner and G. Schultz, *Bot. Acta* **104** (1991) 240.
29. C. Leuschner and G. Schultz, *Phytochemistry* **30** (1991) 2203.
30. S. Hedges and R.B. Herbert, *Phytochemistry* **20** (1981) 2064.
31. Y.-Y. Ford, R.G. Ratcliffe and R.J. Robins, *Phytochemistry* **43** (1996) 115.
32. Y.-Y. Ford, G.G. Fox, R.G. Ratcliffe and R.J. Robins, *Phytochemistry* **36** (1994) 333.
33. M.H. Zenk, *J. Nat. Prod.* **43** (1980), 438.
34. L.-F. Tietze, *Angew. Chem. Int. Ed. Engl.* **22** (1983) 828.
35. R.B. Herbert in *The Monoterpenoid Indole Alkaloids*, ed. J.E. Saxton (J. Wiley and Sons, London, 1994), p.1.
36. R.T. Brown, J. Leonard and S.K. Sleigh, *J.C.S. Chem. Commun.* **1977** 636.
37. J. Stöckigt, H.P. Husson, C. Kan-Fan and M.H. Zenk, *J.C.S. Chem. Commun.* **1977** 164.
38. J. Royer and H.-P. Husson, *Janssen Chimica Acta* **11** (1993) 3.
39. U. Pfitzner and M.H. Zenk, *Planta Medica* **46** (1982) 10.
40. L. Britsch and H. Grisebach, *Eur. J. Biochem.* **156** (1986) 569.
41. M.H. Zenk, B. Ulbrich, J. Busse and J. Stöckigt, *Anal. Biochem.* **101** (1980) 182.
42. D. Strack and H.-P. Mock, in *Methods in Plant Biochemistry, Volume 9. Enzymes of Secondary Metabolism*, ed. P.J. Lea (Academic Press, London, 1993), p. 45.

43. M.J.C. Rhodes in *Regulation of enzyme synthesis and activity in higher plants.*, ed. H. Smith (Academic Press, London, 1977), p. 245.
44. E. Lewinsohn, M. Gijzen and R. Croteau, *Plant Physiol.* **96** (1991) 44.
45. W.R. Alonso and R. Croteau in *Methods in Plant Biochemistry, Volume 9. Enzymes of Secondary Metabolism*, ed. P.J. Lea (Academic Press, London, 1993), p. 239.
46. P. Moesta and C.A. West, *Arch. Biochem. Biophys.* **238** (1995) 325.
47. N.J. Walton, A.C.J. Peerless, R.J. Robins, M.J.C. Rhodes, H.D. Boswell and D.J. Robins, *Planta* **193** (1994) 9.
48. C.A. Mikaliak, F. Karp and R. Croteau in *Methods in Plant Biochemistry, Volume 9. Enzymes of Secondary Metabolism*, ed. P.J. Lea (Academic Press, London, 1993), p. 261.
49. B.A. Halkier, *Phytochemistry* **43** (1996) 1.
50. T. Lange and J.E. Graebe in *Methods in Plant Biochemistry, Volume 9. Enzymes of Secondary Metabolism*, ed. P.J. Lea (Academic Press, London, 1993), p. 403.
51. J. Treimer and M.H. Zenk, *FEBS Lett.* **97** (1979) 159.
52. V. De Luca, in *Methods in Plant Biochemistry, Volume 9. Enzymes of Secondary Metabolism*, ed. P.J. Lea (Academic Press, London, 1993), p. 345.
53. T. Bohlmann, V. De Luca, U. Eilert and W. Martin, *Plant J.* **7** (1995) 495.
54. J.E. Graebe, *Annu. Rev. Plant Physiol.* **38** (1987) 419.
55. A.L. Phillips, *Plant Physiol. Biochem.* **36** (1998) 115.
56. B. Bartel, *Annu. Rev. Plant Physiol. Plant Mol. Biol.* **48** (1997) 51.
57. M. Koornneef, K.M. Léon-Kloosterziel, S.H. Schwartz and A.D. Zeevaart, *Plant Physiol. Biochem.* **36** (1998) 83.
58. M. Szekeres and C. Koncz, *Plant Physiol. Biochem.* **36** (1998) 145.
59. G.M. Reddy and E.H. Coe, *Science* **138** (1962) 149.
60. J.B. Harborne in *The Chemistry of Flavonoid Compounds*, ed. T.A. Geissman (Pergamon Press, Oxford, 1962), p. 593.
61. W. Heller and G. Forkmann in *The Flavonoids: Advances in Research*, ed. J. B. Harborne (Chapman and Hall, London, 1988), p. 399.
62. K. Hahlbrock and D. Scheel, *Annu. Rev. Plant Physiol. Mol. Biol.* **40** (1989) 347.
63. G. Forkmann, *Plant Breeding* **106** (1991) 1.
64. H.K. Dooner, T.P. Robbins and R.A. Jorgensen, *Annu Rev. Genetics* **25** (1991) 173.
65. N.V. Federoff, D.B. Furtek and O.E. Nelson, *Proc. Natl. Acad. Sci. USA* **81** (1984) 3825.
66. T.A. Holton and E.C. Cornish, *Plant Cell* **7** (1995) 1071.
67. A.G.M. Gerats and C. Martin, in *Recent Advances in Phytochemistry, Volume 26. Phenolic Metabolism in Plants*, eds. H.A. Stafford and R.K. Ibrahim (Plenum Press, New York, 1992), p. 165.
68. C. Chapple in *Recent Advances in Phytochemistry, Volume 28. Genetic Engineering of Plant Secondary Metabolism*, eds. B.E. Ellis, G.W. Kuroki and H.A. Stafford (Plenum Press, New York, 1994), p. 251.
69. R. Schmidt, *Plant Physiol. Biochem.* **36** (1998) 1.

70. A.H. Meijer, E. Souer, R. Verpoorte and J.H.C. Hoge, *Plant Mol. Biol.* **34** (1997) 935.
71. A. Slater, M.J. Maunders, K. Edwards, W. Schuch and D. Grierson, *Plant Mol. Biol.* **5** (1985) 137.
72. J.E. Gray, S. Picton, J.J. Giovannoni and D. Grierson, *Plant Cell and Environment* **17** (1994) 557.
73. A.J. Hamilton, G.W. Lycett and D. Grierson, *Nature* **346** (1990), 284.
74. N. Hibi, S. Higashiguchi, T. Hashimoto and Y. Yamada, *Plant Cell* **6** (1994) 723.
75. J.J.M.R. Jacobs, L. Kluijtmans, Z.M.H. Kolen, K. Erckens, R.R.J. Arroo, A.F. Croes and G.J. Wullems, *J. Plant Physiol.* **146** (1995) 418.
76. P.C. Babbitt and J.A. Gerlt, *J. Biol. Chem.* **272** (1997) 30591.
77. R. Welle and J. Schröder, *Arch. Biochem. Biophys.* **293** (1992) 377.
78. T.M. Kutchan, A. Bock and H. Dittrich, *Phytochemistry* **35** (1994) 353.
79. E. Hochuli, H. Dobeli and A. Schacher, *J. Chromatogr.* **411** (1987) 177.
80. K.T.D. DeSilva, G.N. Smith and K.E.H. Warren, *J.C.S. Chem. Commun.* **1971** 905.
81. R.T. Brown, C.L. Chapple and A.G. Lashford, *J.C.S. Chem. Commun.* **1975** 295.
82. T.J.C. Luijendijk, A. Nowak and R. Verpoorte, *Phytochemistry* **41** (1996) 1451.
83. S. Löffler, B. Deus-Neumann and M.H. Zenk, *Phytochemistry* **38** (1995) 1387.
84. T. Frenzel and M.H. Zenk, *Phytochemistry* **29** (1990) 3505.
85. R. Whetten and R. Sederoff, *Plant Cell* **7** (1995) 1001.
86. A.M. Boudet, D.P. Goffner and J. Grima-Pettenati, *C. R. Acad. Sci. Paris* **319** (1996) 317.
87. R. Lenz and M.H. Zenk, *J. Biol. Chem.* **270** (1995) 31091.
88. H.A. Stafford, in *The Biochemistry of Plants, Volume 7,* eds. P.K. Stumpf and E.E. Conn (Academic Press, New York, 1981), p. 117.
89. G. Hrazdina and G.J. Wagner, *Arch. Biochem. Biophys.* **237** (1985) 88.
90. G.Hrazdina, in *Recent Advances in Phytochemistry, Volume 26. Phenolic Metabolism in Plants*, eds. H.A. Stafford and R.K. Ibrahim (Plenum Press, New York, 1992), p. 1.
91. G. Hrazdina and R.A. Jensen, *Annu. Rev. Plant Physiol. Plant Mol. Biol.* **43** (1992) 241.
92. C.G. Smith, M.W. Rodgers, A. Zimmerlin, D. Ferdinando and G.P. Bolwell, *Planta* **192** (1994) 155.
93. J. Gershenzon and R. Croteau, in *Recent Advances in Phytochemistry, Volume 24. Biochemistry of the Mevalonic Acid Pathway to Terpenoids*, eds. G.H.N. Towers and H.A. Stafford (Plenum Press, New York, 1990), p. 99.
94. H. Kleinig, *Annu. Rev. Plant Physiol. Plant Mol. Biol.* **40** (1989) 39.
95. J. Ohlrogge and J. Browse, Plant Cell 7 (1995) 957.
96. K.M. Herrmann, Plant Cell 7 (1995) 907.
97. L.M. Weaver and K.M. Herrmann, *Trends Plant Sci.* **2** (1997) 346.
98. D.A. Hopwood and D.H. Sherman, *Annu. Rev. Genetics* **24** (1990) 37.

99. R. Plater and W.R. Strohl, in *Recent Advances in Phytochemistry, Volume 28. Genetic Engineering of Plant Secondary Metabolism*, eds. B.E. Ellis, G.W. Kuroki and H.A. Stafford (Plenum Press, New York, 1994), p. 61.
100. G. Schröder, J.W.S. Brown and J. Schröder, *Eur. J. Biochem.* **172** (1988) 161.
101. J. Schröder, *Trends Plant Sci.* **2** (1997) 373.
102. Y. Helariutta, M. Koltainen, P. Elomaa, N. Kalakinnen, K. Bremer, T.H. Teeri and V.A. Albert,
Proc. Natl. Acad. Sci. USA **93** (1996) 9033.
103. R. Welle, G. Schröder, E. Schiltz, H. Grisebach and J. Schröder, *Eur. J. Biochem.* **196** (1991) 423.
104. L. B. Davin, N.G. Lewis and T. Umezawa, in *Recent Advances in Phytochemistry, Volume 26. Phenolic Metabolism in Plants*, eds. H.A. Stafford and R.K. Ibrahim (Plenum Press, New York, 1992), p. 325.
105. L.B. Davin, H.-B. Wang, A.L. Crowell, D.L. Bedgar, D.M. Martin, S. Sarkanen and N.G. Lewis, *Science* **275** (1997) 362.
106. R. Mc-C. Lilley, M. Stitt, G. Mader and H.W. Heldt, *Plant Physiol.* **70** (1982) 965.
107. C.G. Smith, M.W. Rodgers, A. Zimmerlin, D. Ferdinando and G.P. Bolwell, *Planta* **192** (1994) 155.
108. C. Robinson and A. Mant, *Trends Plant Sci.* **2** (1997) 431.
109. R. Gerhardt and H.W. Heldt, *Plant Physiol.* **75** (1984) 542.
110. H. Winter, D.G. Robinson and H.W. Heldt, *Planta* **191** (1993) 180.
111. M.H. Zenk, M. Rüffer, T.M. Kutchan and E. Galneder, in *Applications of Plant Cell and Tissue Culture, Ciba Foundation Symposium* **137** (1988) 213.
112. Y. Yamada and N. Okada, *Phytochemistry* **24** (1985) 457.
113. D.J. McGarvey and R. Croteau, *Plant Cell* **7** (1995) 1015.
114. J. Chappell, *Annu. Rev. Plant Physiol. Plant Mol. Biol.* **46** (1995) 521.
115. E. Soler, G. Feron, M. Clastrid, R. Dargent, M. Gleizer and C. Ambid, *Planta* **187** (1992) 171.
116. M. Kuntz, S. Römer, C. Suire, P. Hugueney, J.H. Weil, R. Schantz and B. Camara, *Plant J.* **2** (1992) 25.
117. D. McCaskill and R. Croteau, *Planta* **197** (1995) 49.
118. J.D. Cohen and R.S. Bandurski, *Annu. Rev. Plant Physiol.* **33** (1982) 403.
119. R.A. Dixon and N.L. Paiva, *Plant Cell* **7** (1995) 1085.
120. W. Barz and U. Mackenbrock, *Plant Cell Tiss. Org. Cult.* **38** (1994) 199.
121. V. Sharma and D. Strack, *Planta* **163** (1985) 563.
122. W. Hopp and H.U. Seitz, *Planta* **170** (1995) 74.
123. J.P. Renaudin and J. Guern, *Biochem. Soc. Trans.* **16** (1988) 63.
124. M. Wink, *Adv. Bot. Res.* **25** (1997) 141.
125. S.S.M. Waldhauser and T.W. Baumann, *Phytochemistry* **42** (1996) 985.
126. F. Vazquez-Flota, E. De Carolis, A.-M. Alarco and V. De Luca, *Plant Mol. Biol.* **34** (1997) 935.
127. B.L. Møller and J.E. Poulton, in *Methods in Plant Biochemistry, Volume 9. Enzymes of Secondary Metabolism*, ed. P.J. Lea (Academic Press, London, 1993), p. 183.

128. A.M. Bones and J.T. Rossiter, *Physiol. Plant.* **97** (1996) 194.
129. W.E. Dyer, J.M. Henstrand, A.K. Handa and K.M. Herrmann, *Proc. Natl. Acad. Sci. USA* **86** (1989) 7370.
130. K.F. McKue and E.E. Conn, *Proc. Natl. Acad. Sci. USA* **86** (1989) 7374.
131. J.M. Henstrand, K.F. McKue, K. Brink, A.K. Handa, K.M. Herrmann and E.E. Conn, *Plant Physiol.* **98** (1992) 761.
132. J. Görlach, H.-R. Raesecke, D. Rentsch, M. Regenass, P. Roy, M. Zala, C. Keel, T. Boller, N. Amrhein and J. Schmid, *Proc. Natl. Acad. Sci. USA* **92** (1995) 3166.
133. K.K. Niyogi and G.K. Fink, *Plant Cell* **4** (1992) 721.
134. R.A. Dixon, M.J. Harrison and N.L. Paiva, *Physiol. Plant.* **93** (1995) 385.
135. B.S. Shorrosh, R.A. Dixon and J.B. Ohlrogge, *Proc. Natl. Acad. Sci. USA* **91** (1994) 4323.
136. L.A. Wanner, G. Li, D. Ware, I.E. Sommsich and K.R. Davis, *Plant Mol. Biol.* **27** (1995) 327.
137. X. Liang, M. Dron, C.L. Cramer, R.A. Dixon and C.J. Lamb, *J. Biol. Chem.* **264** (1989) 14486.
138. A. Leyva, X. Liang, J.A. Pinto-Toro, R.A. Dixon and C.J. Lamb, *Plant Cell* **4** (1992) 263.
139. D. Shufflebottom, K. Edwards, W. Schuch and M. Bevan, *Plant J.* **3** (1993) 835.
140. R. Lois, A. Diettrich, K. Hahlbrock and W. Schulz, *EMBO J.* **8** (1989) 1641.
141. S. Ohl, C.A. Hedrick, J. Chory and C.J. Lamb, *Plant Cell* **2** (1990) 837.
142. G.P. Bolwell, J.N. Bell, C.L. Cramer, W. Schuch, C.J. Lamb and R.A. Dixon, *Eur. J. Biochem.* **149** (1985) 411.
143. J. Jorrin and R.A. Dixon, *Plant Physiol.* **92** (1990) 447.
144. C. Appert, E. Logemann, K. Hahlbrock, J. Schmid and N. Amrhein, *Eur. J. Biochem.* **225** (1994) 491.
145. L. Beerhues, H. Robenek, and R. Wiermann, *Planta* **173** (1988) 544.
146. L. Beerhues and R. Wiermann, *Planta* **173** (1988) 532.
147. R.E. Koes, C.E. Spelt and J.N.M. Mol, *Plant Mol. Biol.* **12** (1989) 213.
148. H. Junghans, K. Dalkin and R.A. Dixon, *Plant Mol. Biol.* **22** (1993) 239.
149. P. Facchini and V. De Luca, *J. Biol. Chem.* **269** (1994) 22684.
150. U. Eilert, W.G.W. Kurz and F. Constabel, *J. Plant Physiol.* **119** (1985) 65.
151. C. Grand, A. Boudet and A.M. Boudet, *Planta* **158** (1983) 225.
152. A. Fahn, *Secretory Tissues in Plants* (Academic Press, London, 1979), 302 pp.
153. P. John, *Biosynthesis of the Major Crop Products* (John Wiley & Sons, Chichester, 1992), 154 pp.
154. R.J. Robins, E.G. Bent and M.J.C. Rhodes, *Planta* **185** (1991) 385.
155. W. Jahnen and K. Hahlbrock, *Planta* **173** (1988) 197.
156. E. Schmelzer, W. Janen and K. Hahlbrock, *Proc. Natl. Acad. Sci. USA* **85** (1988) 2989.
157. E. Schmelzer, S. Krüger-Lebus and K. Hahlbrock, *Plant Cell* **1** (1989) 993.
158. S.-C. Wu and K. Hahlbrock, *Z. Natursforsch.* **47C** (1992) 591.

159. T. Hashimoto, A. Hayashi, Y. Amano, I. Kohno, H. Iwanari, S. Usuda and Y. Yamada, *J. Biol. Chem.* **266** (1991) 4648.
160. T. Hartmann, A. Ehmke, U. Eilert, K. von Borstel and C. Theuring, *Planta* **177** (1989) 98.
161. A. Cartayrade, E. Neau, C. Sohier, J.P. Balz, J.P. Carde and J. Walter, *Plant Physiol. Biochem.* **35** (1997) 859.
162. T.D. Sharkey, *Trends Plant Sci.* **1** (1996) 78.
163. N. Dudareva, J.C. D'Auria, K.H. Nam, R.A. Raguso and E. Pichersky, *Plant J.* **14** (1998) 297.
164. J.L. Firmin, K.E. Wilson, I. Rossen and A.W.B. Johnston, *Nature* **324** (1986) 90.
165. C. Wasternack and B. Parthier, *Trends Plant Sci.* **2** (1997) 302.
166. H. Kende, *Annu. Rev. Plant Physiol. Plant Mol. Biol.* **44** (1993) 283.
167. J. Bruin, M.W. Sabelis and M. Dicke, *Trends Ecol. Evolut.* **10** (1995) 167.
168. E.A. Eltayeb and J.G. Roddick, *Phytochemistry* **24** (1985) 253.
169. K. Iwai, T. Suzuki and H. Fujiwake, *Agric. Biol. Chem.* **43** (1979) 2493.
170. Y. Kitamura, R. Yamashita, H. Miura and M.J. Watanabe, *J. Plant Physiol.* **146** (1995) 210.
171. Y. Kitamura, M. Shigehiro and T. Ikenaga, *Phytochemistry* **42** (1996) 1331.
172. I.T. Baldwin and T.E. Ohnneiss, *Oecologia* **98** (1994) 385.
173. R.J. Robins, J.M. Furze and M.J.C. Rhodes, *Phytochemistry* **24** (1985) 709.
174. H. Kacser and J.A. Burns, *Symp. Soc. Exp. Biol.* **27** (1973) 65.
175. A. Cornish-Bowden and J.-H. S. Hofmeyr, *Biochem. J.* **298** (1994) 367.
176. D.A. Fell, *Biochem. J.* **286** (1992) 313.
177. T. ap Rees and S.A. Hill, *Plant, Cell and Environment* **17** (1994) 587.
178. M. Stitt, *Aust. J. Plant Physiol.* **32** (1995) 635.
179. M. Stitt and U. Sonnewald, *Annu. Rev. Plant Physiol. Plant Mol. Biol.* **46** (1995) 361.
180. P. Niederberger, R. Prasad, G. Miozzari and H. Kacser, *Biochem J.* **287** (1992) 473.
181. N.J. Bate, J. Orr, W. Ni, A. Meromi, T. Nadler-Hassar, P.W. Doerner, R.A. Dixon, C.J. Lamb and Y. Elkind, *Proc. Natl. Acad. Sci. USA* **91** (1994) 7608.
182. P.A. Howells, V.J.H. Sewalt, N.L. Paiva, Y. Elkind, N.J. Bate, C. Lamb and R.A. Dixon, *Plant Physiol.* **112** (1996) 1617.
183. D. Burtin and A.J. Michael, *Biochem. J.* **325** (1997) 331.
184. A. J. Lidgett, *Molecular Characterisation of Polyamine Biosynthetic Genes and Effects on Alkaloid Metabolism in Nicotiana tabacum*, PhD thesis, Monash University, Victoria, Australia, 1997.
185. G.P. Bolwell, C.L. Cramer, C.J. Lamb, W. Schuch and R.A. Dixon, *Planta* **169** (1996) 97-107.
186. M. Mavandad, R. Edwards, X. Liang, C.J. Lamb and R.A. Dixon, *Plant Physiol.* **94** (1990) 671.
187. G.P. Bolwell, M. Mavandad, D.J. Millar, K.J. Edwards, W. Schuch and R.A. Dixon, *Phytochemistry* **27** (1988) 2109-2117.

188. G.J. Loake, O. Faktor, C.J. Lamb and R.A. Dixon, *Proc. Natl. Acad. Sci. USA* **89** (1992) 9230.
189. C. Martin, *Current Opinion Biotech.* **7** (1996) 130.
190. F. Katagiri and N.-H. Chua, *Trends Genet.* **8** (1991) 22.
191. B. Weisshaar, G.A. Armstrong, A. Block, O. da Costa e Silva and K. Hahlbrock, *EMBO J.* **10** (1991) 1777.
192. L.M. Yu, C.J. Lamb and R.A. Dixon, *Plant J.* **3** (1993) 805.
193. P. Schulze-Lefert, J.L. Dangl, M. Becker-Andre, K. Hahlbrock and W. Schultz, *EMBO J.* **8** (1989) 651.
194. P. Schulze-Lefert, M. Becker-Andre, W. Schulz, K. Hahlbrock and J.L. Dangl, *Plant Cell* **1** (1989) 707.
195. R.W.M. Sablowski, E. Moyano, F.A. Culianez-Macia, W. Schuch, C. Martin and M. Bevan, *EMBO J.* **13** (1994) 128.
196. S. Ohl, S.A. Hedrick, J. Chory and C.J. Lamb, *Plant Cell* **2** (1990) 837.
197. R.G.K. Donald and A.R. Cashmore, *EMBO J.* **9** (!990) 1717.
198. H. Sommer, U. Bonas and H. Saedler, *Mol. Gen. Genet.* **211** (1988) 49.
199. I.M. van der Meer, C.E. Spelt, J.N.M. Mol and A.R. Stuitje, *Plant. Mol. Biol.* **15** (1990) 95.
200. S.M. Huits, A.G.M. Gerats, M.M. Kreike, J.N.M. Mol and R.E. Koes, *Plant J.* **6** (1994) 295.
201. A.J. van Tunen, S.A. Hartman, L.A. Mur and J.N.M. Mol, *Plant Mol. Biol.* **12** (1989) 539.
202. F. Quattrochio, J.F. Wing, H.T.C. Leppen, J.N.M. Mol and R.E. Koes, *Plant Cell* **5** (1993) 1497.
203. K. Hahlbrock, D. Scheel, E. Logemann, T. Nürnberger, M. Parniske, S. Reinold, W.R.Sacks and E. Schmelzer, *Proc. Natl. Acad. Sci. USA* **92** (1995) 4150.
204. R.A. Dixon and M.J. Harrison, *Adv. Genet.* **28** (1990) 165.
205. J. Mol, G. Jenkins, E. Schafer and D. Weiss, *Crit. Rev. Plant Sci.* **15** (1996) 525.
206. R.E. Koes, F. Quattrochio and J.N.M. Mol, *BioEssays* **16** (1994) 123.
207. R. Lois, A. Diettrich, K. Hahlbrock and W. Schulz, *EMBO J.* **8** (1989) 1641.
208. D. Jackson, K. Roberts and C. Martin, *Plant J.* **2** (1992) 425.
209. C. Martin, A. Prescott, S. Mackay, J. Bartlett and E. Vrijlandt, *Plant J.* 1 (1991) 37.
210. M. Mooney, T. Desnos, K. Harrison, J. Jones, R. Carpenter and E. Coen, *Plant J.* **7** (1995) 333.
211. A.M. Lloyd, V. Walbot and R.W. Davis, *Science* **258** (1992) 1773.
212. E. Moyano, J.F. Martinez-Garcia and C. Martin, *Plant Cell* **8** (1996) 1519.
213. L. Tamagnone, A. Merida, A. Parr, S. Mackay, F.A. Cullianez-Macia, K. Roberts and C. Martin, *Plant Cell* **10** (1998) 135.
214. A. Weigel and O. Nilsson, *Nature* **377** (1995) 495.
215. M.A. Mandel and M.F. Yanofsky, *Nature* **377** (1995) 522.
216. E.S. Coen, *Annu. Rev. Plant Physiol. Plant Mol. Biol.* **42** (1991) 241.
217. H.C. Larkin, D.G. Oppenheimer, A.M. Lloyd, E.T. Paparozzi and M.D. Marks, *Plant Cell* **6** (1994) 1065.

218. R.B. Flavell, *Proc. Natl. Acad. Sci. USA* **91** (1994) 3490.
219. P. Meyer and H. Saedler, *Annu. Rev. Plant Physiol. Plant Mol. Biol.* **47** (1996) 23.
220. P. Meyer in *Transgenic Organisms - Biological and Social Implications*, eds. J. Tomiuk, K. Wöhrmann and A. Sentker (Birkhauser Verlag, Basel, 1996), p. 5.
221. C. Napoli, C. Lemieux and R. Jorgensen, *Plant Cell* **2** (1990) 279.
222. A.R. van der Krol, L.A. Mur, M. Beld, J. Mol and A.R. Stuitje, *Plant Cell* **2** (1990) 291.
223. D. Grierson, R.G. Fray, A.J. Hamilton, C.J.S. Smith and C.F. Watson, *Trends Biotechnol.* **9** (1991) 122.
224. D.C. Baulcombe, *Plant Mol. Biol.* **32** (1996) 79.
225. M.L. Abler and P.J. Green, *Plant Mol. Biol.* **32** (1996) 63.
226. P. Gil and P.J. Green, *EMBO J.* **15** (1996) 1678.
227. M.L. Sullivan and P.J. Green, *Plant Mol. Biol.* **23** (1993) 1091.
228. L.F. Dickey, T.-T. Nguyen, G.C. Allen and W.F. Thompson, *Plant Cell* **6** (1994) 1171.
229. K.D. Jofuku, R.D. Schipper and R.B. Goldberg, *Plant Cell* **1** (1989) 427.
230. T.A. Voelker, J. Moreno and M.J. Chrispeels, *Plant Cell* **2** (1990) 255.
231. R.S. Boston, P.V. Viitanen and E. Vierling, *Plant Mol. Biol.* **32** (1996) 191.
232. S.M. Hemmingsen, C. Woolford, S.M. Van der Vries, K. Tilly, D.T. Dennis, C.P. Georgopoulos, R. Hendrix and R.J. Ellis, *Nature* **333** (1988) 330.
233. R.J. Ellis, *Plant J.* **1** (1991) 9.
234. J.P. Hendrick and F.-U. Hartl, *Annu. Rev. Biochem.* **62** (1993) 349.
235. E. Pedrazzini and A. Vitale, *Plant Physiol. Biochem.* **34** (1996) 207.
236. R.D. Vierstra, *Annu. Rev. Plant Physiol. Plant Mol. Biol.* **44** (1993) 385.
237. R.D. Vierstra, *Plant Mol. Biol.* **32** (1996) 275.
238. P. Spanu, D.G. Grosskopf, G. Felix and T. Boller, *Plant Physiol.* **106** (1994) 529.
239. G.P. Bolwell, *Phytochemistry* **31** (1992) 4081.
240. P.A. Spencer and G.H.N. Towers, *Phytochemistry* **27** (1988) 2781.
241. T.D. Sharkey, *Trends Plant Sci.* **1** (1996) 78.
242. L. Hobbie, C. Timpte and M. Estelle, *Plant Mol. Biol.* **26** (1994) 1499.
243. R. Hoolley, *Plant Mol. Biol.* **26** (1994) 1529.
244. B. Brzobohatý, I. Moore and K. Palme, *Plant Mol. Biol.* **26** (1994) 1483.
245. T.I. Zarembinski and A. Theologis, *Plant Mol. Biol.* **26** (1994) 1579.
246. R. Fluhr and A.K. Mattoo, *Crit. Rev. Plant Sci.* **15** (1996) 479.
247. J. Giraudat, F. Parcy, N. Bertauche, F. Gosti, J. Leung, P.-C. Morris, M. Bouvier-Durand and N. Vartanian, *Plant Mol. Biol.* **26** (1994) 1557.
248. C. Wasternak and B. Parthier, *Trends Plant Sci.* **2** (1997) 302.
249. R.A. Creelman and J.E. Mullet, *Annu. Rev. Plant Physiol. Plant Mol. Biol.* **48** (1997) 355.
250. S. Blechert, W. Brodschelm, S. Hölder, L. Kammerer, T.M. Kutchan, M.J. Mueller, X.-Q. Xia and M.H. Zenk, *Proc. Natl. Acad. Sci. USA* **92** (1995) 4099.
251. E.E. Farmer, *Plant Mol. Biol.* **26** (1994) 1423.
252. J. Durner, J. Shah and D.F. Klessig, *Trends Plant Sci.* **2** (1997) 266.

253. D.F. Klessig and J. Malamy, *Plant Mol. Biol.* **26** (1994) 1439.
254. I. Raskin, *Annu. Rev. Plant Physiol. Plant Mol. Biol.* **43** (1992) 439.
255. H. Lee, J. Léon and I. Raskin, *Proc. Natl. Acad. Sci. USA* **92** (1995) 4076.
256. R.A. Dixon, M.J. Harrison and C.J. Lamb, *Annu. Rev. Phytopathol.* **32** (1994) 479.
257. J. Chory, M. Chatterjee, R.K. Cook, T. Elich, C. Fankhauser, J. Li, P. Nagpal, M. Neff, A. Pepper, D. Poole, J. Reed and V. Vitart, *Proc. Natl. Acad. Sci. USA* **93** (1996) 12066.
258. H. Barbier-Brygoo, J. Joyard, A. Pugin and R. Ranjeva, *Trends Plant Sci.* **2** (1997) 214.
259. N. Wei and X.-W. Deng, *Plant Physiol.* **112** (1996) 871.
260. C.R. Redhead and K. Palme, *Crit. Rev. Plant Sci.* **15** (1996) 425.
261. M.J. Mueller, W. Brodschelm, E. Spannagl and M.H. Zenk, *Proc. Natl. Acad. Sci. USA* **90** (1993) 7490.
262. T.M. Kutchan, *The Plant Cell* **7** (1995) 1059.
263. A. Schaller and C.A. Ryan, *BioEssays* **18** (1995) 27.
264. S. Seo, M. Okamoto, H. Seto, K. Ishizuka, H. Sano and Y.Ohashi, Science **270** (1995) 1988.
265. J. Ryals, K.A. Lawton, T.P. Delaney, L. Friedrich, H. Kessmann, U. Neuenschwander, S. Uknes, B. Vernooij and K. Weymann, *Proc. Natl. Acad. Sci. USA* **92** (1995) 4202.
266. C. Lamb and R.A. Dixon, *Annu. Rev. Plant Physiol. Plant Mol. Biol.* **48** (1997) 251.
267. Z. Chen, J. Malamy, J. Henning, U. Conrath, P. Sánchez-Casas, H. Silva, J. Ricigliano and D.F. Klessig, *Proc. Natl. Acad. Sci. USA* **92** (1995) 4134.
268. E.R. Ward, J.A. Ryals and B.J. Miflin, *Plant Mol. Biol.* **22** (1993) 361.
269. R.G. Birch, *Annu. Rev. Plant Physiol. Plant Mol. Biol.* **48** (1997) 297.
270. J.D. Hoheisel, *Trends Biotechnol.* **15** (1997) 465.
271. M.U. Kopp, A.J. deMello and A. Manz, *Science* **280** (1998) 1046.

Chapter 3

MODERN METHODS OF SECONDARY PRODUCT ISOLATION AND ANALYSIS

TERIS A. VAN BEEK

1. Introduction

1.1. General

Different people investigate different plants for different purposes. Many of these investigations are concerned with secondary metabolites. In this chapter, the methods and rationale behind the extraction, purification and analysis of these compounds will be discussed. The exact purpose of the investigation, the chemical structures of the compound(s) of interest and of impurities, the availability of plant material, chemicals and apparatus, as well as manpower and time schedule, together determine to a large extent the analytical route which should be followed to solve any individual problem. In the next section these factors will be briefly reviewed before proceeding to a discussion of the different analytical techniques. Several abbreviations of common phytochemical or chromatographic terms will be used throughout this chapter. For an explanation of abbreviations, see Section 7.

1.2. Aim of Investigation

Why do plants and their constituents need to be extracted, separated and analysed in the first place? The reasons can be manifold. It may for instance be necessary to extract and separate bulk quantities of a plant with a simultaneous bioassay of each fraction because an initial extract of the plant showed promising antitumour or other pharmacological activity. In such a case only fractions which show activity are further separated and eventually one hopes to finish with one or more active substances. In a separate step the structures of the substances can then be elucidated. The aim of such a bioassay-assisted fractionation is quite different from a chemotaxonomic study of several species of one genus in which all alkaloids need to be identified for comparison and a possible taxonomical reclassification. In contrast with the first example it is known beforehand which class of compounds needs to be isolated, which greatly facilitates the isolation and purification route chosen. On the other hand - because all alkaloids need to be isolated - there are high demands for the quality of the final purification and separation step. Some alkaloids may be very difficult to separate. For these two isolations a 100% yield is desirable but not necessary. In a third example, a large number of plants may need to be analytically screened for their content of a single commercially-important metabolite as part of a plant breeding programme. In this case the compound is known and there is no need for its preparative isolation. But in this example sample throughput and cost aspects, allied to 100% recovery and reproducible, sensitive and selective detection, are important. This aim may be as difficult to achieve as the requirements of the first two examples cited, especially within a short period of time.

Thus for each new investigation it is necessary first to define clearly what the aims of the investigation are and only then to start developing a protocol which fulfils these aims. An important consideration will always be the physical and chemical properties of the compound(s) of interest and those of the unwanted compounds, often denoted by the term “matrix”. One should try to get an idea of the overall polarity of the compounds of interest because this will determine to a large extent their behaviour during extraction, partitioning and chromatography. Another important factor is the presence of acidic or basic groups because ionization of such groups will have a large influence on the solubility. One can think of carboxylic acids (ionized at pH > 6), phenols (ionized at pH > 12) and amino groups (protonated at pH < 7). Secondary metabolites possessing a basic tertiary amino group are collectively indicated by the term “alkaloids” and such compounds can be readily purified by manipulating the pH during partitions. At low pH they are water soluble while at high pH they are soluble in moderately apolar organic solvents, e.g. chloroform or ethyl acetate. For neutral compounds such a quick and efficient clean-up is often not feasible and chromatography may be necessary. Much will also depend on whether one is interested in only one compound or in several from one class. In the latter case the compounds may share a common chemical structure which can be used to advantage to purify or detect them. For instance the neutral parthenolide [1] and other α-methylenebutyrolactones occurring in *Tanacetum parthenium* could be selectively and sensitively detected with HPLC-UV (369 nm) after derivatisation with 9-thiomethylanthracene (1). Other neutral compounds lacking this moiety did not react. To help to define a phytochemical problem in Figure 1 some questions are posed and suggestions are given.

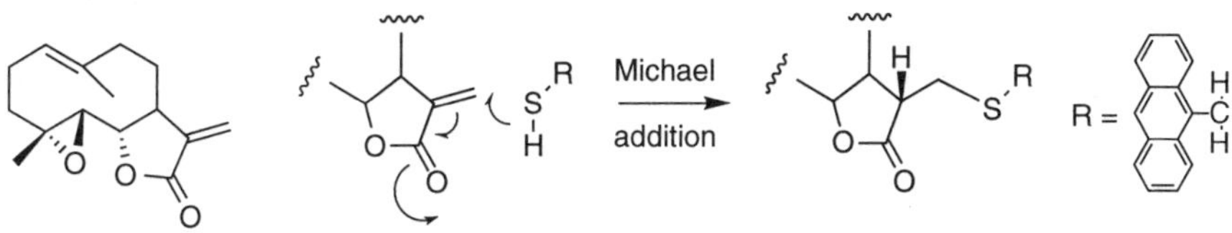

parthenolide [1] α-methylenebutyrolactone Michael adduct

phytochemical analysis

→ preparative separation → known compounds / unknown compounds

→ analytical separation → quantitative / qualitative

known compounds	unknown compounds	quantitative	qualitative
1. one or more?	1. one or more?	1. one or more compounds?	1. one or more compounds?
2. matrix type?	2. re-isolation or first time?	2. matrix type?	2. matrix type?
3. how much?	3. how much?	3. sensitivity?	3. sensitivity?
4. how pure?	4. how pure?	4. reproducibility?	4. speed?
5. how unstable?	Use mild methods.	5. which apparatus?	5. lab or field?
6. which apparatus?	Use partitions initially.	6. how many samples?	6. reliability?
Check literature	Do chemotaxonomic search.	Check literature.	Check literature.
Optimise method, look at all properties of compounds.		Select optimal method and evaluate/optimise.	If necessary, use sample clean-up.
Try out on small scale			

Figure 1. Flow chart for defining a phytochemical problem. For each type of problem different questions have to be answered prior to the start of any practical work.

The presence of labile groups, e.g. certain esters and tertiary alcohols, may also influence the extraction and separation procedure. Whether one is developing a preparative or a quantitative method, decomposition is always unwanted and artefact formation is something to be reckoned with. This aspect is discussed in more detail in Section 5. The physical properties of the compound are another important factor: what is the boiling point and can the compound be steam distilled or run through a gas chromatograph? Does it show good UV absorption at a selective wavelength so that it can easily be monitored after chromatography? All the above questions should also be posed for the matrix and at a later stage for unwanted materials which interfere with the purification or analysis. Only by making optimal use of the chemical and physical differences between compounds will it be possible to solve phytochemical problems. Of course most compounds are not novel and even those that are will resemble other known structures. Thus there is much to be learnt from the literature. If a good method exists the wheel should not be reinvented. On the other hand many poorly validated and sometimes faulty quantitative assays have been and keep being published in the literature and a critical reading of each method is indicated. Any quantitative method lacking recovery data should not be used without further validation. Even if one does not wish to use an earlier method it is good to be aware of the approaches which have been used before. Still other factors that play a role in determining the final analytical method are the availability of manpower, time, equipment, consumables, reference compounds and experience. For instance one may have encountered in the literature for a particular analyte an elegant LC-MS-MS method requiring almost no sample clean-up. If however this piece of equipment unfortunately happens to be absent, it may be a good idea to give an old HPLC a second chance after some additional sample clean-up. And even if all published methods are excellent, not all of them are equally useful and some are sheer analytical overkill. It is good to be aware of this and to reassess frequently the aims of the investigation.

1.3. Phytochemical Literature

There exists a large amount of literature on the methodology and approaches for the isolation, separation and quantitative and qualitative analysis of secondary products from plants. Unfortunately this information is rather dispersed both in time and in source. For instance food chemists working on the aroma compounds of strawberries will publish the phytochemical part of their work in the experimental section of a paper in a food science journal which may not be available in a general chemical library. To retrieve such papers a search on the plant or compounds under investigation in Chemical Abstracts, or to a lesser extent Biological Abstracts, is the method of choice.

Instead of this thorough approach one could scan some journals with a high percentage of articles on the isolation, purification and analysis of secondary metabolites. Phytochemical Analysis is the only journal devoted to the publication of phytochemical methodology. This bimonthly journal also contains a "current awareness in phytochemical analysis" section with ~250 up-to-date abstracts on the analysis of plant constituents. Other useful journals include: Journal of Natural Products, Phytochemistry, Planta Medica, Natural Product Letters and Fitoterapia.

For methodological purposes, the most interesting part of articles published in these journals is the experimental section. For articles on essential oils and other plant volatiles the Journal of Essential Oil Research and Flavour and Fragrance Journal are good sources. Chromatographic journals (Journal of Chromatography, Journal of Liquid Chromatography, Chromatographia) and several more specialised journals (Journal of Chemical Ecology, Journal of Agricultural and Food Chemistry, Chemical and Pharmaceutical Bulletin, Helvetica Chimica Acta) contain on a regular basis articles in which phytochemical analysis plays a role.

Useful books or chapters in books on phytochemical analysis are (2-9). Occasionally some information on the analysis of specific groups of compounds can be found in the series: "Alkaloids: chemical and biological perspectives" (Ed. S.W. Pelletier, Wiley, New York), "The Alkaloids: chemistry and pharmacology" (Ed. G. Cordell, Academic Press, San Diego), "Progress in the chemistry of organic natural products" (Eds. W. Herz, G.W. Kirby, R.E. Moore, W. Steglich and Ch. Tamm, Springer, Vienna), "Proceedings of the Phytochemical Society of Europe" (PSE, Oxford University Press, Oxford), "Recent Advances in Phytochemistry" (Proceedings of the Phytochemical Society of North America, Plenum Press, New York), "ACS Symposium Series" (American Chemical Society, Washington), "Modern methods of plant analysis, new series" (Eds. J.F. Linskens and H.F. Jackson, Springer, Berlin), "Methods in Plant Biochemistry" (Eds. P.M. Dey and J.B. Harborne, Academic Press, London), "Studies in Natural Products Chemistry", Vol. 7, 9, 13, 15 and 17 (Ed. Atta-ur-Rahman, Elsevier, Amsterdam), "Natural Products Reports" (The Royal Society of Chemistry, London), "Journal of Chromatography Library" (Elsevier, Amsterdam) and the new "Medicinal and Aromatic Plants - Industrial Profiles" (Ed. R. Hardman, Harwood, Chur, Switzerland).

2. Extraction Techniques

2.1. Introduction

Only very rarely can plants be analysed directly for one particular compound or a group of constituents without an extraction of some sort. One technique capable of achieving this to a limited extent is near infrared spectroscopy (NIR) (10) which will, however, not be further discussed in this chapter. An example of a spontaneous preparative isolation without an extraction step is the formation of highly pure cedrol crystals on chunks of cedarwood. However, in almost all cases regardless of the aim of the investigation, at least an extraction and usually also a purification step (sample clean-up) are necessary. The extraction serves not only to remove and separate the compounds of interest from the insoluble high-molecular weight parts of the plants but also from other extractables which could interfere with later steps. Cellulose, hemicellulose and lignin, on a weight basis normally constitute the majority of dry plant material. In bark cutin and suberin are of importance too (11). Compounds which are (partially) extractable include many primary metabolites, chlorophyll, pectic substances, starch, (glyco)proteins and salts, in addition to secondary metabolites. The weight percentage of secondary metabolites is strongly dependent on the plant species and plant part (fruits, seeds, stems, bark, wood, flowers, leaves) but is normally below 10%.

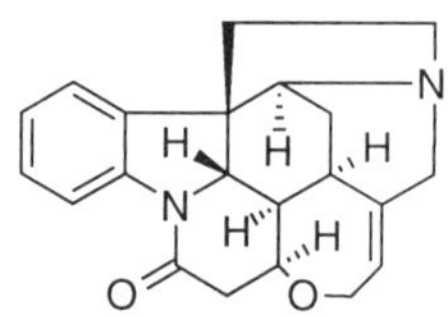

strychnine [2]

Within the plant, secondary metabolites are often bound by, dissolved in or screened off by other compounds present. This makes them difficult to extract. The interactions between compounds of interest and other compounds are collectively called 'matrix' effects. To achieve a quick and complete extraction, the matrix effects have to be broken up. For instance, if one wants to extract the alkaloid strychnine [2] from the hard *Strychnos* seed, dilute sulphuric acid will extract strychnine only very slowly in spite of the fact that this alkaloid is readily soluble in the extraction medium as its sulphate salt. This is caused by the fact that the acid cannot reach the strychnine because together the hard insoluble seed hull and the high percentage of fat form an effective barrier. Dilute sulphuric acid can only be used after first crushing the seeds and then removing the apolar fats with hexane or petroleum ether. Alkaloids are poorly soluble in the highly apolar alkanes which are excellent for the selective removal of the fatty matrix. After drying the crushed defatted remains of the seeds, the alkaloids can then be selectively removed by the highly polar aqueous sulphuric acid. Many compounds of intermediate polarity not extracted by hexane will stay in the seeds because they are equally insoluble in dilute acid. If necessary such compounds can be extracted by solvents of intermediate polarity, such as ethyl acetate, acetone or methanol.

The above example is typical of many extractions of secondary metabolites in which two or three phases can be distinguished (Figure 2). Assuming that a solvent is chosen in which the analyte has sufficient solubility, in the initial phase free analyte molecules on or close to the surface rapidly dissolve. In the next phase solvent molecules enter the plant material and may break up interactions between solute molecules and other molecules in the plant (matrix effects). The solvent may also facilitate the diffusion of analyte out of the plant material into the solvent proper by swelling or by making membranes more permeable to the analyte. In the third and final phase diffusion of free analyte molecules is the limiting factor. A model for spherical particles has shown that with supercritical fluids as the extraction medium this is a logarithmic process so it is impossible to recover 100% (12, 13). Practice confirms this. The same model shows that the two parameters which determine the extraction rate during the diffusion-limited phase are, in order of decreasing importance, the particle size and the diffusion coefficient. However for various reasons the model has its limitations: plant particles are irregular, analytes are not evenly distributed, the plant material does not have a uniform composition, there may be severe matrix effects and solubility problems may exist for some analytes, especially during a maceration. In some cases the solvent may not even be able to reach the bulk of the analyte without a pre-extraction (strychnine from *Strychnos* seeds).

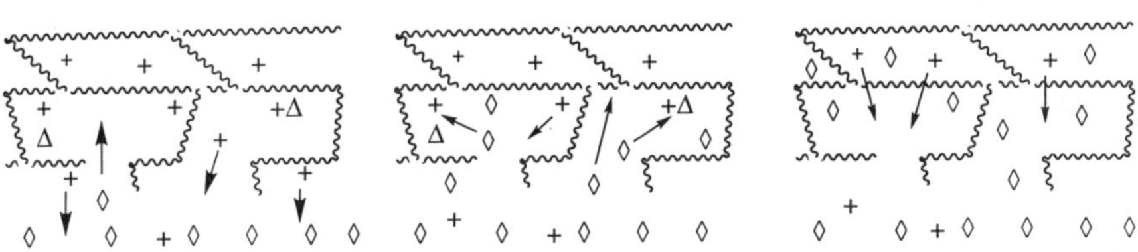

Figure 2. Schematic drawing of the extraction process. In step 1 (left) solvent molecules (◊) enter disrupted membranes, and free analyte (+) dissolves rapidly. In step 2 (middle) solvent enters the plant material, disrupts matrix effects (+Δ), setting analytes free in the process, and makes membranes more permeable. In step 3 (right) free analyte slowly diffuses from the interior.

Therefore it is essential to check the completeness of an extraction, whether analytical or preparative. The technique of spiking (adding pure analyte) is useless for plants because of the diffusion and matrix problems referred to earlier. It is better to construct an extraction curve as given in Figure 3 by analysing subsequent fractions and then to decide in retrospect when one reaches an acceptable recovery. Even then it is wise to re-extract the remaining plant material with a stronger solvent to make sure that a more strongly bound percentage of the analyte is not overlooked (different degrees of matrix effects).

The important role of the particle size on the speed and completeness of any extraction process is clearly shown in Table I where an imaginary piece of wood of original dimensions 10 x 10 x 10 cm is divided into increasingly smaller cubes. As the number of particles gets bigger and bigger, the total surface area where the extraction solvent can come into contact with the plant material also increases while the maximum diffusion length decreases from the initial 5 cm to a mere one eighth of a mm. Additionally many cell walls will be disrupted in the process, further speeding up the initial stages of the extraction. Further diminishing the particle size is not advisable. A maximum diffusion pathway of 0.125 mm is acceptable and smaller particles tend to clog filters.

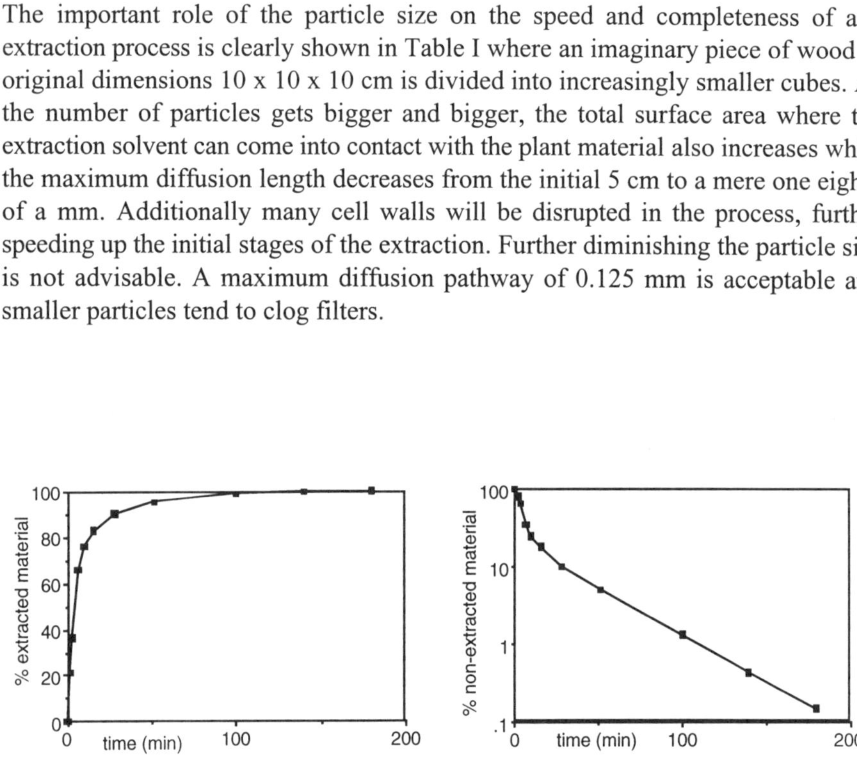

Figure 3a (left) and 3b (right). Typical extraction curve of plant material. Initially there is a rapid extraction (solubility or mass transfer limited), then the speed slowly declines (disruption of matrix effects, diffusion) and finally the extraction becomes diffusion-limited (logarithmic part). This can be seen in the curve in 3b. Thus the extraction never becomes complete.

Table I. Relation between number of particles, particle size, total available surface and maximum diffusion length while keeping the volume and weight constant

Number of particles	Size (cm)	Total surface (dm^2)	Max. diffusion length (mm)
1	10x10x10	6	50
1000	1x1x1	60	5
1,000,000	0.1x0.1x0.1	600	0.5
64,000,000	0.025x0.025x0.025	2400	0.125

In principle four different types of extraction medium can be distinguished: (1) liquids, (2) supercritical fluids, (3) steam, and (4) gases. The first two types of medium can really physically dissolve analytes while the last two types function more as a transportation medium to remove selectively more volatile analytes. In selected cases (oil from seeds, juice from fruits) plant materials can be pressed to obtain the fraction of interest. Still another possibility is to liquefy a plant part. Certain fungal enzyme mixtures can "dissolve" cell walls and after a certain period a soup remains containing the analytes. This is an effective way to disrupt strong matrix effects. An example is the liquefaction of chicory roots. After chopping 8 kg of roots into 3.5 mm pieces, a 2 h treatment with 8 g of crude enzymes at pH=4.5 and 40°C, followed by centrifugation, gave 5 l clear supernatant. Extraction with ethyl acetate gave the moderately apolar bitter sesquiterpene lactones (14). More apolar constituents did not dissolve and were removed by the centrifugation step.

Each of the four main types of extraction has advantages and disadvantages and many problems later on during the purification or quantitation of analytes can be circumvented if the proper extraction medium and the right conditions are selected. It is clear that no matter how sophisticated the equipment for the final purification or quantitation, if 25% is lost due to a poor extraction this cannot be remedied. In the remainder of Section 2, the four different types of extraction medium which are used for the isolation of secondary metabolites from plants are discussed in more detail.

2.2. Solvent Extraction

This is by far the most common and universal extraction technique. Two types of extraction can be distinguished: (I) a pre-extraction either to remove impurities from the plant material or to make the analytes available for further extraction (see *Strychnos* seeds in Section 2.1), or (II) extraction of the compounds of interest. In a type I extraction the solvent should not dissolve the analyte to an appreciable extent while the impurities should be readily soluble. Examples are hexane or petroleum ether for seeds or leaves to remove large amounts of fats or chlorophyll respectively. Limited information about solubilities can be found in the Merck Index (15) or the Handbook of Chemistry and Physics (16). As a rule of thumb, analytes which contain fewer carbon atoms than 4 times the combined number of oxygen and nitrogen atoms are poorly soluble in hexane or petroleum ether. For more polar analytes like glycosides, instead of hydrocarbons, dichloromethane or preferably petroleum ether / ethyl acetate mixtures can be used which dissolve more unwanted compounds. An example is the removal of carotenes, essential oils and other apolar impurities from orange peel with either chloroform or a hexane / ethyl acetate mixture prior to the extraction of the flavonoid glycoside, hesperidin [3], with methanol.

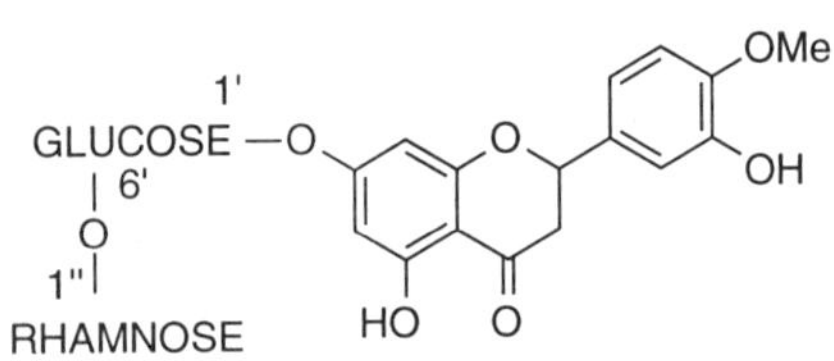

hesperidin [3]

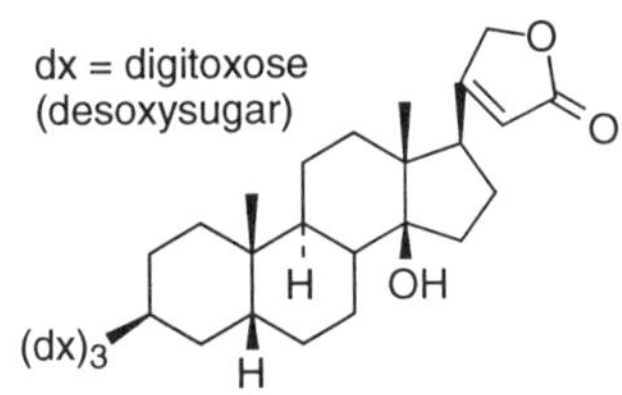

digitoxin [4]

In contrast, in a type II extraction the solvent should be sufficiently "strong" to dissolve all metabolites of interest and to disrupt any matrix effects. On the other hand selectivity is equally important. If a quick and complete extraction can be achieved either with hexane /ether = 95 / 5 or with pure ether, the former mixture is superior because fewer compound dissolve in it, meaning also fewer impurities. Generally speaking the selectivity decreases from hexane (a "weak" solvent) to methanol (a "strong" solvent). Until methanol the listing of solvents from weak (syn. selective) to strong (syn. non-selective) closely follows the eluotropic ranking of solvents mentioned inSection 3.5. A strong increase in selectivity is again seen in going from methanol to water. Many apolar impurities like chlorophyll are insoluble in water or water / methanol mixtures. An extreme example of this is the initial extraction used in the quantitative determination of *Digitalis purpurea* leaves according to the 8th edition of the Dutch Pharmacopeia (17). The main active constituent is digitoxin [4], a cardiac glycoside whose solubility in water according to the same Pharmacopeia is less than 1 part in 10,000. Nevertheless the extraction is carried out by a 1 h maceration with pure water. The explanation for this is that only 1 mg of glycoside is present while 50 ml of water is used. Thus the solubility is not exceeded and few of the many apolar impurities present in the leaf show up in the aqueous extract. After precipitation of some phenols with lead(II) ions, the sugars are removed by an acidic hydrolysis. After extraction with chloroform, the resulting aglycone is determined colorimetrically. Had a less selective, but more effective, solvent than water been used in the initial extraction, many apolar impurities would have been present in the chloroform extract, possibly interfering with the final colorimetric reaction. Conclusion: the "best" solvent is usually not the solvent in which the analytes are most soluble but the solvent which still gives a good extraction of the analytes while at the same time co-extracting a minimum of impurities.

Technically, three different types of solvent extraction can be distinguished. During a *maceration* the plant material is soaked in the solvent. The extraction time can vary from several minutes up to weeks. Other variables include the amount of agitation, plant - solvent ratio, moisture content, temperature and the number of times that the extraction is repeated. A well known maceration is the preparation of a cup of tea. Also the dramatic influence of stirring and the temperature on the extraction rate is well illustrated by this example. In a *percolation*-type extraction the plant material is continuously flushed with fresh solvent. The extraction is continued until sufficient compound is extracted. If necessary the same material can be re-extracted with a second solvent. A coffee machine is an example of this type of. extraction. In a *Soxhlet* extraction - similar to a percolation - the plant material is continuously flushed with fresh solvent (Figure 4). This fresh solvent is formed by boiling the solvent containing the extracted analytes. Thus in contrast to a percolation the total amount of solvent is limited. In spite of what is sometimes thought, a Soxhlet extraction can be far from complete due to channelling or the presence of air in the semi-permeable thimble containing the plant material. The strong and weak points of the three main types of extraction have been summarised in Table II. As a general guideline one can say that a maceration is most useful for the extraction of compounds occurring in low concentrations which are strongly bound and/or diffuse slowly. A percolation is more useful for extracting loosely bound compounds which occur in high concentrations or have a low solubility in the solvent. A Soxhlet is useful for certain quantitative extractions of thermally stable compounds.

A recently-introduced fully-automated new variant of percolation-type extraction is the "Accelerated Solvent Extractor". This extractor is equipped with an autosampler for 24 steel extraction vials (11-33 ml) and can extract with any solvent up to 200° and 200 atm. Due to the high pressure the solvent does not boil whilst the high temperature gives a more efficient extraction. Claimed advantages are automation of

Figure 4. Schematic drawing of a Soxhlet extractor.

Table II. Advantages and disadvantages of maceration, percolation and Soxhlet types of extraction

Advantages	Disadvantages
Maceration	
Simple and cheap	Analytes must be sufficiently soluble without stirring and/or heating slow
Limited use of solvents	
Can give good and selective extraction	
Percolation	
Exhaustive extraction possible	Some supervision necessary
Mild technique	Selectivity can be a problem
	Can be slow
	Large amounts of solvent necessary
Soxhlet	
Low solvent consumption	Harsh technique, thermal decomposition
No supervision necessary	Water and electricity needed
	Extraction not as good as sometimes thought
	Less suitable for large scale operation

the extraction process, time gain and low solvent consumption (18, 19). No reports in the field of natural products have been made so far but potentially there are plenty of applications for routine analysis of thermally stable secondary metabolites.

2.3. Supercritical Fluid Extraction (SFE)

Supercritical fluids have properties which are intermediate between those of gases and liquids. Substances become supercritical when both their pressure and temperature exceed the supercritical values (T_c and P_c). For CO_2, which is the most commonly used fluid, these values are 31° and 73 atm respectively. Supercritical fluids possess a near-liquid-like density which gives them almost liquid-like solvating power. On the other hand the diffusion constant is more gas-like, which leads to rapid mass transfer. The viscosity is considerably less than that of liquids, which is especially advantageous for supercritical fluid chromatography (see Section 3.7), though less so for SFE. Finally, they have very low surface tensions and hence easily penetrate different matrices. Thus, in a manner of speaking, supercritical fluids combine the best of both worlds and have interesting properties for extracting plant constituents. Only in recent years has SFE gained use as a laboratory tool for the small-scale extraction of plant parts, mainly for sample preparation purposes. Previously only large-scale industrial applications were known, such as the decaffeination of coffee beans and the extraction of hops. Reviews of the application of SFE in the natural products field are given in (20-22).

Table III. Advantages and disadvantages of Supercritical Fluid Extraction (SFE)

Advantages	Disadvantages
Mild technique	Less suitable for very polar products
Ideally no organic solvents needed, i.e. environmentally friendly and cheap	Rather complex apparatus needed
Can give fast extractions	Involves high pressures
Extractions can be automated	Less suitable for the extraction of leaves
With 100% CO_2 no solvent removal step	Trapping can be troublesome
Solvating power variable via pressure	Scale-up is difficult
	Fresh plant materials are difficult to extract

The properties of supercritical fluids mentioned above lead to a number of advantages (see Table III) of SFE over traditional extractions with solvents (Section 2.2). SFE extractions are fast, environmentally-friendly and sometimes more selective. This is primarily due to the fact that the pressure and temperature have a strong influence on the density, and thus on the solvating power. One can fractionate simply by increasing the pressure. However supercritical fluids have no magical properties and have also disadvantages (Table III). In particular the solvating power of pure CO_2 is weak (≈freon). It can extract only apolar materials like fats, certain triterpenes and essential oils. The solvating power can be increased by adding methanol to the fluid (in pre-mixed tanks, with a second pump or by prior addition to the extraction cell). However SFE then loses one of its biggest advantages: it is no longer solvent-free. When large amounts of uninteresting apolar materials are present, like chlorophyll in leaves, SFE is also not very selective. Such impurities are co-extracted as in a solvent extraction. Finally preparative uses outside the industrial field are difficult to realise due to the high cost of the apparatus, preparative vessels and the CO2 itself. In conclusion: SFE seems most valuable for the small-scale extraction of thermally labile apolar secondary metabolites from dry plant materials.

The SFE process can, if necessary, be monitored on-line by various spectroscopic techniques, like UV, FT-IR or NMR. Piperine [5] and caffeine [6] could be readilydetected during the extraction of black pepper and roasted coffee respectively with the latter technique (23). Detailed information about developing an SFE method is given in (24). More basic information on SFE can be found in (25, 26).

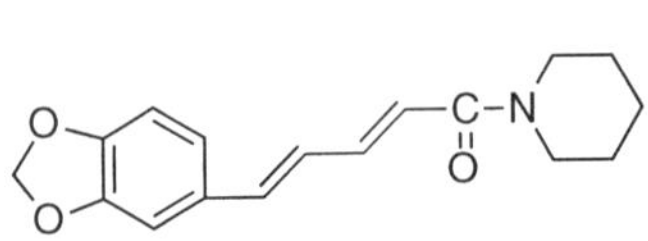
piperine [5]

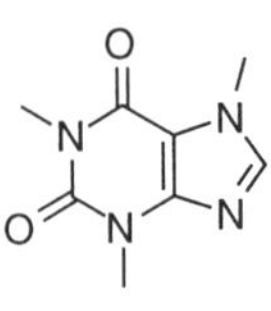
caffeine [6]

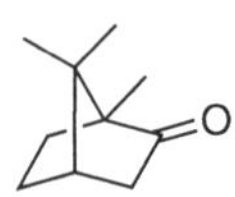
camphor [7]

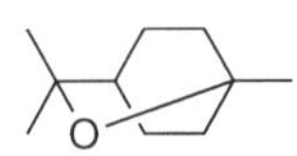
1,8-cineole [8]

A schematic drawing of an SF extractor is depicted in Figure 5. A CO_2 cylinder supplies a modified HPLC pump with liquid CO_2 which, upon entering the oven, becomes a supercritical fluid. If necessary, a second modifier pump can add a small percentage of methanol or another polar solvent to the CO_2 for increasing the solubility and extraction rate of more polar secondary metabolites. The extraction takes place in a thick-walled stainless steel thimble (1-100 ml) which contains the sample. The sample can be pre-mixed with inert sand or celite to increase the extraction surface or to avoid a physical displacement of small sample particles by the fluid. The extraction can be either static, dynamic or a combination of the two. A static extraction is comparable to a maceration and is most useful when matrix effects and diffusion are important. A dynamic extraction is comparable to a percolation and is most useful when solubility in the fluid is a problem. Of course a dynamic step is also necessary for removing any free analytes. The temperature and pressure in the extraction vessel can be regulated and are usually between 35°-150°C and 120-680 atm respectively.

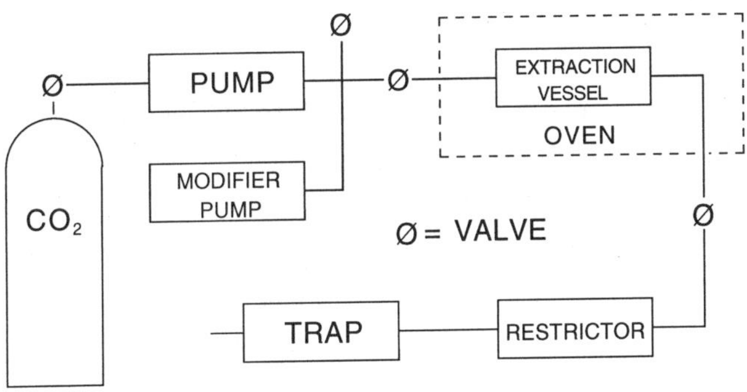

Figure 5. Schematic drawing of a Supercritical Fluid Extractor. The extraction vessel is situated in an oven so that the temperature can be regulated. The trap can be either a liquid or solid trap.

After leaving the cell the fluid is decompressed to atmospheric pressure with a restrictor. This can be a piece of narrow-bore tubing. Clogging can be a problem. Finally the extracted analytes are separated from the gaseous CO_2 in a trap. Two types of trap exist. In a liquid trap the end of the restrictor is placed in a small volume of a volatile solvent. The analytes dissolve in the solvent and can be analysed afterwards. Disadvantages of this type of trapping are losses due to aerosol formation and loss of solvent during trapping. With a solid trap the gas and analytes pass through (cooled) steel balls, silica or C18 material which retains the analytes. Afterwards the trap can be eluted with a small volume of a suitable organic solvent to yield an analysable extract.

An example of SFE is the extraction of 1 g *Rosmarinus officinalis* leaves with pure

CO_2 (27). The highest recoveries were obtained with relative high flow rates of 4 ml/min during 20 min at 333 atm and 40°C. Trapping was carried out on C18 material at 30°C. After eluting with dichloromethane the extract was investigated by means of GC. The extract was also investigated by means of SDE (see Section 2.4). Although qualitatively the same terpenes were extracted, there were significant quantitative differences between the SFE and SDE results. Camphor [7] and 1,8-cineole [8] were the major constituents and are good examples of compounds which can be extracted with 100% CO_2. On average the relative standard deviations were smaller for the SDE method. The SDE extraction time was 3 times as long as the SFE extraction time. For the extraction of more polar compounds the addition of a polar modifier such as methanol is needed. Ginkgolides A-J [9-12] could be extracted from crude ginkgo extracts with 10% methanol in CO_2 at 360 atm and 45°C if the extract dissolved in methanol was premixed with silica. The addition of methanol was more important for breaking up matrix effects than for dissolving the ginkgolides in the fluid (28).

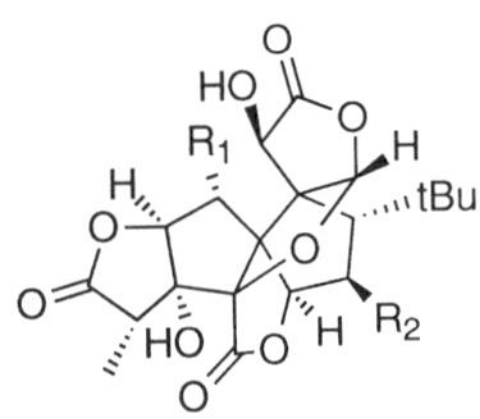

ginkgolide A [9] $R_1=R_2=H$
ginkgolide B [10]$R_1=OH$, $R_2=H$
ginkgolide C [11]$R_1=R_2=OH$
ginkgolide D [12] $R_1=H$, $R_2=OH$

2.4. Steam Distillation

Steam distillation is a very selective, simple and clean isolation technique for the isolation of water-insoluble volatile constituents from various matrices, including plant materials. It is selective because only volatiles (b.p. 100° - 350°C) are isolated. As volatiles rarely constitute more than 5%, and usually less than 1%, of the total weight an enrichment factor of 100 is not uncommon and is difficult to achieve by either liquid extraction or SFE. Simple because no expensive apparatus or special steps prior to the extraction are necessary. Clean because colourless or light yellow oils are obtained consisting only of volatiles. No fats or other apolar materials are present in the extract. The advantages and disadvantages are summarised in Table IV.

The principle of steam distillation is based on Dalton's Law: the total vapour pressure of a system consisting of two immiscible solvents (water and essential oil) is equal to the sum of their partial vapour pressures. As one of the components is water, the boiling point of the mixture under atmospheric pressure will be slightly less than 100° (see Figure 6). Theory predicts further - due to the relatively high boiling point of water in relation to its molecular weight - that any distillate will contain a relatively high weight percentage of essential oil constituents. Thus even constituents having boiling points of over 300° (sesquiterpene alcohols) can be isolated below 100°C by means of steam distillation.

Table IV. Advantages and disadvantages of the steam distillation technique

Advantages	Disadvantages
Selective for volatiles, e.g.essential oils	Only suitable for apolar volatiles
Very simple and cheap technique	Possible decomposition due to the presence of water and high temperatures
Only water needed as solvent	Unsuitable for mg scale operation
Well suitable for preparative scale extractions	Quantitative distillation is time-consuming
Very clean extracts	Distillate varies during distillation
High enrichment factor	Can lose more polar trace constituents

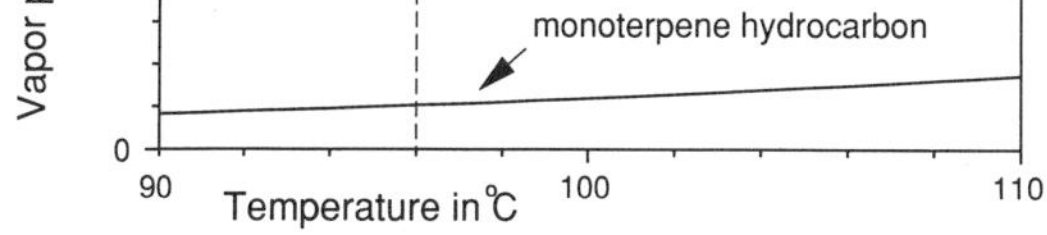

Figure 6. Part of the vapour pressure curves of water, a monoterpene hydrocarbon and a mixture of the two. The mixture boils at 96° under atmospheric pressure (dashed lines) although the two individual constituents have boiling points of 100°C and 160°C respectively. The distillate will consist of about equal amounts of both.

Two - not in principle different - forms of steam distillation can be distinguished: (1) a true steam distillation in which steam prepared in a separate steam generator is blown through the plant material, and (2) a hydrodistillation where the plant material is suspended in boiling water and where steam is generated *in situ*. The latter type is more popular for the laboratory scale preparation of essential oils. A special continuous form of hydrodistillation named a Clevenger apparatus after its inventor is used for the quantitative determination of essential oils according to the European Pharmacopeia (Figure 7a). True steam distillation is more commonly used for the industrial preparation of essential oils. A thorough discussion of the steam distillation of essential oils can be found in (29). Steam distillation should be clearly distinguished from fractional distillation (Section 3.9). The latter is a separation technique while the former is an isolation technique. With the steam distillation process all volatiles are isolated as a mixture.

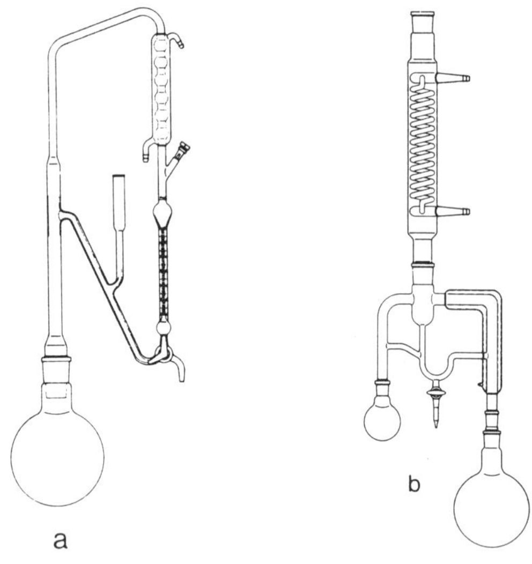

Figure 7. A (left): Clevenger-type apparatus for the continuous distillation of essential oils. Oil which is lighter than water is collected in the conical bulb while water returns to the distillation flask. **B (right)**: A Likens and Nickerson-type continuous steam distillation extraction (SDE) apparatus. Volatiles move via a hydrodistillation and a vapour phase extraction from the larger water-containing roundbottom flask to the smaller organic-solvent-containing bulb while water and organic solvent run back to their respective containers.

Matrix effects and diffusion also play an important role during the distillation process. When aromatic plants where all the oil is present on the surface, e.g. glandular hairs, are distilled, diffusion hardly plays a role and the distillation time is short. All oil components are quickly removed. In contrast, when the oil is present deeper in seeds or woody material, the distillation rate is determined by matrix effects and diffusion. Thus hydrocarbons distil more slowly than corresponding alcohols in spite of the fact that the former have lower boiling points. This is caused by their much lower water solubility: they remain more associated with other constituents, especially fatty materials and they dissolve much less in the aqueous medium, thus inhibiting a rapid diffusion to the surface (30).

If very small amounts of volatiles have to be isolated or if volatiles which have a significant solubility in water are present (e.g. hexenal), a special isolation technique which combines a hydrodistillation and an extraction can be used (Figure 7b). This technique is called Steam Distillation Extraction (SDE). In the vapour phase an extraction of the hydrodistillate containing the plant volatiles with a highly volatile organic solvent (e.g. pentane / ether = 1 / 1) takes place. This effectively strips the steam of any plant volatiles. Next both the steam and the organic solvent condense and due to the difference in density and the ingenious glass construction in Figure 7b the two liquid phases each return to their own roundbottom flask. Once these volatiles have entered the small flask containing the organic solvent they are trapped because their boiling points are 100° - 300° higher than that of the organic solvent. After the end of the extraction the organic solvent can be removed and the concentrated volatiles can be further investigated, e.g. with GC.

2.5. Headspace Techniques

The headspace of an object, e.g. a flower, is the atmosphere directly surrounding it. The headspace contains different volatiles in varying amounts which originate either from the object or the environment. The headspace can be collected and analysed. Two types of headspace analysis can be distinguished:
(I) equilibrium or static headspace - used for analytical purposes
(II) dynamic headspace - used for analytical and preparative purposes.

With technique (I) a plant part is placed in a closed vial which, if needed, can be heated. After a certain amount of time ($\approx$30 min) an equilibrium is reached between the concentration of volatiles in the plant and the surrounding headspace. The concentration ratio will be dependent on the boiling point, the temperature and matrix effects. Although the vapour composition can be interesting in its own right, one is usually more interested in the amount of volatiles in the original plant part. In principle this could be calculated from the vapour phase composition; however in practice several problems arise:

- It is difficult to transfer accurately a known and representative volume of headspace to a GC by hand. Due to the excess pressure, headspace can leak away from a syringe and condensation of high-boiling components can occur. Specialised automated headspace samplers exist which solve these problems.

- Splitless injection of a 1 ml volume of headspace into a capillary GC is not possible unless some additional focussing takes place.

- For intact plant parts, no suitable reference mixtures can be prepared because the matrix effects cannot be quantified. To solve this problem the Multiple Headspace Extraction (MHE) technique was developed. The MHE technique is based on a repeated extraction of volatiles from a sample. As long as equilibrium exists before the volatiles are removed, the concentration in the original material declines in a logarithmic fashion. By calculation, a value for the total amount originally present can in principle be obtained from two measurements.

- When very small amounts of volatiles are present sensitivity problems can occur. One may resort to dynamic headspace sampling which is more sensitive.

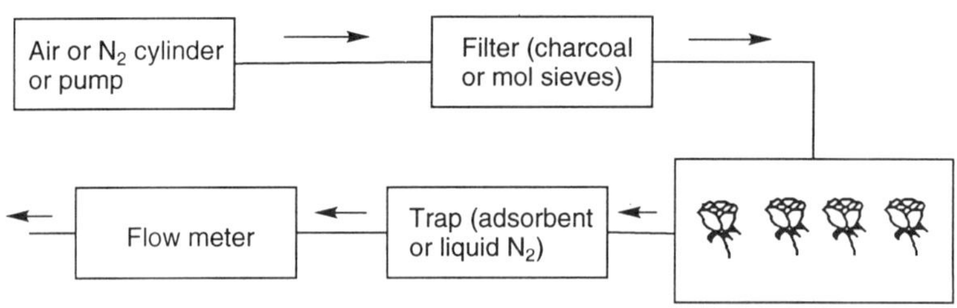

Figure 8. Schematic representation of dynamic headspace sampling equipment.

With technique (II) clean air or nitrogen is first led over the sample and then either over a trap containing a suitable adsorbant or over a cold trap (Figure 8). Clean air can be obtained by prior filtering over charcoal, molecular sieves and/or adsorbants. This step is essential because standard air or nitrogen contains too high a percentage of environmental pollutants. All the tubing should preferably be of Teflon® because other materials release impurities (phthalates, siloxanes). The volatiles released by the plant material are carried along with the gas and then focussed at the trap. Adsorbent traps are more common and contain materials with a high surface area and a high affinity for apolar compounds. Most common are synthetic polymers such as Tenax®, Porapak® or Amberlite® XAD, activated charcoal or C18-derivatised silica. If volatile concentrations are low and the sampling container large it is better to connect the trap only after the container has been thoroughly flushed with purified gas. Otherwise impurities originally present in the container can still interfere with the analysis. The set-up in Figure 8 (sample in closed container) is not suitable for sampling larger plants under natural conditions. Then one has to resort to non-closed sampling. This increases the risk of contamination with other volatiles not produced by the sample, especially with long sampling times. An example of this type of sampling is the analysis of many different orchids (31).

After a certain time (1 min to 24 hr) in which sufficient volatiles are collected, the trap can be thermally or chemically desorbed. Thermal desorption is carried out in a specially-constructed oven connected on-line to a GC. Because of the large volume of carrier gas necessary to desorb all volatiles, a direct connection to a capillary column is not possible due to excessive band broadening. Therefore refocussing, either by a smaller trap with adsorbant, or by low temperatures (-75° to -150°C) is necessary. After focussing, the secondary trap is rapidly heated giving ideally sharp peaks. Advantages are automation and high sensitivity; disadvantages are possible thermal decomposition and non-repeatability. This desorption technique was used for the analysis of the headspace of *Brassica napus* flowers (32).

With chemical desorption, the trap is flushed with a small quantity of solvent (pentane, Et_2O or CS_2), concentrated and injected splitless onto a normal GC. This approach is much simpler but contamination and sensitivity can be a problem. This desorption technique was used for the analysis of apple headspace (33). A preparative application of this technique was the isolation of ≈0.25 mg of each of the kairomone constituents 4,8-dimethyl-1,3E,7-nonatriene and 4,8,12-trimethyl-1,3E,7E,11-tridecatetraene (34). For this purpose 5 kg of bean leaves infested with spider mites was flushed with approximately 1500 l of purified air. The volatiles were trapped on Tenax®. After flushing the Tenax® with pentane / acetone = 4 / 1 and concentration to 0.6 ml, the extract was separated by preparative capillary GC. Enough was isolated for NMR measurements which resulted in elucidation of the structures.

Both techniques (I) and (II) are very mild and selective for the isolation of small amounts of volatiles from plants. The headspace composition of an aromatic plant or a flower can differ considerably both qualitatively and quantitatively from an essential oil obtained from the same plant or flower because of the much harsher conditions during a steam distillation and because of matrix effects. This is dramatically illustrated by a comparison of the volatile composition of an SDE extract and a dynamic headspace extract of kiwi fruit flowers (35). In Table V all the components which occur in concentrations > 5% in either of the extracts are tabulated. In particular, the qualitative differences are hard to explain. Nonetheless, especially with the proper equipment, equilibrium headspace can be an elegant, fast and economic way to analyse aromatic plants. Some advantages and disadvantages of headspace sampling are summarised in Table VI. A review on the application of headspace sampling for the analysis of aromatic plants and flowers is given in (36).

Table V. Major constituents in % of SDE and headspace (HS) extract of the same kiwi fruit flowers

Compound	SDE	HS	Compound	SDE	HS
Hexyl acetate	0.03	5.47	Unknown $C_{17}H_{34}$	8.95	nd
(Z)-3-Hexenyl acetate	0.12	16.52	Unknown $C_{17}H_{32}$	nd	8.57
Heptanal	9.07	nd	α-Farnesene	7.42	29.70
Nonanal	12.57	nd	Unknown	4.82	6.92
Pentadecane	4.49	9.06	hydrocarbon		

Data taken from: (35) nd: not detected

Recently a simple alternative for headspace extraction called solid phase microextraction (SPME) has become commercially available. It consists of a 1 cm length of fused silica fibre, coated with a 0.1 mm layer of stationary phase, e.g. polydimethylsiloxane, which is bonded to a stainless steel plunger. The plunger can be retracted into a hollow needle. The needle can pierce the septum of a container with the sample, after which the plunger is lowered. Once equilibrium is reached, the needle is transferred to a GC and in the injection port the adsorbed volatiles are thermally desorbed from the stationary phase. SPME has been used for investigating the headspace of rancid corn oil (37) and crushed strawberries (38). The authors of the latter investigation conclude that SPME would be suitable for the screening of

Table VI. Advantages and disadvantages of the Headspace (HS) sampling technique

Advantages	Disadvantages
Selective for volatiles	Many variables in dynamic HS sampling
Very mild technique	Dynamic HS quantitatively less suitable
Can be fast, no clean-up necessary	Impurities can be a problem in dynamic HS
Living plants can be sampled	Static HS apparatus complex
	Introduction of sample on GC difficult
	Less suitable for preparative purposes

less volatile aroma compounds in a plant breeding programme. For highly volatile substances, like E-2-hexenal the reproducibility was poor. An even more simple technique than SPME, is solvent microextraction: the needle coated with immobilised stationary phase is replaced by a Teflon® rod containing a few μl of organic solvent containing an internal standard. After being held in a stirred aqueous solution or the headspace of the sample for a certain time, the solvent is sampled with a microsyringe and injected into a GC for analysis (39). Claimed advantages are the simpler construction and faster vaporization in the GC injector.

3. Separation and Purification Techniques

3.1. Introduction

Having obtained an extract which contains as much as possible of the compounds of interest and as little as possible of any interfering substances, some additional separation or sample preparation step is usually necessary. This step serves to obtain the desired compound(s) in purified form so that they can be used for further measurements. One can think of commercial isolations, biological tests or quantitative measurements (see Section 4 for various detectors). In the following an overview is given of the most important techniques for this purpose, some suggestions as to the type of applications for which they are most useful and some examples.

3.2. Solvent Partition

The most widely used instrument for separating certain classes of secondary metabolites, e.g. alkaloids from other classes like flavonoids or impurities like chlorophyll, has been and remains the separatory funnel. Some advantages and disadvantages of this technique are given in Table VII. After obtaining the initial extract of interest (see Section 2 for more details) partitioning is almost always the next step, especially in preparative applications. On an analytical scale, partitioning is gradually being displaced by SPE (solid phase extraction). Crude extracts are not directly suitable for chromatography because they contain compounds of widely varying polarities and thus do not totally dissolve in the initial weak solvent. It is much better first to fractionate the crude extract into different portions each consisting of compounds of roughly similar polarity. In addition, cost considerations play an important role: a preliminary partition step is carried out faster than a chromatographic step, consumes less solvent, reduces the amount to be separated, and does not use any expensive stationary phase. As mentioned in Section 1.2 tertiary alkaloids are easily purified by means of partitioning with a separatory funnel because their polarity and thus solubility change dramatically upon changing the pH of the aqueous phase. This is well illustrated by the data in Table VIII. The partition coefficient K changes approximately 4 orders of magnitude upon lowering the pH from 7 to 1.

Table VII. Advantages and disadvantages of solvent partitioning in a separatory funnel

Advantages	Disadvantages
No special apparatus necessary	Limited separation power
High capacity	Troublesome emulsions form frequently
Scale-up is easy	Not easily automated
Solvents can be adjusted to compounds	
No irreversible adsorption	

Also, certain neutral compounds can be purified by a proper choice of solvent system. An example of the selective partitioning of glycosides can be found in the paper by Kuwajima *et al.* (40). The extraction scheme is given in Figure 9. The initial crude MeOH extract was first treated with water which dissolved only the more polar

Table VIII. Solubilities of quinine, strychnine and their sulphates in water and chloroform

Compound	Water	Chloroform
Quinine [13]	1 : 1900	1 : 1.2
Quinine bisulphate	1 : 9	1 : 625
Strychnine [2]	1: 6400	1 : 5
Strychnine sulphate	1 : 35	1 : 220

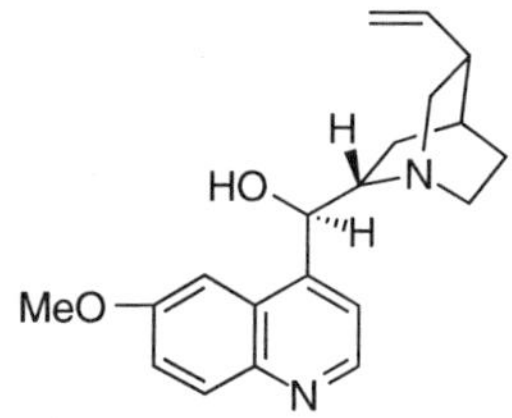

quinine [13]

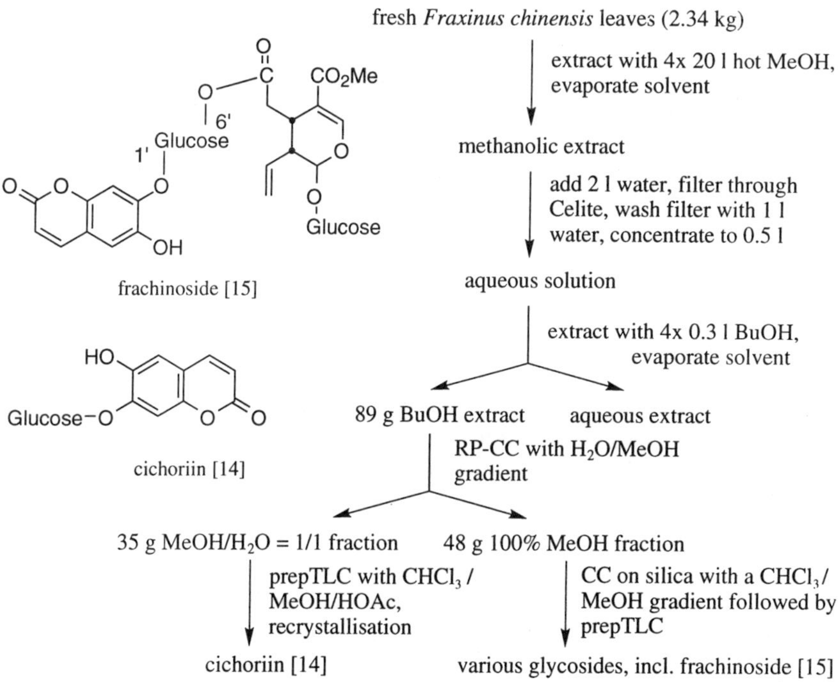

Figure 9. Fractionation scheme for the isolation of various coumarins from *Fraxinus chinensis* leaves (40). In step 2 the glycosides were dissolved in the selective solvent water. In the next step, partition between butanol and water, they partitioned into the organic phase. Many impurities were lost in the process. RP-CC stands for Reversed Phase Column Chromatography.

compounds, including the glycosides. The aqueous extract was subsequently extracted with *n*-butanol. A characteristic property of many glycosides is that they partition into water when the organic phase is chloroform or diethyl ether, but partition into butanol in a butanol / water system. Water and butanol are only just immiscible and thus form a more selective partition system than water / chloroform. Several additional chromatographic steps were necessary before any pure compounds like cichoriin [14] and frachinoside [15] could be obtained. For a more selective partitioning of very apolar compounds (e.g. steriods) one can use non-aqueous systems like methanol/hexane.

3.3. Chromatography

Chromatography is a separation method in which the analytes are distributed between two phases, one of which is stationary while the other moves. Chromatography is the most useful technique for the separation of phytochemicals. With the exception of gravity column chromatography, all of the chromatographic techniques used today, and which are discussed in Sections 3.4 - 3.9, were introduced only after 1950. Due to space limitations only some basic chromatographic parameters will be mentioned and defined here. For general chromatographic theories the reader is referred to (41-43). Important parameters are the retention time of a solute (t_r) and an unretained compound (t_m), the corrected time of a solute ($t_{r'}$), the peak width at the baseline (w) and halfway of the peak ($w_{0.5}$), the capacity factor (K') of a solute which is a measure of the retention relative to the solvent, the effective number of plates of a chromatographic column (N_{eff}), the height of a theoretical plate (H), the selectivity factor (α) which is representative of the relative retention of two compounds and the resolution (R_s) between two different solutes. All of them can be easily calculated from a chromatogram (Figure 10) by means of the formulae given below.

Especially noteworthy is the formula in which the resolution is expressed as the product of the plate number (N), the selectivity factor and the capacity factor. This formula clearly shows that if one has a good column (N_{eff} sufficient) and optimal retention (K' between 2 and 10) the only way remaining to improve a poor separation is to change the selectivity factor a. A modest 5% increase in a, e.g. from 1.05 to 1.10 results in a dramatic 50% (!) increase in resolution. It is thus better to invest some time in choosing the right stationary phase and solvent than in buying expensive columns with high plate numbers.

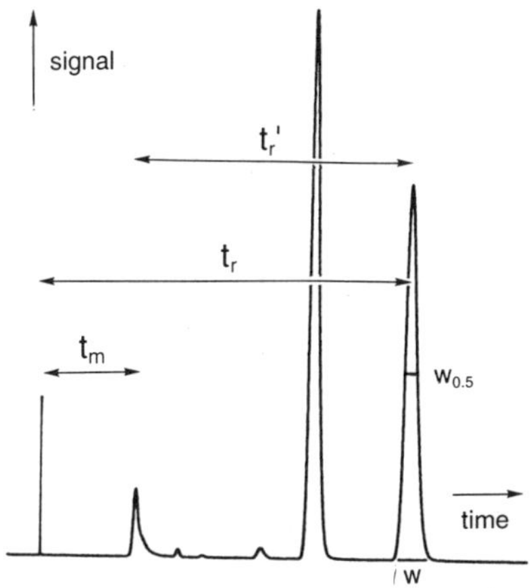

Figure 10. Example chromatogram showing t_r, $t_{r'}$, t_m, $w_{0.5}$ and w. From the measured parameters the following values could be calculated: $K'_A = 1.83$, $K'_B = 2.85$, $\alpha = 1.56$, $N_{eff} = 2500$ and $R_s = 3.3$.

$t_{r'} = t_r - t_m \qquad K' = (t_r - t_m)/t_m$

$\alpha = K'_B / K'_A \quad N_{eff} = 5.54 \,.\, (t_r / w_{0.5})^2$

$H = L / N_{eff}$ (L = column length)

$R_s = 2(t_r^B - t_r^A) / (w_A + w_B)$

$R_s = {}^1/_4 \,.\, (N_{eff})^{0.5} \,.\, \{(\alpha - 1) / \alpha\} \,.\, \{K' / (K' + 1)\}$

3.4. Counter-Current Techniques

Counter-current chromatography can be thought of as a large number of separatory funnels connected to one another and filled with an immiscible solvent system in which one phase remains stationary while the other phase moves. The first (Craig) apparatus truly worked according to this principle (44). Counter-current chromatography is thus pure liquid/liquid partition chromatography and lacks any solid supports. An example of counter-current distribution is the purification of a flavonoid glycoside from Roman chamomile (45). A crude butanol fraction (29.5 g) was separated in 120 fractions with the solvent system ethyl acetate / propanol / water = 2 / 1 / 2 on an apparatus with 120 elements of 25 ml each. The flavonoid of interest occurred in eight fractions and was further purified by column chromatography on Sephadex® LH-20 and selective precipitation (45).

As the Craig apparatus was too large, since 1970 a number of smaller and more efficient techniques have been introduced. One such technique, which was quite popular in the eighties but less so now, is droplet counter-current chromatography (DCCC). It consists of ≈300 vertical silanised glass tubes (40 cm length and i.d. 2-4 mm) connected by narrow-bore Teflon® tubing. The tubes are filled with a stationary phase by means of a pump, expelling all air in the process. The stationary phase consists of either the upper or lower layer of a ternary or quaternary solvent system. These usually consist of mixtures of chloroform, propanol, methanol and water. The smaller the difference in polarity between the two phases, the more selective the separation. Once the tubes have been filled, pumping of mobile phase starts. If this is the lower layer small droplets will form at the top of the tubes and will fall down. When all tubes contain mobile phase droplets, no more stationary phase will be displaced. Using a valve, dissolved sample can be introduced, starting the separation. Due to the small droplet size and their turbulence in the tubes, an effective partitioning of analyte will take place. It is also possible by arranging the upper phase to enter at the last tube to let the apparatus run in the "reversed" or "ascending" mode, i.e. with chloroform / methanol as stationary phase and methanol / water as mobile phase. In this case the mobile phase droplets ascend in the tubes and compounds will elute in reverse order.

Although non-aqueous solvent systems for the separation of apolar products, such as essential oil constituents, do exist, most applications of DCCC are for the separation of moderately polar products such as saponins, many glycosides, alkaloids and tannins. A large number of applications and references can be found in (4, 5). Advantages and disadvantages of DCCC and other counter-current chromatographic techniques can be found in Table IX. It is likely that DCCC will gradually decrease in importance because of the frequent necessity for the use of chloroform, an undesirable solvent on environmental and health grounds, and the development of faster counter-current techniques.

There are now several new counter-current instruments on the market which give both a better and faster separation than DCCC. This is achieved by pumping mobile phase through quickly-rotating narrow-bore Teflon® tubing filled with stationary

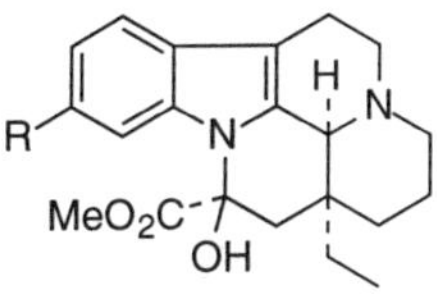

vincamine [16]: R=H
vincine [17]: R=OMe

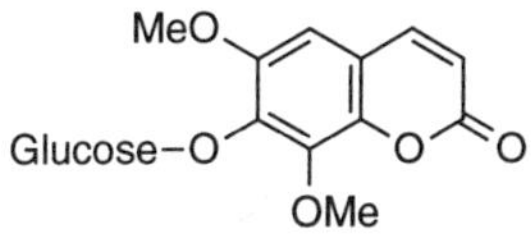

eleutheroside B_1 [18]

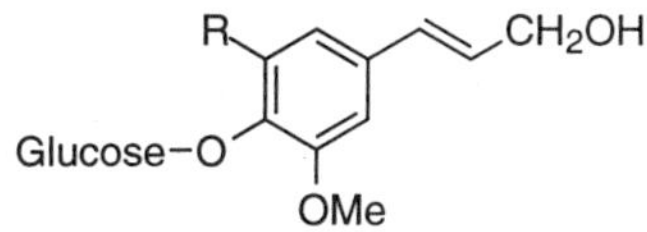

eleutheroside B: [19] R=OMe
coniferin [20]: R=H

Table IX. Advantages and disadvantages of counter-current chromatographic methods

Advantages	Disadvantages
No solid support necessary	Low resolution (DCCC and RLCC)
High capacity, low solvent composition	Slow (DCCC and RLCC)
No irreversible adsorption	Droplet formation necessary (DCCC)
Total recovery of sample	Scale-up is very difficult
Solvents can be adjusted to compounds	Detection more difficult than with HPLC
No pH limitation	Not easily automated, mainly preparative
No tailing	Gradients difficult

phase and wrapped round a spool. Due to the centripetal force, stationary phase is retained, while there is also a much more efficient contact between the phases. There are no special requirements for the solvent system with regard to droplet formation. Approximately 2 mg of vincine [16] and vincamine [17] could be baseline separated in 75 min with a multi-layer coiled column design (46). The effective cycle time could be reduced to only 42 min because no other compounds were present in the mixture. The solvent was hexane / ethanol / water = 6 / 5 / 5 and the lower layer was used as the mobile phase with a flow rate of 2.6 ml/min. Centrifugal partition chromatography (CPCC) was also capable of separating eleutheroside B_1 [18], eleutheroside B [19] and coniferin [20] from Siberian ginseng on a preparative scale (47). The lower phase of chloroform / methanol / water = 7 / 13 / 8 was used as the mobile phase at 3 ml/min, with tubing of 2.6 mm i.d. wrapped around a coil rotating at 700 rpm (content: 360 ml). Detection was by UV at 254 nm (47). Reviews on the various forms of centrifugal counter-current chromatography together with applications can be found elsewhere (4, 5, 48-51).

3.5. Liquid Chromatography

This is the most universal and frequently used technique for the separation and purification of secondary metabolites. By a proper choice of the stationary phase, the mobile phase, and various other conditions such as particle size, flow speed, injection volume, sample concentration and temperature, essentially any two compounds can be separated on both an analytical and preparative scale. A number of separation mechanisms and corresponding stationary phases can be distinguished (Table X).

Table X. Separation mechanisms distinguished in LC, with examples of stationary phases

Separation mechanism	Example phases
Adsorption chromatography	Silica, alumina, Florisil®, polyamide
Partition chromatography	Reversed-phase materials (C18), cellulose
Ion-pair chromatography	Reversed-phase materials (C18)
Ion-exchange chromatography	Cation- and anion-exchangers
Complexation chromatography	Cation-exchangers loaded with Ag(I)
Chiral chromatography	Chiral phases or chiral mobile phase additives
Gel filtration chromatography	Various gels, Sephadex® LH-20, Sephasorb®

In the phytochemical laboratory, adsorption chromatography on silica is by far the most frequently employed method of separation in the initial stages of purification. Silica is cheap and versatile and a wealth of experience exists with its use. Under normal circumstances, the polar phases used for adsorption chromatography are most suitable for separating compounds which show a difference in polar groups. These polar groups can interact with polar groups (e.g. silanols) on the stationary phase surface. It will be easy to separate a sesquiterpene alcohol (e.g. farnesol [21]) from a sesquiterpene diol (e.g. alloaromadendrane-7α,11α-diol [22]) because the latter has two polar groups and will thus be more strongly retained. Adsorption chromatography - or normal- or straight-phase chromatography as it also called - is less suitable for separating compounds with the same polar groups which differ only in their apolar part (e.g. farnesol [21] and geraniol [23]). The solutes can be displaced from the active sites by solvent molecules which contain polar groups. The more polar the solvents, the better they can displace polar solutes. A ranking of solvents in order of increasing solvent strength for normal-phase chromatography is called the

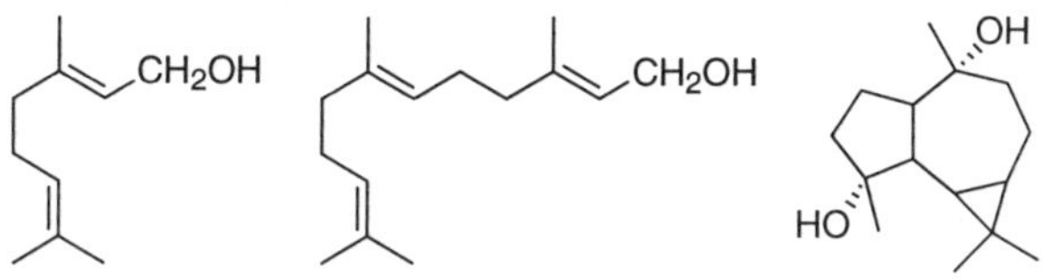

geraniol [23] farnesol [21] alloaromadendrane-7α,11α-diol [22]

eluotropic series: alkanes (hexane, petroleum ether), toluene, dibutyl ether, dichloromethane, chloroform, *tert*-butyl methyl ether, diethyl ether, ethyl acetate, *n*-butanol, methyl ethyl ketone, acetone, propanol, ethanol, acetonitrile, methanol, water. The solvent strength should be adjusted in such a manner that K' values between 1 and 15 are obtained. This can be achieved by mixing appropriate proportions of stronger and weaker solvents, e.g. hexane and ethyl acetate, or ethyl acetate and methanol. For selectivity reasons other solvents are also frequently used. Silica is weakly acidic and thus there is an additional interaction with basic secondary metabolites such as alkaloids. Tailing can be the result. To avoid tailing a small percentage of base can be added to the solvent (ammonia, diethylamine). If no separation on silica can be achieved, alumina may be suitable since this has a different selectivity. Alumina is available in acidic, neutral and basic forms. Florisil® has been used for the purification of homoerythrina alkaloids in combination with a chloroform / methanol gradient (52). Polyamide is mostly employed for the separation of polyphenols like flavonoids and anthocyanins; for an example see (53).

Partition chromatography used to take place on water-saturated cellulose with organic solvents. A problem was the gradual removal of the aqueous stationary phase because it could partially dissolve in the organic phase. Nowadays partition chromatography takes place on silica derivatised with apolar side chains, such as octadecyl (C18) or octyl (C8) silica. Because these groups are chemically bound to the surface they cannot be removed. Chromatography on such phases has been named reversed-phase (RP) chromatography because in contrast to normal-phase chromatography the stationary phase is apolar and the mobile phase (methanol / water mixtures) polar. Compounds partition between the apolar alkyl chains and the aqueous phase, although adsorption effects by residual silanols also play a small role. Such phases are most useful for the separation of compounds which differ in their fat solubility/water solubility ratio. Thus - in contrast to adsorption chromatography - changes in the apolar part of the molecule can result in a different retention. Geraniol [23] and farnesol [21] are easy to separate on RP phases, but geraniol [23] and the diol [22] may be more difficult to separate because the influence of the second hydroxyl group may be similar to the effect of the additional five carbon atoms. RP phases without mobile-phase modifiers are poorly suitable for the separation of highly polar compounds such as sugars and small ionized molecules like glucosinolates.

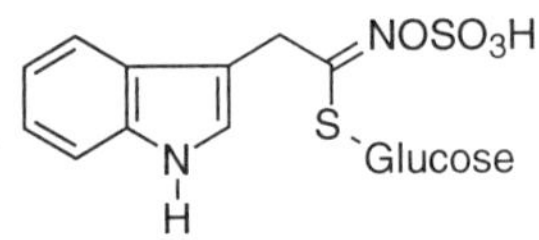

glucobrassicin [24]

Such small ionized molecules can be better separated by the addition of an ion-pair reagent (54). Glucobrassicin [24] could be well separated from other glucosinolates on C18 phases after the addition of tetrabutylammonium sulphate (TAS) to the solvent (55). TAS forms a stable ion pair with [24]. This neutral species with several alkyl chains can be retained by C18 phases (55). Similarly, alkaloids can be analysed in their ionized form with acidic ion-pair reagents, e.g. octane sulphonate or perchloric acid (see Section 3.5.5).

Ion-exchange chromatography is sometimes used for HPLC separations (Section 3.5.5) of organic acids. Its main use lies in SPE sample clean-up for HPLC or GC or for the removal of certain reagents used during the extraction or purification, e.g. the iodomercurate ion to precipitate alkaloids (see Section 3.10) or the TAS used in preparative ion-pair HPLC. An example of the former use is the purification of some neuroexcitatory amino acids occurring in algae by a strongly basic anion exchanger (56). The column extracted the amino acids from a neutral 80% alcohol solution. After removing many polar impurities with pure water, the analytes were eluted with 2 M acetic acid.

Complexation chromatography is also infrequently employed. The separation of olefins by complexation with silver(I) is an example (57, 58). Phenylboronic phases have been used in the SPE clean-up of vicinal diols.

Although there is lately much interest in chiral LC, the attention is not focussed on secondary metabolites from plants, in contrast to chiral GC. Chiral separations can be achieved either by using a chiral stationary phase or by adding a chiral additive to the mobile phase which interacts with the solute. More general information on chiral LC can be found in (59, 60).

Because of the rather small molecular weights of secondary metabolites, the small mass difference between compounds from the same class, the limited separating power and the high costs of size-exclusion columns, this type of chromatography is only sporadically used in phytochemical separations. Thus it is not further discussed here with the exception of Sephadex® LH-20, a hydroxypropylated dextran gel. With C18 materials, this gel is - after silica - the most frequently employed stationary phase for the preparative separation of secondary metabolites. Depending on the conditions this material shows size-exclusion, adsorption, partition or mixed-mode effects (see also Section 3.5.1).

In the following paragraphs the technically different forms of column chromatography (CC) are described. They are ordered according to increasing pressures of operation and complexity and decreasing particle size of the stationary phase. In terms of stationary phases which can be used or selectivity there are no essential differences between them.

3.5.1. Gravity liquid chromatography

This is the oldest and most simple form of chromatography. Most often normal phase materials (silica, alumina) and large particle sizes (60 - 200 μm) are used. For small-scale separations (sample size < 1 g) this technique is gradually being replaced by flash chromatography (see next Section). Gravity liquid chromatography is still practised as a first step in large-scale preparative separations (sample size 5 - 100 g and stationary phase weight 1 - 10 kg) because of its low cost. As cheap RP phases, XAD resins can be used (61). Although resolution is far from optimal in such separations under overload conditions, usually the main purpose is only to reduce the sample weight for the next chromatographic step(s). If the sample cannot be dissolved in the initial solvent (e.g. petroleum ether / ethyl acetate) the sample can first be dissolved in a strong solvent (ethyl acetate / methanol) after which silica (same weight as the sample) is added. After pouring the slurry in a roundbottom flask the solvent can be removed under vacuum. When the silica/sample mixture is dry, it can be put on top of the column. An example of CC is the intermediate purification of some antimicrobial abietane diterpenoids on a silica column (25 x 3 cm) with isocratic chloroform as solvent. Initial purification was with VLC while the final purification took place with pTLC (62).

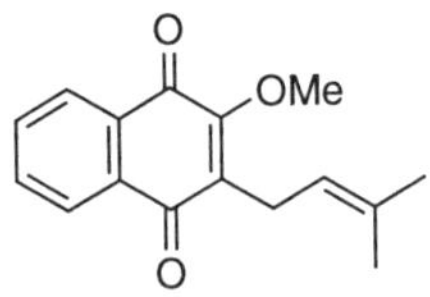

lapachol methylether [25]

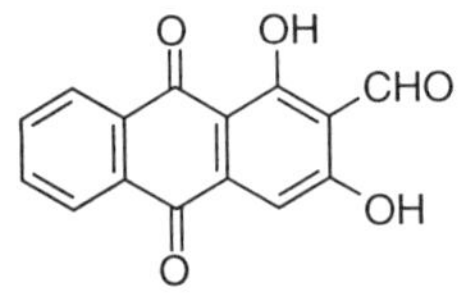

nordamnacantal [26]

A mild stationary phase which is frequently used for a final purification of almost pure substances is Sephadex® LH-20. Because this phase is non-compressible it is eluted by force of gravity. An example of its use is the separation of lapachol methylether [25] and nordamnacantal [26]. As the molecular weights of these compound are similar in this case the separation must be due to polar interactions (63).

3.5.2. Flash chromatography

The only difference between flash chromatography and gravity column chromatography is the pressurization of the air (or nitrogen) above the solvent. Due to this higher pressure, higher flow rates are achieved and/or smaller particle sizes (40-60 μm) can be used which in turn results in faster separations and/or better separations. The pressure can be applied by pressurised air or nitrogen and a needle valve or the simpler and safer construction in Figure 11. An example of flash chromatography is the separation of three dihydroanthracenones from a 15 g sample of a methanolic stem and leaf extract of *Gasteria bicolor* on silica with, as solvent, ethyl acetate / chloroform = 1 / 1 (64). More examples of the use of flash chromatography in plant sciences can be found in (4, 5).

3.5.3. Vacuum Liquid Chromatography (VLC)

VLC is essentially a hybrid technique of column chromatography and preparative

TLC with multiple developments. TLC-grade stationary phase (mostly silica or alumina) is tightly packed under vacuum to a height of 4-5 cm on a glass filter.After a washing step with petroleum ether and drying by applying the vacuum, the sample is put on top of the stationary phase either as a solution in the most apolar solvent possible or as a dry powder adsorbed onto the stationary phase. The filter is then eluted with 10-25 small portions of solvent of increasing strength (e.g. a gradient from hexane / ethyl acetate via pure ethyl acetate via ethyl acetate / methanol to - finally - pure methanol). After each solvent the column is sucked dry. The technique can be used for both small and large samples by varying the diameter of the glass filter. For 0.5 - 1 g samples a 2.5 cm i.d. suffices. Claimed advantages are that it is simpler, faster and cheaper than open column chromatography or preparative TLC while giving superior resolution (65-67). Channelling and air-oxidation constitute problems, however, and today VLC is not nearly as much used as preparative TLC and open-column or flash chromatography.

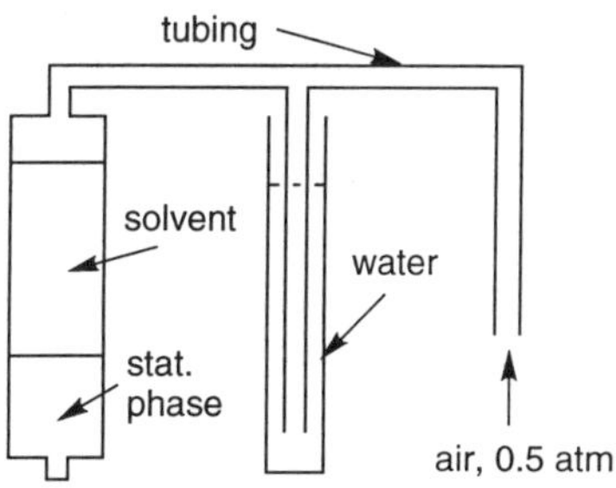

Figure 11. Simple and safe apparatus for pressurising flash chromatography columns. Excess pressurised air escapes via the open end of the tubing at the bottom of the water column. If a 1 m water column is used, the pressure above the solvent is a constant 0.1 atm.

An example of the use of VLC is in the isolation of alkaloids from *Zanthoxylum budrunga* roots. After a pre-extraction with hexane to remove apolar impurities, the alkaloids were extracted with ethanol. The ethanol fraction was partially purified by an acid-base partition, after which a fraction was separated on a silica VLC column with a benzene-chloroform gradient. The final purification took place by means of centrifugal TLC on silica with a hexane / ether gradient (see Section 3.5.7) (68). VLC was also used for the separation of iridoid glycosides from *Plantago media*. Fresh aerial parts were twice extracted with methanol. After concentration and the addition of water, the aqueous layer was partitioned against ether, ethyl acetate and n-butanol. As many glycosides are poorly soluble in organic solvents except butanol (see

Section 3.2), the iridoids were partitioned finally into the butanol layer. The iridoids were adsorbed on charcoal, washed with water / alcohol mixtures and eluted with alcohol / acetone. This 330 mg fraction was separated on 16 g silica with VLC. Fraction 3 (50 ml $CHCl_3$ / EtOH = 3 / 1), 5 (50 ml $CHCl_3$ / EtOH = 3 / 1) and 6 (100 ml EtOH) contained 35 mg [27], 62 mg [28] and [29] respectively, all pure (67).

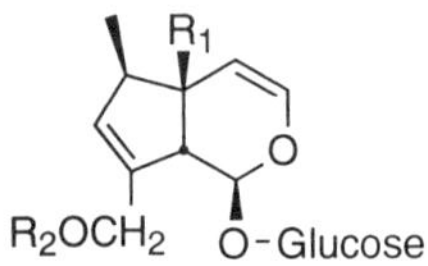

[27]: R_1=OH, R_2=Ac
[28]: R_1=R_2=H
[29]: R_1=OH, R_2=H

3.5.4. Low and Medium Pressure Liquid Chromatography (LPLC and MPLC)

LPLC and MPLC are similar to HPLC with respect to the fact that the solvent itself is pressurised. LPLC operates at pressures from <1 to 6 atm and uses simple membrane pumps. An example is the ready-made Lobar system from Merck. The particle size is similar to that of flash chromatography. Different column dimensions and stationary phases, including RP-types, are available. Fusaric acid [30] and 9,10-dehydrofusaric acid [31], isolated from culture filtrates from *Fusarium nygamai*, were separated on an RP-column (MeCN / H_2O = 1 / 1, 3 atm) with the Lobar system (69).

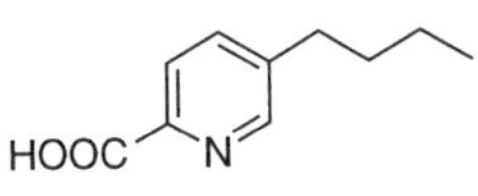

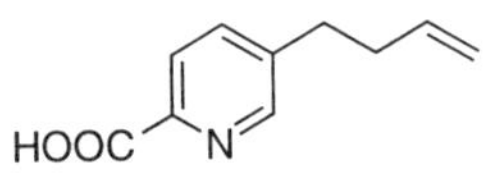

fusaric acid [30] 9,10-dehydrofusaric acid [31]

MPLC uses more sophisticated apparatus and smaller particles (10-40 μm) but is otherwise similar to LPLC. Standard HPLC pumps can be used but are not strictly necessary. Pressures typically vary from 3-50 atm. MPLC is flexible because the columns can be filled by the user with any stationary phase. One MPLC system is commercially produced by Büchi (system B-680). Other systems use axial or radial compression to improve the packing of the stationary phase. MPLC on a silica (15-25 μm) column of dimensions 46 x 2.6 cm and a benzene / ethyl acetate gradient at 40 ml/min was employed for the final purification of 300 mg 2-methoxy-eudesma-1,4,6-trien-3-one, an antifungal sesquiterpene from *Parthenium* resin (70). A Büchi system was successful for the intermediate separation of various flavonoid glycosides from wild garlic (71). Both silica and RP-C18 stationary phases were used with ethyl acetate / methanol and H_2O / methanol gradients respectively. Many examples of the application of LPLC and MPLC in the period 1980 - 1985 can be found in (4).

3.5.5. High Pressure Liquid Chromatography (HPLC)

High Pressure Liquid Chromatography is the most powerful liquid chromatographic technique currently available. It is characterised by the use of small particles - typically in the 3-10 μm range - with a narrow size distribution. Due to the small particle size flow resistance is large which requires high pressures (100-300 atm). The separations are usually efficient and therefore HPLC also stands for High Performance Liquid Chromatography. A good 25 cm column has around 20,000 theoretical plates. Because of this high efficiency HPLC is frequently used for quantitative purposes, in sharp contrast to the LC techniques discussed above which are exclusively used for preparative purposes. Other distinctions are the much smaller dimensions of HPLC, the use of on-line detectors and the large use of RP-materials.

The high pressures are achieved with special high pressure pumps. In the analytical mode with 4.6 mm i.d. columns a fixed volume of 5-50 μl sample solution can be injected by means of a 6-way valve (Figure 12). For routine analysis this step can be automated with an autosampler. The compounds are then separated on the column. RP phases such as C18 are most often used. Normal-phase columns (e.g. silica) are not popular because (I) they tend to degrade rapidly due to adsorption of highly polar impurities, and (II) retention times on such columns are irreproducible due the varying percentage of water on the column and in the solvent. These considerations are less important for the lower-pressure, preparative, normal-phase columns discussed in Section 3.5.2 - 3.5.4 because these columns are much cheaper, the material is often discarded after use and the separation is only carried out a few times. The RP materials are more robust and retention times are much more reproducible because humidity does not play a role (the solvents already contain a significant percentage of water). After eluting from the column the compounds are detected on-line (Section 4.2). Sometimes more than one detector is used. When quantitation is desired an integrator has to be connected to the detector. Usually no internal standardisation is used with HPLC, because - in contrast to GC - the fixed loop volume allows for reproducible injection volumes. Two basic books on HPLC are (72, 73).

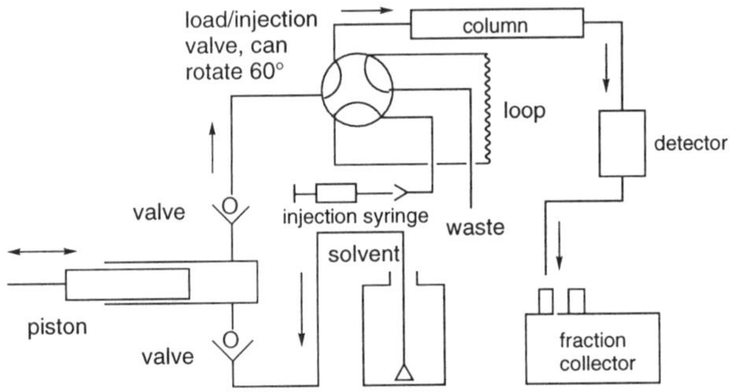

Figure 12. Schematic drawing of an HPLC apparatus in the isocratic mode. The solvent is pumped via a high-pressure injection valve through the column and the detector. The turnable valve (depicted in the "load" position) and the loop make it possible to inject repeatedly a constant volume of sample. The fraction collector is optional and only used in the preparative mode.

Many things can go wrong with HPLC: air in the pumps or in the detector, clogging of columns due to particles, leaks, poor pulse dampening, voids in the column, spikes, drifting baselines, poor peak shape, irreproducible retention times, ghost peaks, etc. To help novices a book on HPLC troubleshooting has appeared (74). The same author also has a useful troubleshooting column in the journal LC-GC. HPLC can also be used in the semipreparative (i.d. 8-20 mm) or preparative mode (i.d. 2.5 cm and larger) for the actual collection of compounds. Because of the extreme costs of such columns and the large quantities of expensive, highly pure solvents preparative HPLC is only used as a last step in the purification of plant constituents. A useful book on preparative HPLC is (75).

Separations can be optimised by changing one or more of many variables. If the separation is excellent but takes too much time, one could try a gradient run. During the run the solvent gradually changes and becomes stronger which causes a quicker and sharper elution of later-eluting constituents. A disadvantage is that the apparatus becomes more complex, retention times become less reproducible, solvents cannot be recycled and an equilibration step is necessary after each run. If the selectivity and retention are satisfactory but the peaks are too broad one could mount a better or a longer column. If the peaks are too close to the solvent one could increase the amount of water in the solvent (RP columns). If the peak shape is poor (e.g. tailing peaks of acidic analytes) a small amount of modifier could be added to the solvent, e.g. 0.1% trifluoroacetic acid. If the selectivity is poor one could (I) change one or more of the organic constituents of the solvent, e.g. MeCN instead of MeOH (76), (II) change the column, e.g. C8 or phenyl instead of C18, (III) change the column temperature, or (IV) change the separation mechanism, see Table X.

An example of what is currently achievable by state-of-the-art HPLC is given in Figure 13. This chromatogram shows the separation of 49 flavonoids and anthocyanins by gradient RP-HPLC (53). The compounds were extracted from *Rosa* petals by refluxing with methanol / alcohol under nitrogen (to avoid decomposition of labile anthocyanins). After partitioning against petroleum ether to remove apolar impurities, evaporation, dissolving in methanol / water and filtration, the extracts were analysed by HPLC. By fingerprint comparison the chromatograms could be used for distinguishing between 44 different cultivars (53).

Preparative ion-pair HPLC was used for the isolation of some aporphine alkaloids from the medicinal plant *Artabotrys monteiroae* (77). As the ion-pair reagent, sodium perchlorate under acidic conditions was used. On a 15 cm C18 column, with a gradient from 0.2 M aqueous perchlorate and 20% acetonitrile to aqueous perchlorate and 35% acetonitrile, 21 peaks were visible with diode-array detection at 280 nm. For preparative applications the perchlorate ion as ion-pair reagent was found to be much more pleasant to work with than dodecyl sulphate because it did not foam (77).

An example of normal phase HPLC is the separation of toxic norditerpenoid alkaloids from larkspur (*Delphinium barbeyi*) on neutral alumina (Spherisorb®, 5 μm). With the solvent hexane / iso-propyl alcohol / water = 85 / 14.25 / 0.75, this column gave a good and reproducible resolution of several alkaloids with fine peak shape. Because alumina is - in contrast to silica - not acidic in nature it could be used without the addition of basic tail reducers. The small amount of water present in the mobile phase strongly diminished the dependence of the retention times on the relative humidity. An advantage of this system was that the scale-up on alumina could be well predicted (78).

A complexation mechanism was responsible for the separation of several apolar sesquiterpene hydrocarbons occurring in tolu balsam (79). Normally hydrocarbons cannot be separated at all on silica because they are not retained ($K' = 0$ or $R_f = 1$). Impregnation of the silica with silver salts (e.g. AgClO4) gives a much better selectivity because silver(I) ions form complexes with olefinic double bonds and most terpenes contain 1 to 3 double bonds. Compounds are thus separated according

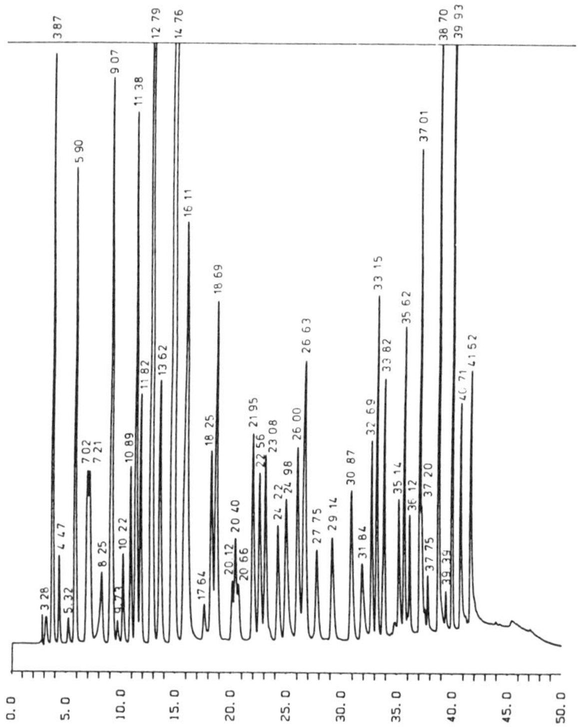

Figure 13. Separation of a mixture of 49 flavonoids and related phenolic compounds on a Hibar pre-packed guard column (30 mm x 4 mm i.d.); LiChrosorb RP-18; 7 μm) in combination with a Hibar pre-packed analytical column (250 mm x 4 mm i.d. Lichrospher 100 CH-18/2; 5 μm). A non-linear gradient from 15% B in A to 95% B in A was used. A: HCO_2H : H_2O (5:95), B: MeCN: MeOH (5:95) was used. Detection by a photodiode array detector at 350 nm. Reprinted from Phytochem. Anal., Vol. 4, C. Van Sumere *et al.*, Improved extraction and reversed-phase high performance liquid chromatographic separation of flavonoids and the identification of *Rosa* cultivars, pp. 279-292 (1993) with kind permission of John Wiley & Sons, Ltd., Chichester, U.K.

to the type and number of double bonds and the presence of polar groups. γ-Gurjunene [32] and 4-epi-γ-gurjunene [33] could be easily separated in such a system (t_r 23 and 19 min respectively) (79).

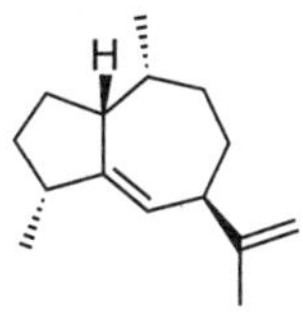

γ-gurjunene [32]

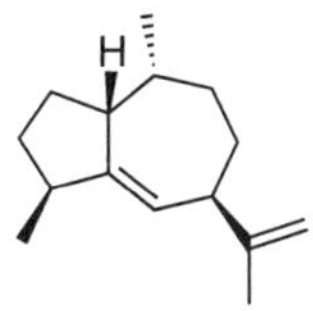

4-epi-γ-gurjunene [33]

3.5.6. Solid Phase Extraction (SPE)

Solid Phase Extraction or SPE in short has rapidly become popular in recent years for purposes of sample clean-up for GC or HPLC and it is gradually replacing the separatory funnel. In principle it is nothing other than a small chromatography column (i.d. 9-12 mm) filled with 100 - 1000 mg of 40 μm stationary phase. Normally the polypropylene columns are bought ready-made and virtually any stationary phase can be chosen. Usually 4-6 steps can be distinguished during an SPE sample clean-up procedure (Figure 14):

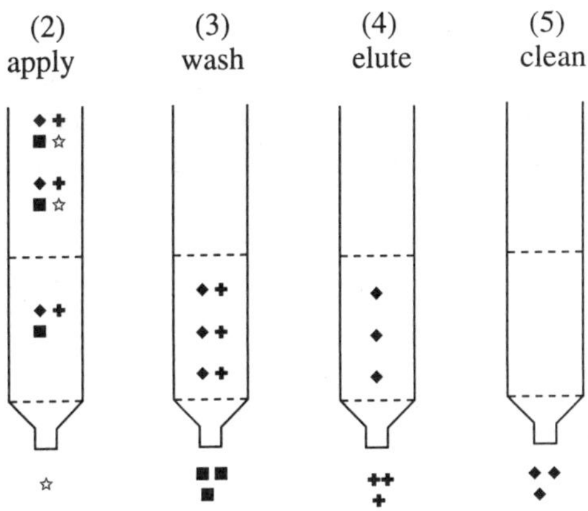

Figure 14. Schematic drawing of an SPE sample clean-up. After cleaning and conditioning the column (step 1, not depicted), the sample is applied (II). Some components (I) are not retained. A wash step (III) follows which removes some of the impurities (■). An elution step (IV) follows which removes the compound of interest (✚). If necessary, the column can be cleaned (V) removing components still retained (◆).

(I) conditioning of the column with a solvent similar to the one in which the sample is dissolved (5 ml for a 500 mg column). When there is doubt about the cleanliness of the column a wash step with a strong solvent (e.g. 100% methanol for a C18 column) could precede step (I).

(II) application of the sample; depending on the strength of the solvent, the size of the column and the concentration of the solute the sample volume could be anywhere between 0.5 ml and 50 ml. For quantitative recoveries the breakthrough volume should not be exceeded. This is the sample volume at which solute starts to exit the column. It occurs when the SPE column is grossly overloaded or when the sample solvent is too strong. In the latter case dilution with water may help. During step (II) unretained impurities will elute.

(III) wash step(s) with small volumes of solvents which gradually increase in elution strength. During this step impurities which are less strongly bound than the compound(s) of interest are eluted. Sometimes this step is omitted.

(IV) elution step with a small volume (typically $\leq$ 2 ml for a 500 mg column) of a solvent which is strong in relation to the solute. The compounds of interest should elute completely during this step. This fraction can be used for further analysis. If the SPE is not reused it can be discarded.

(V) cleaning of the column with a very strong solvent if it is to be reused at a later stage. Whether this is feasible depends on the nature and amounts of impurities, the nature of the stationary phase, cost and procedural regulations. Bare silica is difficult to recycle.

An example of SPE in sample preparation is the purification of polygodial [34] in hexane / ethyl acetate = 8 / 2 extracts of water-pepper (*Polygonum hydropiper*) leaves over a C2 column. The dried extract was dissolved in 20% aqueous acetonitrile and applied to the column. Washing, elution and cleaning occurred with 30% and 60% aqueous acetonitrile and pure acetonitrile respectively. A clean extract resulted which could easily be analysed by RP-HPLC with UV detection at 233 nm (80). An illustrative book about SPE in general and SPE method development is (81).

One more material which deserves mentioning here is Extrelut® (Merck). This is a type of wide-pore diatomaceous earth which is sold in columns similar to SPE columns. It is however almost pure partition chromatography and acts more as a replacement for the separatory funnel in analytical sample preparation. Aqueous

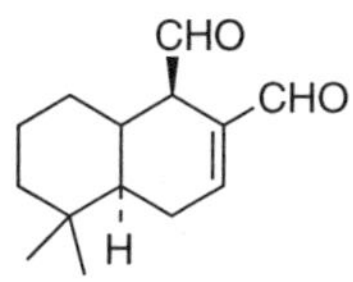

polygodial [34]

extracts are first loaded onto the column. After 20 min, elution with a suitable water-immiscible solvent takes place which can be further analysed. The water from the sample remains on the Extrelut®, which is an essential difference from the SPE techniques discussed above. Claimed advantages are the absence of emulsion formation, less solvent consumption, time gain and the absence of a drying step. A disadvantage in comparison with the traditional funnel is the increased possibility of adsorption on active sites.

Table XI. Advantages of Solid Phase Extraction (SPE) over partitioning with a separatory funnel

Advantages	Disadvantages
Much more flexible and selective	Less suitable for large samples
Less solvent consumption	Can irreversibly adsorb or degrade samples
Time gain	Separations not always reproducible
Less diluted samples	Method development more time consuming
No emulsions	
Better recoveries	
Removes insoluble particles	

3.5.7. Thin Layer Chromatography (TLC)

TLC is the most universal and widely used chromatographic technique available for the analysis of plant constituents. Basically it consists of a thin layer of stationary phase, e.g. silica, on a solid support. The inert support can be either glass, plastic or aluminium. To prevent the stationary phase from falling from the support, a binder, e.g. gypsum, is added. After spotting the 1-10 μl sample solution onto the plate, the plate is placed in a closed tank with a 1 cm layer of solvent at the bottom. Due to the capillary action of the stationary phase particles, the solvent is transported upwards over the plate and a separation can take place similar to other forms of liquid chromatography. Essential differences from most non-planar types of liquid chromatography are (I) that part of the stationary phase is open to the atmosphere, and (II) that the stationary phase is dry when the development starts. TLC has essentially replaced paper chromatography because it can achieve better separations in much less time. More detailed information and some applications on paper chromatography can be found in (3).

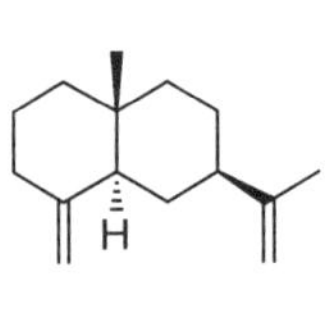

β-selinene [35]

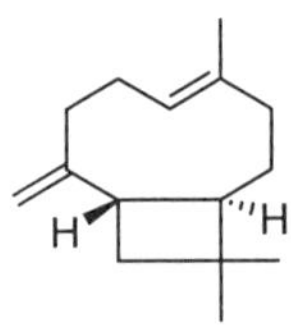

caryophyllene [36]

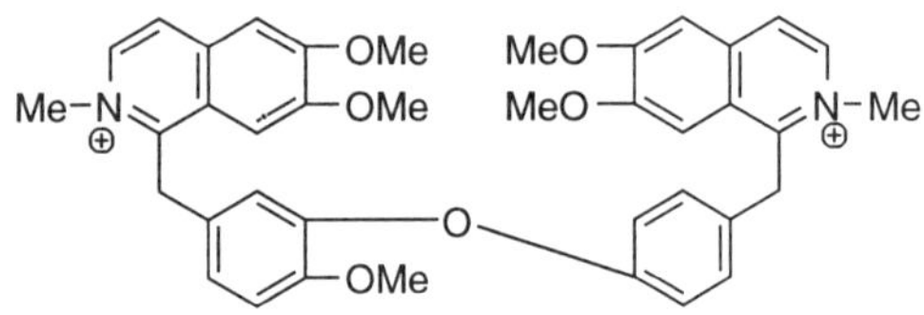

phaeantharine [37]

Although ready-made TLC plates with all kinds of different stationary phases are nowadays commercially available, silica is by far the most commonly used. This is because silica plates are much cheaper (even cheaper when prepared by the user) than other types of plates, because a wealth of experience is available for separations of plant constituents on silica (82, 83) and because silica under the right conditions (selective solvent, suitable mobile phase modifier, or impregnation of stationary phase) can separate compounds ranging from very apolar sesquiterpene hydrocarbons like β-selinene [35] and caryophyllene [36] to highly polar diquaternary alkaloids like phaeantharine [37]. The latter compound has an R_f value of 0.37 on silica with the solvent methanol / 4 M NH_4OH / 1 M NH_4NO_3 = 7 / 2 / 1 (84) while [35] and [36] have R_f values of 0.80 and 0.42 respectively on silica impregnated with 15% $AgNO_3$ with the solvent benzene / acetone = 19 / 1 (85). For higher resolution, so-called High Performance TLC (HPTLC) plates can be used. The higher resolution is due to the smaller average particle size and narrower size distribution (86).

A technique which enhances the resolution of separation is 2-dimensional TLC. After developing in one direction and drying, the plate is turned through 90° and developed a second time with a solvent which has a different selectivity. A disadvantage of

2-dimensional TLC is that only one spot per plate can be applied. Govindarajan used even 3-directional TLC, a more refined form of 2-dimensional TLC for distinguishing between various ginger oils (87). After the normal development in 2 directions the plate was dried and developed a third time, opposite to the second direction, with hexane at -15°. The mono- and sesquiterpene hydrocarbons which constitute an important part of ginger oils cannot be separated by the first two solvents because all hydrocarbons have $R_f \approx 0.95$ in these solvents. In pure hexane at low tem-peratures they are separated, however. All the other oxygenated compounds have $R_f \approx 0$ in hexane so these spots remain stationary during the third development. An example of this technique can be found in Figure 15. This technique of an additional development with a weaker solvent has been further refined: Automated Multiple Development (AMD).

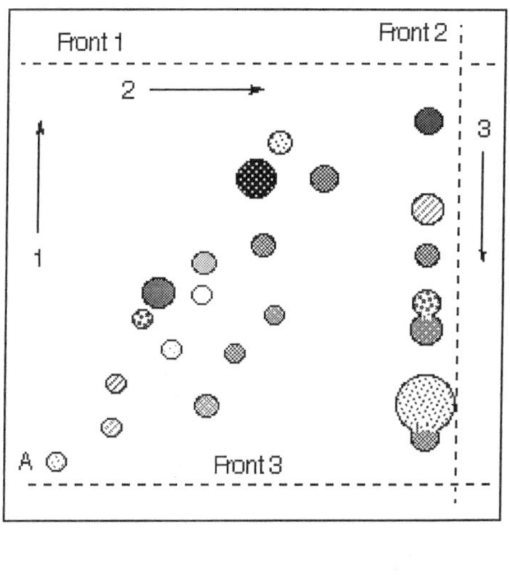

Figure 15. Example of 3-directional TLC after detection. See text for details. A = site of application of sample.

The speed with which a solvent moves along the TLC plate gradually declines. Thus after the solvent has migrated a certain distance the separation can actually get worse because the loss of resolution due to band broadening becomes more important than the gain in resolution due to an increased migration distance. To counter this effect various techniques were developed to achieve a constant speed of the solvent during the separation. Two different principles can be distinguished:

(I) a plate (generally circular) can be rotated while the solvent and solute enter in the centre. Due to centrifugal forces solvent moves from the centre outwards in all directions. If necessary the solvent and solute can be collected. Because such devices can be operated continuously they are especially suitable for preparative applications. This technique is known under the names centrifugal layer chromatography or Rotation Planar Chromatography (RPC). Many applications in the field of natural products are given in (4).

(II) the plate is covered with an elastic membrane pressurised from the outside. The solvent is then pumped through the plate. Because no vapour exists above the plate this technique is more similar to HPLC than to other TLC techniques. This technique is known as Overpressured Layer Chromatography (OPLC). OPLC can be used both for analytical and preparative purposes (88).

Although AMD and these latter techniques have a high separation power it is unlikely that they will ever become as popular as "ordinary" TLC because they lack the one big advantage of TLC: simplicity. A review of more modern TLC techniques is given in (89). More information can be found in (86).

When separation is poor on standard silica, modification of the mobile phase or impregnation of the stationary phase may hold the solution. An example is the impregnation of silica with sodium acetate for the separation of ginkgo terpenetrilactones. On standard silica ginkgolides A [9] and B[10], and J [12] and C [11] cannot be separated because of internal hydrogen bonding. For example in ginkgolide B there is a strong interaction between the 1-OH and 10-OH which makes this compound just as polar as ginkgolide A in spite of the presence of an extra hydroxyl function. Sodium acetate interferes with this internal hydrogen bonding and the terpenes then elute in order of increasing polarity with excellent resolution. R_f values of 0.58, 0.41, 0.25, 0.12 and 0.06 for bilobalide [38] and ginkgolide A, B, J and C respectively were reported (90).

Modified stationary phases based on silica are commercially available. Most reversed phases which are encountered with HPLC can be used for TLC too; however a big disadvantage is their high cost, especially for preparative purposes. A way to avoid these high costs is to impregnate standard silica plates with paraffin. Plates are first developed with a 5-6% solution of liquid paraffin or dimethylpolysiloxane in petroleum ether. After drying, the plates are then developed with an aqueous-organic

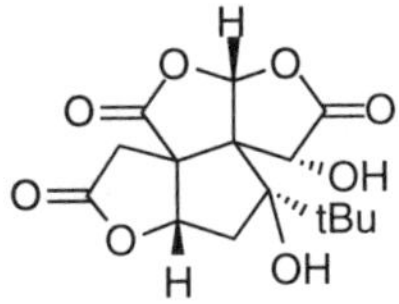

bilobalide [38]

phase similar to RP-HPLC (91, 92). For four phenolic fatty esters from *Buddleja* species a better resolution was seen on the reversed phase plates than on bare silica. Although the reproducibility and resolution of the self-made plates were not as high as those of the commercial RP plates they were acceptable for preparative work (92).

When TLC is used for preparative purposes, normally silica plates of 20 x 20 cm and 0.5 - 2 mm thickness are used. On one such plate an approximately 100 mg sample can be separated. Instead of circular spots or small bands one large band is applied over the entire width of the plate. After development and detection the bands are scraped from the plate and the stationary phase is extracted with a stronger solvent in a small chromatography column. Detection can be a complicating factor when the bands of interest are not visible under UV 254 or 366 nm. The best thing to do then is to cover the plate with another glass plate of approximately 16 x 20 cm so that only two small areas on both sides of the plate remain uncovered. A chromogenic reagent is then applied and the invisible band(s) between the coloured ends of each band are carefully removed. Especially when only small quantities are purified every effort should be made to work as cleanly and quickly as possible without direct sunlight. Tanks should be cleaned beforehand and absolutely no grease should be used. Some advantages and disadvantages of prep-TLC are given in Table XII.

3.6. Capillary Electrophoresis (CE)

This is the only really new and innovative technique which has been introduced for the separation of phytochemicals in recent years. The technique was originally only used for the separation of "biochemical" substances such as proteins, peptides and DNA but following some adaptations it can now be used for the separation of small charged and neutral molecules such as plant secondary metabolites as well. In its most simple form (capillary zone electrophoresis or CZE) it consists of a small piece of fused silica capillary (typical dimensions are 50 cm length and 50 - 150 μm id)

Table XII. Advantages and disadvantages of preparative Thin Layer Chromatography

Advantages	Disadvantages
Very simple, no apparatus needed	Impurities are easily introduced
Limited scale-up (≈1 g product) easy	Large scale-up (≈100 g) very difficult
Analytical TLC excellent pilot technique	Decomposition occurs easily (oxidation)
Slow running bands are visible	Application of sample is time consuming
Plates used only once → no clean-up	R_f values not totally reproducible
Selective detection possible	Sensitive to draft and leakages of chamber
Easy to learn	Stationary phase other than silica expensive
Resolution often better than column chromatography	Silica dust harmful to health (cancer risk)
Faster than column chromatography	Less suitable below 1 mg

filled with a buffer. Both ends of the tubing are placed in a buffer reservoir (see Figure 16A) and a high voltage (30 kV) is applied. As the electrolyte composition is the same along the capillary during the entire run, the current (typically 10 - 100 μA) will be constant. Because at pH > 3 the silanol groups on the walls are negatively charged in the presence of the buffer electrolyte, there is a layer of cations near the wall. These cations and associated water molecules migrate towards the cathode. This flow is called the electroosmotic flow (EOF). Because the speed of this flow is higher than that of the electrophoretic flow (EPF) of negatively charged ions present in the sample, all injected substances eventually arrive at the cathode where they can be detected. Cations with a high charge to mass ratio will move fastest, followed by cations with a smaller ratio, then by all neutral compounds (which move all with the same speed as the EOF), then by anions with a low charge to mass ratio and finally by anions with a high charge to mass ratio (93-96). One of the great advantages of CE over conventional liquid chromatographic techniques is the very flat flow profile and the small dimensions, both of which lead to a minimum of band broadening and thus very high plate numbers (over a million is possible). Two problems are the reproducible injection of the tiny sample volumes (nl) without discrimination of certain analytes and the low concentration sensitivity. Detection most often occurs by UV. Electrochemical detection or MS is also possible.

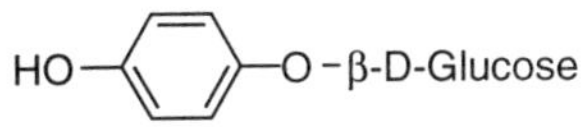

arbutin [39]

Because CZE in its basic form is limited to charged molecules (simple amines, alkaloids, acids), modifications were sought to widen the scope. One possibility is to add an agent which forms a charged complex with the solute, for instance the borate ion, which can form negatively charged complexes with sugars and certain phenols. An example is the CZE separation of phenols in an aqueous extract of *Uvae ursi* leaves (97). As buffer a 0.1 mol/l sodium borate buffer of pH = 9.5 was used. The dimensions of the fused silica capillary were 75 μm I.D. and 50 cm length while the voltage and current were 17.5 kV and 70 μA respectively. Before each run the capillary had to be rinsed with base and buffer solution. Detection occurred with UV at 214 nm. The main component, arbutin [39], was well separated from other components and could be quantified with resorcinol as internal standard. Flavonols can also be nicely separated under similar conditions (98).

A more generally usable modification for the separation of neutral molecules is micellar electrokinetic capillary chromatography (abbreviated as MEKC or MECC). In this technique, a surfactant, e.g. sodium dodecyl sulphate (SDS) is added to the buffer. Above the critical micelle concentration of the surfactant, negatively charged micelles are formed which move electrophoretically towards the anode. Because the EOF is larger than the EPF, there is a continuous stream of micelles towards the cathode (Figure 16B) (99). The hydrophobic tails of the micelles form an apolar phase in which the solutes can partition. This is similar to partition chromatography. Thus more apolar compounds arrive later at the detector than polar compounds because the net movement of the micelles is slower than that of the EOF. There are many different variables in MEKC which influence elution order, resolution and plate number: voltage, micelle type and concentration, electrolyte type and concentration, percentage organic solvent, pH, temperature, pre-run rinsing and the addition of various modifiers.

An example of MEKC is the separation of nine coumarins occurring in *Angelica* species. The conditions of the separation are given in the legend of Figure 16. Compared to gradient RP-HPLC the MEKC method gave a shorter run time (22 min

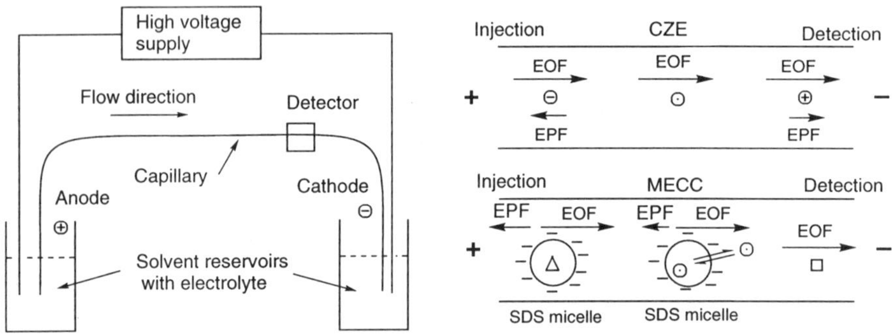

Figure 16. A (left) Schematic drawing of CE equipment. **B** (right) Schematic representation of the relative influence of the Electroosmotic flow (EOF) and the Electrophoretic flow (EPF) in (CZE, upper) and Micellar Electrokinetic Capillary Chromatography (MECC, lower). Legend: upper: ⊕ = positively charged species, ⊙ = neutral species, ⊖ = negatively charged species; lower: **Δ** =apolar species, ⊙ = polar-apolar species, □ = very polar species

excl. 12 min washsteps compared to 47 min excl. stabilisation for HPLC), a better resolution of psoralen and xanthotoxin, a lower resolution of osthol and columbianadin, a noisier baseline and a ten times higher detection limit (HPLC: ≈0.25μg/ml, MEKC: ≈2.5 μg/ml). Because of the four different wash steps the MEKC method needs to be automated. Both methods gave comparable precision and accuracy. Plate numbers for the nine coumarins varied from $4 \cdot 10^4$ to $3 \cdot 10^5$ (100). MEKC was also successfully applied in the determination of illicit drug substances, e.g. psilocybin, morphine, mescaline, cocaine and Δ^9-THC. MEKC resolved at least twice as many peaks as HPLC in less time with superior resolution. Due to this, and the fact that all peaks must elute before the micellar aggregate migration time, MEKC is well suited for general drug screening (101). The present status of capillary electrophoresis is comparable with that of HPLC in the seventies. Undoubtedly equipment will become more sophisticated in the coming years and new applications for natural products will be published. It will develop as an additonal analytical technique complementary to HPLC. For some very difficult separations CE may be better while for more simple separations HPLC will remain the method of choice. In many cases the choice will be dictated by available expertise and equipment. Some advantages and disadvantages of CE relative to HPLC are given in Table XIII (95, 96).

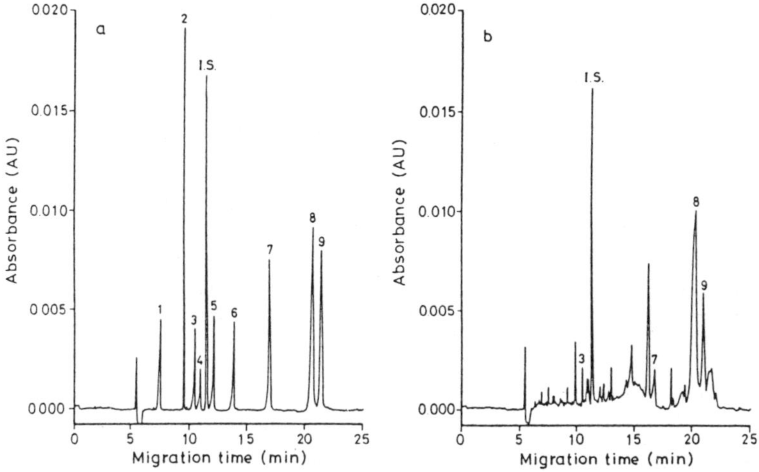

Figure 17. Capillary electropherograms of (a) a mixture of nine authentic coumarins and (b) an extract of *Angelica* root. Injection: 2 sec hydrodynamic of a methanolic extract. Electrophoretic medium: buffer (20 mM SDS, 15 mM sodium borate, sodium dihydrogenphosphate (pH 8.26)) / MeCN = 96 / 4. Capillary: 67.5 cm (59.5 cm to detector) x 50 µm i.d. Voltage: 20 kV. Temp. 30°. Detection: UV at 200 nm. For peak assignment see structures below. Internal Standard (I.S.) = 4-hydroxybenzoic acid. Reprinted from J. Chromatogr., Vol. 710A, C.-T. Chen and S.-J. Sheu, Separation of coumarins by micellar electrokinetic capillary chromatography, pp. 323-329 (1995) with kind permission of Elsevier Science - NL, Sara Burgerhartstraat 25, 1055 KV Amsterdam, The Netherlands.

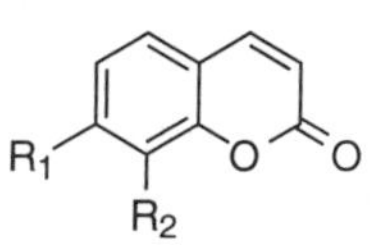

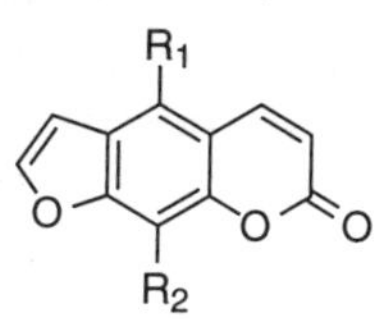

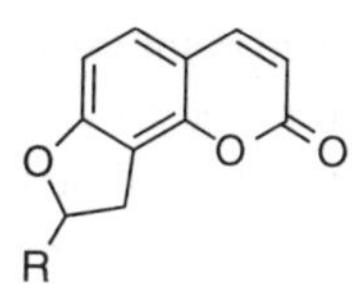

1 coumarin $R_1 = R_2 = H$
2 umbelliferone $R_1 = OH$, $R_2 = H$
8 osthol $R_1 = OMe$, $R_2 =$ CH_2-CH=C(Me)$_2$

4 psoralen $R_1 = R_2 = H$
5 xanthotoxin $R_1 = H$, $R_2 = OMe$
6 bergapten $R_1 = OMe$, $R_2 = H$

3 columbianetin $R = C(OH)(Me)_2$
7 columbianetin acetate $R = C(OAc)(Me)_2$
9 columbianadin $R = C(O_2CC(Me){=}CHMe)(Me)_2$

Table XIII. Advantages and disadvantages of Capillary Electrophoresis in comparison with HPLC

Advantages	Disadvantages
Much scope for future development	Not a mature technique
Much higher plate numbers (> 105)	Fewer published separations for natural products
Lower (organic) solvent consumption	Lower concentration sensitivity
Higher mass sensitivity	Unsuitable for preparative work, no scale-up
Lower reproducibility	Injection more complicated
Less sample required	Less suitable for very apolar compounds
Cheaper columns	

3.7. Supercritical Fluid Chromatography (SFC)

The use of a supercritical fluid as the mobile phase in chromatography has not found such widespread use as GC or HPLC but this may be more a matter of limited experience, published applications and financial limitations than theoretical and practical problems. SFC has a position intermediate between GC and HPLC because supercritical fluids have properties intermediate between those of gases and liquids (see Section 2.3 for more information on supercritical fluids). In Table XIV some advantages of SFC are given. As with SFE, by far the most common fluid is CO_2. To increase the solvation and eluting power 1-20% of organic solvent, e.g. methanol, can be added. To the modifier small amounts of additives such as trifluoroacetic acid or triethylamine can be added to improve the peak shape or to change the selectivity. The selectivity can also be altered by changes in the temperature, pressure and the proportion of modifier. Both open tubular capillary columns and packed HPLC columns filled with either normal- or reversed-phase materials have been used, but packed columns filled with normal-phase materials (bare silica, diol, aminopropyl or cyanopropyl) are most promising. SFC is thus most valuable for replacing both analytical and preparative separations which used to be carried out by normal phase HPLC. Such SFC separations are much more reproducible and faster than the equivalent HPLC separations. It is further useful for separating moderately apolar to very apolar compounds which cannot be separated by GC (boiling point too high or thermally unstable) and for chiral separations. A general SFC review with some natural product applications is given in (102). A more exhaustive review of applications in the field of natural products is (20). Three books devoted to SFC are (103-105).

Because of the intermediate nature of supercritical fluids both a number of LC and GC detectors are usable in SFC. This can be an important consideration for choosing SFC. LC detectors which have been commonly used include UV, fluorescence, MS, NMR and ELSD (see Section 4.2.4). MS is easier to couple with SFC than with HPLC. NMR is interesting because pure CO_2 has no protons and thus gives no signal. GC detectors include FID, ECD, MS and FT-IR. FID can only be used with pure CO_2 or CO_2 modified with formic acid. Dual detection is possible.

Table XIV. Advantages and disadvantages of Supercritical Fluid Chromatography

Advantages	Disadvantages
Suitable for volatile and involatile samples	Lower mobile phase flexibility than HPLC
Mild temperatures (40°), less decomposition	Not suitable for very polar substances
Can be combined with LC and GC detectors	Fewer published applications than HPLC
Standard HPLC columns can be used	Apparatus more expensive than HPLC
Higher optimum flow rates than with HPLC	Separations less well understood
Density (i.e. pressure) is an extra variable	Slower than GC
Lower solvent consumption, i.e. less waste	
Inflammable mobile phase (safety aspects)	

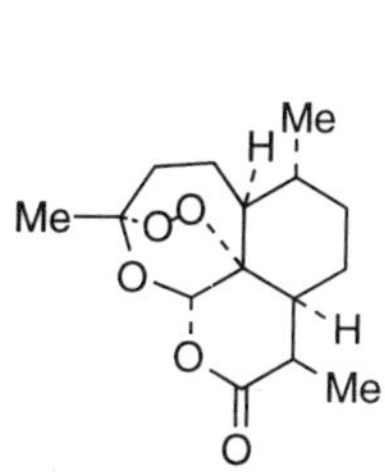

artemisinin [40]

A compound which can be eluted (t_r = 8.2 min) with 100% carbon dioxide at 17.2 MPa and 80° on a packed column (200 x 1 mm) with 5 μm cyanopropyl phase is the antimalarial artemisinin [40] (qinghaosu). The analyte was detected by means of electron capture (Section 4.4.3). Apparently the peroxide moiety confers sufficient electron capture properties to [40] for a sensitive detection. Concentrations of ≈ 100 ng/ml in plasma could be quantified (106). Probably the method could be used for the determination of [40] in plants. Other recent examples of SFC of plant products are the analysis of hydroxy- and methoxy-substituted flavonols on capillary polydimethylsiloxane columns with FT-IR and FI detection (107) and of iridal triterpenoids on packed C-18 or cyanopropyl columns with UV detection (108).

3.8 Gas Chromatography (GC)

Gas chromatography (GC) is based on the partition of substances between a liquid stationary phase and a gaseous mobile phase. For transportation by the mobile phase the solute needs to spend some time in the vapour phase and thus a significant vapour pressure is necessary at the temperature selected for the separation. As 350° should be considered the practical limit (because of oven maximum and stability of stationary phase and analyte) the application of gas chromatography is limited to those secondary metabolites which have boiling points below 450° and which are stable at the temperature of the separation. Thus GC is less universal than HPLC. An example of a high-boiling stable plant compound which can still be separated by GC around 300° is β-sitosterol [41]. All essential oil compounds have lower boiling points than [41] and are as a group eminently suitable for GC analysis. Derivatisation (see below) can extend the range of compounds analysable by GC.

A GC-separation between two compounds is primarily based on a difference in boiling point, and to a smaller extent on different interactions with the stationary phase. Both apolar (e.g. polydimethylsiloxane) and polar stationary phases (e.g. polyethylene glycol) exist. Apolar phases are more durable and separate compounds mainly acccording to their volatility. Essential oil constituents elute in the order: monoterpene hydrocarbons, oxygenated monoterpenes, sesquiterpene hydrocarbons and finally oxygenated sesquiterpenes. Polar phases retain compounds with polar groups more strongly (e.g. alcohols) so that essential oil constituents elute now in the order: monoterpene hydrocarbons, oxygenated monoterpenes and sesquiterpene hydrocarbons (overlapping), and finally oxygenated sesquiterpenes. This disadvantage is partially offset by the fact that all groups elute over a longer period of time, i.e. with polar columns more compounds can be resolved. If a particular pair of compounds cannot be resolved on one column, changing to a column with a different selectivity will often produce a separation.

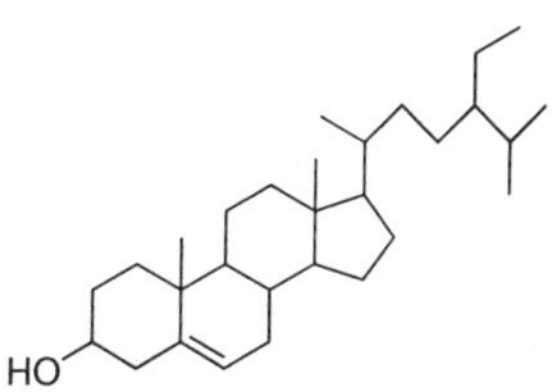

β-sitosterol [41]

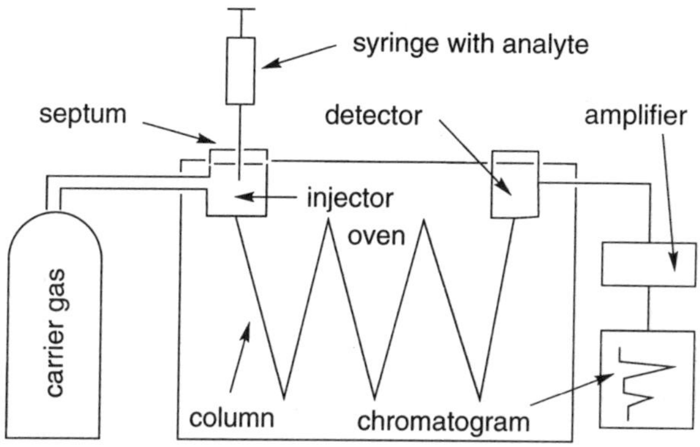

Figure 18. Schematic representation of a gas chromatograph.

A schematic drawing of a gas chromatograph is given in Figure 18. Compounds to be investigated are dissolved in a volatile organic solvent (e.g hexane, acetone) in concentrations between 0.005 and 0.5% and ≈1 μl is injected with a syringe equipped with a sharp needle (an HPLC syringe is unsuitable). The solvent and the solute quickly vaporise in the injector (≈250°) and are removed by the carrier gas stream from the injector into the column. After retracting the syringe the pierced septum which covers the top of the injector seals again. In the large majority of cases the carrier gas is hydrogen, helium or nitrogen. Hydrogen with the highest diffusion coefficient and the lowest viscosity is chromatographically speaking the best carrier gas. Unfortunately it is explosive. Helium is a good - albeit expensive - second choice. For ECD (Section 4.4.3) nitrogen is necessary.

GC columns exist in two varieties: packed or capillary. Packed columns are becoming less and less common. They are between 1 and 4 m long, have internal diameters from 2-6 mm and are filled with an inert, high specific surface material (e.g Chromosorb®) coated with 3-20% stationary phase. Many different stationary phases exist for fine-tuning the selectivity. This is necessary because the separation power is limited. Capillary columns (0.2 - 0.5 mm i.d.) are nowadays the standard. They are made of highly inert fused silica which has been coated on the inside with a thin film (0.1 - 5 μm) of stationary phase and are 10 - 100 m long. A standard 25 m x 0.25 mm i.d. column with a 0.25 µm film typically generates 100,000 theoretical plates. Because of these high plate numbers selectivity becomes less important.

Table XV. Advantages and disadvantages of GC in comparison with HPLC

Advantages	Disadvantages
Much higher plate numbers (> 10^5); fast	Without derivatisation, only for volatiles
Separation based on volatility and polarity	No mobile phase manipulation possible
No solvent consumption	Supply of high purity gases needed
Sensitive universal detector available	Only small amount of sample injectable
Columns very durable	Poorly suitable for preparative work

In one run more than 200 compounds can be routinely baseline-separated and this is a distinct advantage over HPLC (Table XV). If a mixture which contains compounds with widely different boiling points (e.g. an essential oil) has to be separated, temperature programming (e.g. from 50° to 250°) will result in the elution of all compounds within a reasonable time. The effect is comparable to gradient elution in HPLC. After the separation the compounds have to be detected. This is discussed separately in Section 4.4. For more basic information on GC see (43) or (42).

If thermally stable volatile plant constituents have to be analysed analytical GC is strongly preferred over HPLC or SFC because its separation power per unit time is much higher. An additional advantage of GC is that its separation is based both on differences in volatility and polarity. Many volatile terpenes have similar polarities and are therefore difficult to separate by LC. Still another advantage over HPLC is that a sensitive and universal detector exists for secondary metabolites (FID, see Section 4.4.1). Safranal [42] in saffron could be readily quantified by GC-FID with β-cyclocitral as internal standard. The isolation occurred selectively by heating ≈0.1 g saffron at 70° during 8 min and refocussing with a Tenax trap held at -30°C. Injection occurred by heating the trap to 200°C. A good separation was obtained with a 50 m capillary column (109).

Compounds with several hydroxyl, amino or carboxylic acid functionalities are non-volatile and cannot be separated by GC in their native form. However after

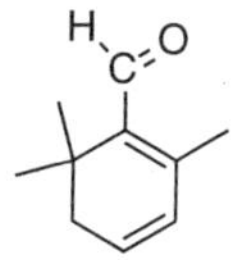

safranal [42]

derivatisation, e.g. silylation, acetylation or methylation, many compounds become volatile enough for GC analysis. For instance flavonoid aglycones are polar, high-melting involatile compounds which would - if introduced into a GC - never leave the injection chamber. However after silylation, in which every hydroxyl group is converted to a trimethylsilyl ether function, they can be analysed by GC (110). Halogen-containing derivatisation agents, e.g. heptafluorobutyric anhydride, give derivatives which can be selectively and very sensitively analysed by GC-ECD. A disadvantage of any derivatisation is that it is an extra step which can cause problems, e.g. incomplete reaction, more than one product, unpleasant reagents, time loss. In some cases HPLC may be the preferred technique for such compounds. More information on derivatisation, reagents and precautions can be found in (111).

Another example of analytical GC is the chiral evaluation of aroma compounds in essential oils and fruits. This can be a sensitive method to detect adulteration. Frequently, but not always, aroma constituents occur in nature almost exclusively as one of the enantiomers. Chirality can also influence the odour. The enantiomeric ratio of (+)- and (-)-linalool was determined by 2-dimensional GC. In the first oven a "normal" separation took place on an apolar column (DB-5, 30 m x 0.25 mm, 0.25 μm d_f). When linalool (+, -, or a mixture) eluted, this peak was transferred via a heated transfer line to a second oven equipped with a chiral alkylated cyclodextrin-phase GC column (Lipodex C, 25 m x 0.25 mm). On this column (+)- and (-)-linalool were resolved. Detection took place by means of mass spectrometry (112).

Analytical GC is also frequently used for qualitative purposes. Two different approaches can be pursued: (I) identification on the basis of retention times, preferably on two different columns, and (II) identification by comparison of a mass spectrum with that of many spectra in a mass spectral library. Mass spectra of volatile compounds can be obtained by GC-MS (see Section 4.4.6). Neither identification method is foolproof and many erroneous identifications exist in the literature,

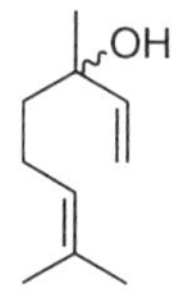

linalool [43]

especially in the essential oil field. A good practice is to combine both methods and to check if the proposed MS-identification fits with the retention indices on different columns. Even better is to perform a co-injection of the sample with an equal amount of an authentic reference compound on two diferent columns which should give exactly coinciding peaks (spiking experiment). Unfortunately reference substances are not always available, in which case identification remains tentative. If compounds cannot be identified, preparative isolation and subsequent identification by NMR is a possibility. This can be done by preparative LC, SFC and GC. LC suffers from poor resolution and analyte losses during evaporation of the mobile phase while GC has as disadvantage the very low "loadability" of the columns. This is particularly true for capillary columns and thus packed GC columns still find some use for preparative applications because of their somewhat higher "loadability". In (113) more information can be found on preparative GC with packed columns.

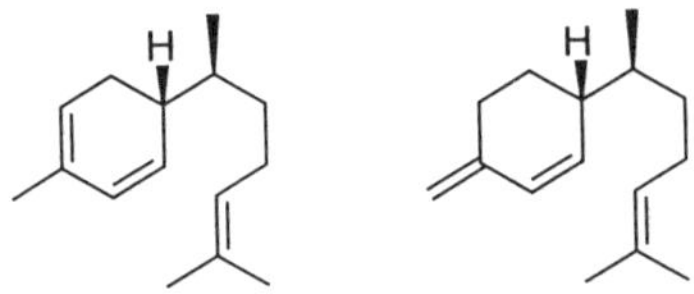

α-zingeberene [44] β-sesquiphellandrene [45]

An example of the use of preparative capillary gas chromatography is the separation of the sesquiterpene hydrocarbons α-zingiberene [44] and β-sesquiphellandrene [45] which occur in ginger oil. These two very apolar compounds differ only in the position of one double bond. They can neither be separated with normal phase HPLC nor with RP-HPLC. However they can be well resolved with GC. Retention indices on polydimethylsiloxane and polyethylene glycol capillary columns are 1489/1730 and 1518/1781 for [44] and [45] respectively (retention indices can be calculated by comparing the retention time of an analyte with those of the two n-alkanes eluting just before and after the analyte). Thus on a 30 m megabore (0.53 mm i.d., 3 μm film thickness) polydimethylsiloxane column ≈ 400 μg of the mixture could be baseline separated (see Figure 19). To minimize the cycle time a second sample was injected before elution of the first sample without adverse effects on the resolution. Trapping was carried out in glass tubes at room temperature. Within a few hours 5-10 mg quantities of the pure substances could be obtained which were enough for recording proton- and carbon NMR spectra (114). Nowadays fully automated equipment exists for carrying out such separations.

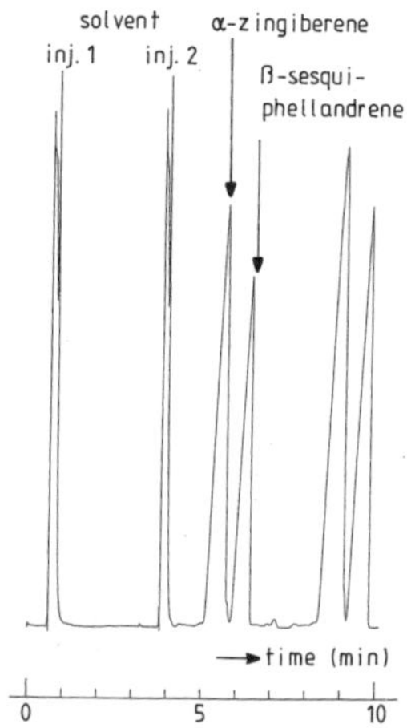

Figure 19. Preparative capillary separation of the sesquiterpene hydrocarbons α-zingiberene [44] and β-sesquiphellandrene [45]. Experi-mental conditions: column DB-1, 30 m x 0.53 mm, 3 μm film thickness, 0.8 μl splitless injection of a 50% solution in pentane; temp. 175°. Reprinted from Phytochem. Anal., Vol. 2, T.A. van Beek and G.P. Lelyveld, Isolation and identification of the five major sesquiterpene hydrocarbons of ginger, pp. 26-34 (1991), with kind permission of John Wiley & Sons, Ltd.

3.9. Fractional (Vacuum) Distillation

This formerly important technique has now lost much of its attractiveness in the laboratory purification of natural products. This is both due to the smaller scale on which (new) natural products are isolated and the advent of more powerful chromatographic methods to separate volatiles. The separation is based on a difference in boiling points. The best equipment available is capable of separating two compounds differing by 3°C in their boiling point. By lowering the pressure, natural products with boiling points at atmospheric pressure as high as 350°C can still be separated at temperatures as low as 100°C, thus minimising potential thermal decomposition. For a more complete description of the technique the reader is referred to (44). Nowadays fractional vacuum distillation is still occasionally used for the prefractionation of essential oils (115, 116).

An example is the distillation of 20 g of the neutral part of the essential oil of *Pulicaria gnaphalodes* with a 30 cm Vigreux column up to 45°/0.02 Torr. With this procedure the majority of the oil (65% monoterpenes) was separated from the oxygenated sesquiterpenes. To avoid decomposition a trace of the antioxidant BHT was added to the distillation flask. By careful multistep flash chromatography on silica with various petroleum ether 40/60 / diethyl ether gradients several sesquiterpene alcohols could be isolated and identi-fied (116). Some advantages and disadvantages of the technique are presented in Table XVI.

Table XVI. Advantages and disadvantages of fractional vacuum distillation

Advantages	Disadvantages
Technique is simple and apparatus durable	Only suitable for volatiles (b.p. < 350°)
No solvent consumption	Thermal decomposition and oxidation
Excellent for large samples	Minimum sample size is 4 ml
Scale-up is possible	Significant hold-up volume, i.e. losses
Separation mechanism different from HPLC	Separation power much lower than GC
	Fraction collection somewhat complicated
	Good equipment expensive
	Separations are slow (≈ 1 day)

3.10. Precipitation

Precipitation can be used for the selective removal of either impurities or analytes from an extract or solution. It is followed by decanting, filtration or centrifugation. It is a cheap and fast method in comparison to chromatography and can accomplish purifications which are difficult to realise by other means. An example of the precipitation of analytes is the addition of a mercuric iodide complex (Mayer's or Valser's reagent) to an acidic solution of a tertiary or quaternary alkaloid. A heavy whitish precipitate is formed which can be easily separated. After collection the precipitate can be dissolved in acetone / methanol / water = 6 / 2 / 1 and passed over a strong anion exchange column (Cl^-). This exchanges the mercuric iodide complex for a chloride and the crude alkaloid chlorides remain. This is one of the best ways to purify quaternary alkaloids which cannot be separated by conventional purification techniques for alkaloids because they are charged irrespective of the pH. An example is the isolation of luxandrine [46], a quaternary benzylisoquinoline from *Pseudoxandra sclerocarpa* (117).

An example of the removal of interfering substances is the precipitation of phenolic substances with lead(II) ions (see Section 2.2, clean-up of the initial aqueous extract of *Digitalis purpurea* leaves) or - more environmentally friendly - polyvinylpolypyrrolidone (PVPP).

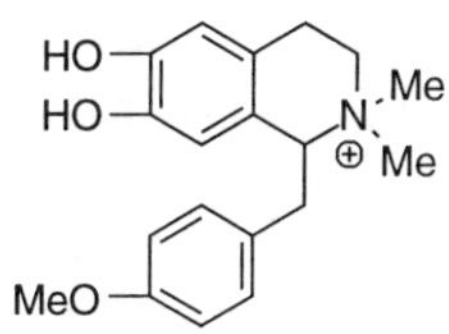

luxandrine [46]

Activated charcoal is often used in the final stages of preparative natural product purifications to remove undefined brownish trace impurities from already chromatographically pure substances. It can be difficult to remove such minor impurities by chromatography or crystallisation. Care should be taken that the substance which is to be purified is not itself adsorbed by the charcoal. Okabe *et al.* used such an adsorption effect to advantage in the isolation of ginkgolides A [9], B [10] and C [11] from *Ginkgo biloba* leaves (118). When the aqueous leaf extract was treated with charcoal the ginkgolides were adsorbed by the charcoal. After filtration the ginkgolides could be recovered from the charcoal with acetone. Here the charcoal was used on a preparative scale as a cheap reversed phase material.

Compounds can also be selectively precipitated by adding a solvent in which they are poorly soluble. An example is the clean-up of certain flavanone glycosides (e.g. hesperidin [2]) which are extremely insoluble in water but moderately soluble in the equally polar formamide. When water is added to a crude solution of these glycosides in formamide they immediately precipitate, after which they can be removed by centrifugation or filtration. Many impurities of similar polarity do not exhibit such a big difference in their solubility in water or formamide, remain in solution and are thus in a simple fashion separated from [2].

3.11. Crystallisation

As most new natural products are nowadays isolated in mg amounts as oils or amorphous solids after several chromatographic steps, crystallisation is no longer a necessarily technique preceding an identification. It remains however the technique of choice for the final purification of known natural products in amounts as low as 100 mg (laboratory scale, e.g. re-isolation of a promising new antitumour compound for initial tests) to as large as many tons (bulk industrial scale, e.g. saccharose). Finding the proper solvent for crystallisation is a matter of trial and error. Usually one chooses a solvent or solvent mixture in which the compound of interest is less soluble than the remaining impurities although the alternative approach - by analogy with the precipitation in Section 3.10 - is also possible. For more practical information about crystallising compounds the reader is referred to (44). Sometimes crystallisation can remove unwanted compounds with similar structures which are difficult - and thus expensive - to separate by chromatographic means. Crystallisation from tetrachloromethane was the key step in an improved isolation procedure for azadirachtin, discussed in Section 6 (119). In conclusion one can say that in spite of all the chromatography crystallisation remains a quick, selective and cheap final purification step.

4. Detection Techniques

4.1. Introduction

As the separation or purification of plant extracts is not a complete goal in itself, some other final step will take place afterwards. When an analytical separation is concerned, this step is usually a qualitative screening or a quantitative determination. If it is a (semi-) preparative separation the aim is to obtain one or more of the compounds of interest in a pure form for further biological or chemical testing. However before this is possible one needs to locate the compound(s) in the various fractions. There exist numerous techniques to accomplish both the analytical and preparative steps. A distinction should be made between so called "on-line" and "off-line" techniques. With "on-line" techniques, compounds are measured directly after leaving the chromatographic system (closed system). "Off-line" techniques are those where there is at least some transfer, and usually also some manipulation of the sample either by humans or a machine before the final measurement takes place. Some advantages and disadvantages of on-line techniques are given in Table XVII. Some spectroscopic techniques like UV spectroscopy can be used both on-line, e.g. HPLC-UV and off-line, e.g. scanning of column fractions in a UV spectro-photometer. In general, on-line techniques are associated with analytical separations whilst off-line techniques are usually associated with preparative applications.

Table XVII. Advantages and disadvantages of on-line techniques in comparison with off-line techniques

Advantages	Disadvantages
Less chance of introducing impurities	Often more expensive
Less chance of decomposition	Less flexible
Faster	Technically more complicated
Less laborious	Impossibility of connecting some detectors
Higher sensitivity	Limited dynamic range

In the following section, various techniques available for the detection of plant compounds after HPLC, TLC or GC are discussed, and some applications are given. Some of the techniques, like mass spectrometry (MS) or NMR spectroscopy, cannot only be used for the qualitative or quantitative detection of (known) compounds but also for their structure elucidation. This specific aspect will not be discussed here but will be the subject of a separate chapter in this volume: see chapter 5 - "Structure elucidation of plant secondary products". Possible detectors for capillary electrophoresis or SFC have already been briefly mentioned in Section 3.6 and

Section 3.7.

4.2. Detection Techniques for Liquid Chromatography

4.2.1. UV spectroscopy

This is the most widely-used detector for liquid chromatography. Its popularity is due to its robustness, relative low cost, good sensitivity (1-10 ng range for many compounds), tunable selectivity, good baseline stability (relatively insensitive to temperature fluctuations and refractive index effects) and large dynamic range. It can be used with gradients. Disadvantages are the need for solvents to be transparent at the detection wavelength and the sometimes large (> 1000-fold) variations in relative response factors at a given wavelength.

With UV spectroscopy, double bonds are detected. Thus UV detection is not the method of choice for compounds lacking double bonds, such as saturated hydrocarbons and sugars. When only isolated double bonds or carbonyls are present in a molecule (e.g. many essential oil constituents), detection has to occur at low wavelengths: 210 - 220 nm and the sensitivity can be relatively poor (see Section 4.2.2 for an example). The choice of solvents is in this case restricted to water, acetonitrile, methanol and other alcohols, and saturated hydrocarbons like hexane. When conjugated double bonds (either dienes or α,β-unsaturated carbonyls) or aromatic rings, are present the wavelength shifts to higher values and the sensitivity increases. Many more solvents can then be used.

In general, detection takes place at the wavelength which gives the best ratio in absorption between compounds of interest and impurities. Thus when a compound shows a maximum at 225 nm and a slightly lower maximum at 285 nm it may be better to measure at 285 nm since there will be less interference from impurities. When UV detection is used for monitoring the eluate in preparative applications there are two possibilities to overcome saturation of the detector: one can install a cell with a shorter path length (e.g. 0.2 instead of 5 mm) or, more cheaply, one can measure at a wavelength different from the absorption maximum (e.g. 320 nm instead of 280 nm).

A UV detector which makes it possible to record UV spectra on-line in less than 1 sec is the Diode Array Detector (DAD). Thus there is no need to stop the flow. Contrary to a normal UV detector polychromatic light passes first through the detection cell and is then dispersed by a grating. The absorbance is measured by a large number of photodiodes, each being responsible for a small section of the spectrum (1-3 nm). By combining the signals of all the diodes the spectrum can be constructed. With the earliest models the sensitivity was rather poor but modern diode array detectors can record good spectra of peaks as small as 0.001 AU. The spectra are useful for compound identification and for checking peak purity. For this reason it has also been called "a poor man's LC-MS". HPLC-DAD is of course most useful for distinguishing between compounds which differ slightly in their UV spectra and less so for compounds showing only an end absorption around 210 nm like monounsaturated olefins. The software allows one to build personal libraries which make automatic identifications possible in a manner similar to GC-MS. An example of a group of compounds where DAD is useful is the indole alkaloids. There are more than 1000 indole alkaloids and at least eight different chromophores occur. An indole (Figure 20, left structure) has a spectrum very different from a dihydroindole moiety (Figure 20, right structure). Thus UV allows one to differentiate considerably between a number of structures. Within a certain chromophore various aromatic substituents (R_{1-4} = H, OH or OMe) at the four different positions lead to slightly different spectra which can be recognised by the DAD software. A small review on DAD and its application in the analysis of carotenoids appeared in (120).

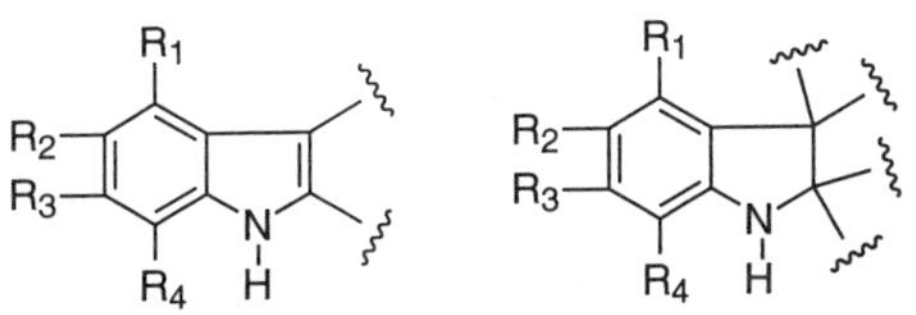

UV: 225, 285, 293 nm UV: 243, 291 nm

Figure 20. Basic structure of indole (left) and dihydro-indole (right) with their UV spectra. Depending on the substitution pattern maxima may shift slightly.

4.2.2. Refractive index detection

This is the most universal detector available for LC. It is based on a difference in refractive index (RI) between solvent and solute, and solvent alone. Because it has several significant disadvantages, it has never gained the wide popularity of the UV detector. Some advantages and disadvantages are given in Table XVIII. It is mainly used for non-volatile compounds which do not absorb in the UV above 200 nm (e.g. sugars) or which have very low extinction coefficients (ε) relative to co-eluting impurities (e.g. carbonyl-containing compounds without additional olefinic bonds). In recent years some applications which used to be performed with RI detection are now carried out with ELS detection (see Section 4.2.4).

An application where RI detection is obviously superior to UV detection is the quantitation of ginkgo terpenetrilactones [9-12, 38] (121). These compounds of intermediate polarity can be extracted from the leaves of *Ginkgo biloba* by boiling in water. After a two-column SPE clean-up, the purified extract still contains some impurities in low concentrations. However, because of the small extinction coefficient and low absorption maximum of these terpenes ($\varepsilon \approx 300$ l/mol . cm at 219 nm), they are poorly visible in the RP-HPLC chromatogram (Figure 21). In contrast, the same extract but now detected with RI shows essentially only the peaks of the solvent and the compounds of interest because on a weight basis they are the major constituents and the relative response factors do not vary nearly as much with RI detection as with UV detection (121).

Table XVIII. Advantages and disadvantages of the Refractive Index detector

Advantages	Disadvantages
Detects all compounds	Poor sensitivity
Can in principle be used with all solvents	Unstable baseline due to extreme sensitivity to temperature and pressure variations
Relative response factors do not vary much	Cannot be combined with gradient elution
Some models usable for preparative work	No changeable selectivity like UV
Relatively cheap	

4.2.3. Fluorescent detection

This is a highly sensitive and selective detector. Only those secondary metabolites which emit light upon irradiation with light of a different (lower) wavelength can be measured. Most of these molecules show extended conjugation. The standard example is the antimalarial alkaloid quinine [13] from *Cinchona* bark which is added to tonic water because of its bitterness. If trace compounds which do show fluorescence have to be quantitatively analysed and such a detector is available its use should be considered because sample clean-up can be less stringent. Another possibility is to derivatise selectively compounds with lack UV absorption so that they become fluorescent. An example is the precolumn derivatisation of non-protein amino acids occurring in *Acacia* seeds with 9-fluoroenylmethyl chloroformate (FMOC-Cl) (122). Excitation and emission wavelengths were 260 and 313 nm respectively.

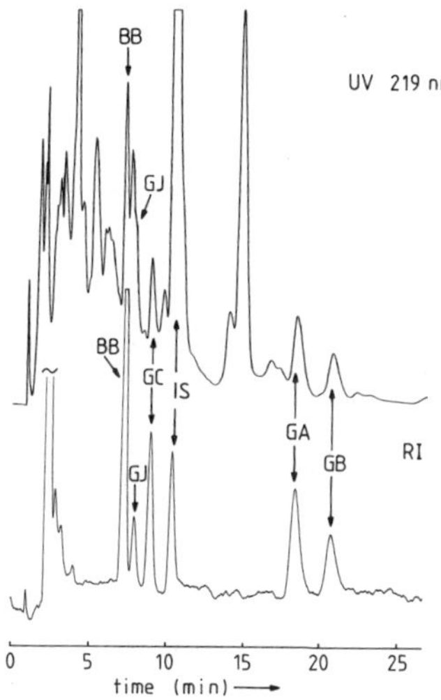

Figure 21. HPLC traces of a phytopharma-ceutical based on an extract of *Ginkgo biloba* leaves after SPE clean-up. Column: C18, solvent: water : methanol = 67 : 33 at 1.0 ml/min. Upper trace, UV detection at 219 nm. Lower trace, RI detection. BB = bilobalide, IS = internal standard (benzyl alcohol), GJ, GC, GA and GB = ginkgolide J, C, A and B respectively. Reprinted from J. Chromatogr., Vol. 543, T.A. van Beek *et al.*, Determination of ginkgolides and bilobalide in *Ginkgo biloba* leaves and phyto-pharmaceuticals, pp. 375-387 (1991), with kind permission of Elsevier Science - NL, Sara Burger-hartstraat 25, 1055 KV Amsterdam, The Netherlands.

4.2.4. Evaporative Light Scattering Detection (ELSD)

The principle of this detector is based on nebulization of the HPLC flow. The small droplets are carried along a heated drift tube with nitrogen gas. After evaporation of the volatile solvent analyte particles remain which scatter a light beam in the detector cell proper. The scattering is dependent on the mass flow (123). This detector has recently gained some popularity as an alternative for the RI detector due to its higher sensitivity, more stable baseline and possible use with gradients. It is most useful for involatile secondary metabolites with lack a suitable UV chromophore and which can be separated in volatile solvents (max. 75% water), e.g. seed oil triacylglycerols (124). Some advantages and disadvantages are given in Table XIX. A nice application of ELSD is the detection of ginkgo terpene trilactones (125). These compounds cannot be analysed by UV and the sensitivity of RI is rather low. Because they are non-volatile and thermally stable ELSD is a logical choice and HPLC-ELSD is indeed clearly superior to HPLC-UV or HPLC-RI for these compounds. A recent review on ELSD appeared in (126).

Table XIX. Advantages and disadvantages of the Evaporative Light Scattering Detector

Advantages	Disadvantages
Detects all involatile compounds	Not suitable for compounds with b.p. <200°
Can be used with all volatile solvents	Involatile modifiers (salts) cannot be used
Can be used with gradient elution	Droplet formation can be troublesome
More sensitive and stable than RI detector	Response not linear with mass flow
No solvent peak	No changeable selectivity like UV
Can also be used as an SFC detector	Only usable for preparative work with split
Sample size and flow rate not critical	Some thermal stress during solvent removal
Apparatus is rather simple	Plugging of nebulizer can occur

4.2.5. Electrochemical detection

This technique is based on the reduction or oxidation of a substance on an electrode. In its most basic form, the detector consists of a small cavity with a working electrode, a counter or auxillary electrode and a reference electrode. At a preset constant voltage between the working- and reference electrodes the current can be measured (controlled potential amperometry). Depending on the standard redox potential, certain compounds will be either reduced or oxidised at the working electrode. Many compounds are less electrochemically active than the solvent or supporting electrolyte i.e. the latter are more easily oxidised or reduced than the solute, and such compounds cannot be determined. Some selectivity can be obtained for the electrochemically active com-pounds by a proper choice of the operating conditions (voltage, mobile phase composition, electrode material). In principle, some compounds can be analysed at the pg level. A review on electrochemical detection of biomolecules appeared in (127).

In spite of this selectivity and sensitivity electrochemical detection has never attained high popularity in phytochemical analysis. Plant compounds which can be analysed include phenols, peroxides, quinones, aromatic amines and ascorbic acid. A nice example is the selective electrochemical detection of the antimalarial compound artemisinin (128). Artemisinin [40] is a sesquiterpene peroxide which can undergo an irreversible two-electron reduction. It is the only constituent in the extract showing this behaviour and therefore selective electrochemical detection is possible. After RP-HPLC separation with a solvent consisting of acetonitrile / 0.1 N ammonium acetate + 500 mg Na_2EDTA/l = 45 / 55, artemisinin was reduced at a gold-mercury cell at -0.8 V. Quantitation was possible with benzoyl peroxide as internal standard. The sensitivity and selectivity were excellent. The clean-up consisted only of an extraction with petroleum ether, evaporation and treatment of the residue with acetonitrile. Three ng was readily detected which meant that only 25 mg of air-dried plant material containing 0.02% [40] was sufficient for a quantitative determination. A severe disadvantage of the method was that oxygen interfered with the detection. All oxygen had to be rigorously removed from the solvents (128).

Another example is the pulsed amperometric detection of some sugar-like indolizidine alkaloids like castanospermine [47] (129). After HPLC separation under acidic conditions on a cation-exchange column, NaOH was added. This partially ionized the hydroxyl groups after which oxidation took place on a gold working electrode. The method was rapid reproducible, selective and sensitive. Linear calibration curves were observed from 10 ng/ml to 20 μg/ml (50 μl injection volume). Catecholamines from *Crataegus* could also be determined by RP-HPLC with amperometric detection (130).

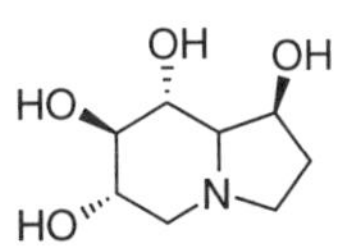

castanospermine [47]

4.2.6. Mass Spectrometry (MS)

The mass spectrometric detector is a highly sensitive and specific universal detector. When used on-line with a chromatographic separation technique (GC-MS, HPLC-MS, SFC-MS, CE-MS) the result is not only a chromatogram but also mass spectra are obtained for each peak. Thus - depending on the ionization technique - variable additional spectral information is available which is useful for the identification of a compound.The mass spectrometer is a rather complex apparatus and ion separation and detection require high vacuum operating conditions. The combination of HPLC and MS is complicated because (I) even a low flow of mobile phase is not compatible with this vacuum condition, and (II) the often low volatility of the analyte.

Table XX. Advantages and disadvantages of the Mass Spectrometric detector

Advantages	Disadvantages
Truly universal detector	Highly complex apparatus
Very sensitive (pg range)	Very expensive (except for tabletop GC-MS)
By choosing a proper interface suitable for GC, HPLC, SFC and CE	Less suitable for dirty samples
Yields molecular weight of analyte	Works under high vacuum
Yields detailed structural information	Only with split usable for preparative work
A large database with mass spectra exists	Structural data vary with ionization technique
Deuterated analyte as ideal internal standard	With HPLC-MS mobile phase problems: in-compatible flow rate, modifiers or solvent
No clean-up with tandem MS-MS systems	

A large number of LC-MS interfaces have been developed during the last two decades, e.g. moving-belt, heated nebulizer, particle beam (131), frit FAB (132), continuous-flow FAB, thermospray, electrospray (ESI) and atmospheric pressure chemical ionization (APCI)). Although one ideal interface is not yet available, currently the latter two interfaces appear to be among the most generally applicable.

For a review see (133). By various techniques (which are also applicable to GC-MS) more selective or sensitive measurements are possible. By a technique called "Selective Ion Monitoring" a chromatogram can be recorded which shows only the eluting compounds which possess in their mass spectrum a peak at the m/z value specified. This increases both the sensitivity and selectivity (see Figure 22).

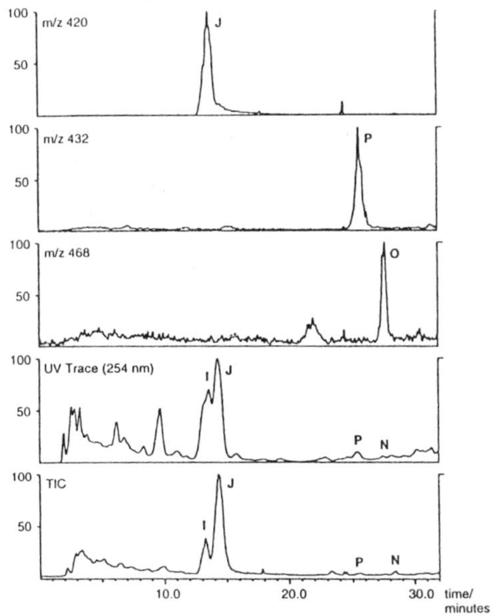

Figure 22. HPLC traces of a mycelial extract of *Armillaria tabescens*. The UV trace at 254 nm and the TIC trace are displayed, in addition to selected-ion traces at *m/z* 420 (armillyl orsellinate (J)), *m/z* 432 (unknown (P)) and *m/z* 468 (arnamiol (O)). Horizontal axis: time in min. Reprinted from J. Chromatogr., Vol. 710A, P. Cremin *et al.*, Liquid chromatographic-thermospray mass spectrometric analysis of sesquiterpenes of *Armillaria* (Eumycota: Basidiomycotina) species, pp. 273-285 (1995), with kind permission of Elsevier Science - NL, Sara Burgerhartstraat 25, 1055 KV Amsterdam, The Netherlands.

In combination with diode array detection (HPLC-DAD-MS) LC-MS forms an exceptionally powerful instrument for a first indication of the composition of crude plant extracts. In combination with library spectra and a good literature survey of the species under investigation many compounds can be reliably identified at the low μg scale without any preparative isolation. As the preparative fractionation and purification of unknown secondary metabolites followed by the off-line recording of IR-, NMR- and mass spectra is very labour-intensive, much time and money can thus be saved. An example of HPLC- thermospray MS is the analysis of sesquiterpenes in *Armillaria* mycelia (134). In Figure 22 the LC-UV trace at 254 nm, the Total Ion Current (TIC) trace and traces at *m/z* ratios specific for individual compounds are given, showing the versatility of this technique. Reviews on the use of various forms of MS for the analysis of glycosides, on HPLC with thermospray MS, and on LC-MS of alkaloids appeared in (135-137) respectively.

4.2.7. NMR spectroscopy

This very expensive detector is potentially the most powerful detector available for LC. Unfortunately it is also a very insensitive detector. Nevertheless in recent years significant progress has been made with regard to sensitivity problems and solvent suppression and several nice applications have been demonstrated although so far not many in the natural products field. A review can be found in (138). Five vitamin A acetate isomers could be identified by NMR after separation on a cyanopropyl column with heptane containing 1% benzene-d6 (139). Spectra were recorded with a flow of 0.2 ml/min. In the stopped flow mode the detection limit was 1.2 μg. LC-NMR was also useful for the characterization of a photodegradation product of azadirachtin [64] (140). It is certain that within the next ten years HPLC-NMR and SFC-NMR will become more widely available and, in particular, the structure elucidation of secondary metabolites in plant extracts should benefit enormously from this development.

4.3. Detection Techniques for Thin Layer Chromatography (TLC)

4.3.1. Qualitative techniques

Most TLC plates are developed for qualitative or semi-quantitative purposes: is the isolated compound identical to the reference or does a fraction consist of > 90% of the compound of interest? Except in the case of coloured compounds (e.g. anthocyanins), some visualisation is necessary prior to answering such questions.

The most simple visualisation method makes use of the fluorescent indicator which is added to most TLC stationary phases. Under UV light of 254 nm, this indicator gives a light green appearance to the silica or alumina (light blue for Reversed Phase C18-TLC plates). When compounds are present that absorb UV light around of 254 nm (aromatic compounds, α,β-unsaturated carbonyls, di- or trienes), the light no longer reaches the indicator and the stationary phase looks dark purple. Thus such compounds are selectively and sensitively (1 μg) revealed. Another simple detection technique makes use of the intrinsic fluorescence of some compounds under UV 366 nm light (see also Section 4.2.3). Because only compounds with an extended conjugated system show such behaviour, this is not a universal technique. If one deals with fluorescent compounds, however, one should use it because of the high selectivity and sensitivity (0.1 μg). A compound which is frequently encountered as an impurity when dealing with plant extracts is chlorophyll. It is easily recognised on TLC by its characteristic red fluorescence. Because of the high sensitivity of this detection technique, one should be careful not to attach too much significance to fluorescent spots. If the same spot cannot be seen under UV 254 nm, it is most likely a minor impurity.

After the preliminary viewing under UV light of 254 and 366 nm, a whole range of spraying reagents is at the disposal of the investigator for further visualisation. These range from universal reagents (e.g. concentrated sulphuric acid) via group-specific reagents (e.g. Dragendorff reagent for alkaloids) to highly specific reagents for one functionality (Ehrlich reagent = 4-dimethyl aminobenzaldehyde/HCl for a primary aromatic amine). Some examples are given in Table XXI. More information can be found in references in two books on TLC (20-22). Even if compounds have already been detected at 254 nm, spraying with a chromogenic reagent can be useful for distinguishing between compounds with similar Rf values. An example of this is the

Table XXI. Some spray reagents for the detection of various natural products on TLC plates

Name of reagent	Characteristic for
Anisaldehyde-sulphuric acid	essential oils, many terpenes, aromatics
Anthrone-sulphuric acid	sugars, glycosides
Chloramine-trichloroacetic acid	cardenolides (cardiac glycosides)
Dragendorff	alkaloids (give brown-orange spots)
Fastblue B	phenolics
Ninhydrin	amino acids
Potassium hydroxide	anthraquinones
Permanganate - soda	olefins
Sulphuric acid (concentrated)	universal reagent for organic compounds
Vanillin-sulphuric acid	essential oils, many terpenes, aromatics

characterisation of some structurally similar dimeric indole alkaloids which occur in the genus *Tabernaemontana*.They all show extinction under UV 254 nm and have similar Rf values. Some of them are even difficult to distinguish with MS or ^{1}H NMR. However, spraying with a solution of 3.25% $FeCl_3$ in 35% $HClO_4$ after development gives for each alkaloid specific colours (Table XXII) during heating

Table XXII. Colours of various bisindole alkaloids with ferric chloride / perchloric acid spray reagent

Name of reagent	Colour changes upon heating*
Conoduramine [48]	gold (Δ0) ⇨intense gold (Δ1, Δ2) ⇨ green gold (Δ3)
3-Hydroxy-conoduramine [49]	gold (Δ1) ⇨ dark blue (Δ2) ⇨ grey-black (Δ3)
Conodurine [50]	dark green (Δ0) ⇨ blue-green (Δ1) ⇨ vague brown-green (Δ2) ⇨ vague gold (Δ3)
3-Hydroxy-conodurine [51]	light blue (Δ0, Δ1), blue (Δ2), grey (Δ3)
Voacamidine [52]	gold-brown (Δ0) ⇨ brown-green-grey (Δ1) ⇨ dark blue (Δ2, Δ3)
Voacamine [53]	gold (Δ2) ⇨ blue-green ⇨ green centre with brown rim (Δ3)
3-Hydroxy-voacamine [54]	blue-grey (Δ2) ⇨ grey (Δ3)

* Δ0 = no heating, Δ1 = 5-10 sec heating, Δ2 = 15-20 sec heating, Δ3 = 40-50 sec heating

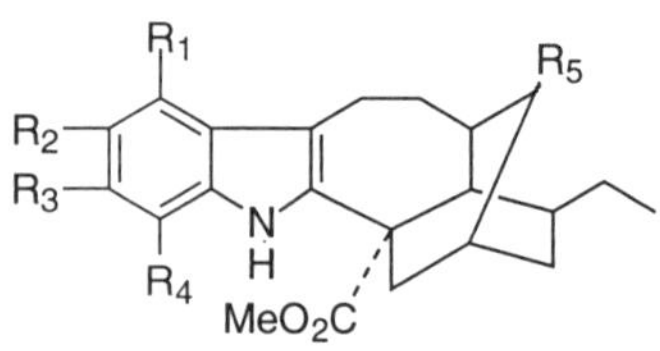

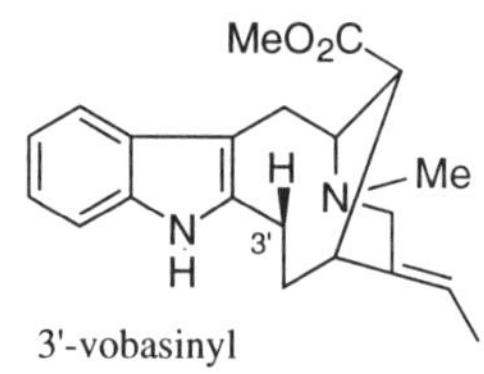

3'-vobasinyl

[48] Conoduramine: R1 = R4 = R5 = H, R2 = 3'-vobasinyl, R3 = OMe
[49] 3-Hydroxy-conoduramine: R1 = R4 = H, R2 = 3'-vobasinyl, R3 = OMe, R5 = OH
[50] Conodurine: R1 = R2 = R5 = H, R3 = OMe, R4 = 3'-vobasinyl
[51] 3-Hydroxy-conodurine: R1 = R2 = H, R3 = OMe, R4 = 3'-vobasinyl, R5 = OH
[52] Voacamidine: R1 = 3'-vobasinyl, R2 = OMe, R3 = R4 = R5 = H
[53] Voacamine: R1 = R4 = R5 = H, R2 = OMe, R3 = 3'-vobasinyl
[54] 3-Hydroxy-voacamine: R1 = 3'-vobasinyl, R2 = OMe, R3 = R4 = H, R5 = OH

with a hot air gun (141). Thus this very simple, fast and cheap chromogenic reaction can achieve results which can be realised only with difficulty or not at all with far more expensive techniques. Sometimes it can be useful to spray with two different spray reagents, e.g. a general reagent and a specific one. A simple way to do this is to apply the sample (S) and the reference (R) twice on the same plate in the order S-R-S-R. After development and drying, the left half of the plate is first covered with a piece of glass and spraying of one S-R pair with reagent 1 takes place. Then the right half is covered and reagent 2 is applied on the second pair. If necessary the plate can be heated afterwards.

4.3.2. Quantitative techniques

After development, spots on a TLC plate can be quantitatively scanned with a densitometer. A densitometer is an instrument which can measure the transmittance, reflectance or fluorescence of spots on TLC plates. A prerequisite is that the spots either show some absorbance at a particular wavelength or some fluorescence. If neither is the case, visualisation with a reagent (see Section 4.3.1 above) is necessary prior to the densitometer scanning. To obtain optimal, reproducible results, application of the sample is best done automatically on high-performance TLC (HPTLC) plates.

If carried out properly, quantitative TLC can be as accurate as HPLC and much faster because sample clean-up can be simple and many samples can be analysed in one run. On a single 10x20 cm plate, if developed horizontally over 4.5 cm and simultaneously from both sides, 50 samples can be analysed within ten minutes. An example is the analysis of the alkaloids ajmalicine and serpentine in plant tissues of *Catharanthus roseus* and hyoscyamine and scopolamine in root tissues of *Datura stramonium* (142).

4.4. Detectors for Gas Chromatography

4.4.1. Flame Ionization Detector (FID)

The FID is by far the most widely-used GC detector for the detection of organic substances, including secondary metabolites, on account of a large number of advantages (Table XXIII). Its principle is based on combustion of the analytes coming from the GC column in a hot hydrogen-air flame. In the process ions are generated which can be sensitively measured.

Table XXIII. Advantages and disadvantages of the Flame Ionization Detector (FID)

Advantages	Disadvantages
Universal for secondary metabolites	Destructive
Sensitive (25 pg detection limit)	Air and hydrogen supply needed
Simple and robust	For capillary columns make-up gas needed
Relatively inexpensive	
No response to carrier gases, H_2O and CO_2	
Very stable baseline	
Most response factors vary between 0.5 and 2	
Mass sensitive, peak area independent of flow rate	
Good linearity, wide dynamic range	

4.4.2. Thermal Conductivity Detector (TCD)

The TCD has not much to offer that the FID cannot do better. It measures a difference in thermal conductivity between the carrier gas alone and the carrier gas with eluting analytes. It is less sensitive and has a poorer baseline stability than the FID. Its two advantages over the FID are (I) that it is a non-destructive detector, which can be useful for preparative (packed column) GC, and (II) that no additional gases apart from the carrier gas (He or H_2) are needed for operation.

4.4.3. Electron Capture Detector (ECD)

This detector can detect only compounds which have a high electron affinity, such as halogenated compounds, compounds possessing nitro groups and compounds having many oxygens. The occurrence of the former two groups is - with the exception of some marine plants - rare in the plant kingdom. An example of its use in selectively detecting oxygenated plant compounds is the detection of artemisinin [40] after an SFC separation discussed in Section 3.7 (106). Also ginkgolides [9-12] can be selectively and sensitively detected by ECD after being made volatile enough for GC analysis by silylation (143). Some advantages and disadvantages of ECD have been summarised in Table XXIV.

Table XXIV. Advantages and disadvantages of the Electron Capture Detector (ECD)

Advantages	Disadvantages
Highly selective	Only usable for few plant constituents
High sensitivity (< 1 pg detection limit)	Radioactive detector
Non-destructive	For capillary columns make-up gas needed
Can also be used for SFC	Small linear range
	For packed columns only N_2 or Ar usable as carrier gas
	Response factors vary considerably

4.4.4. Nitrogen Phosphorus Detector (NPD)

This detector is useful for the selective and sensitive detection of alkaloids and other nitrogen-containing compounds which are volatile by themselves or which have been made volatile after derivatisation. A comparison of NPD with FID for the quantitation of tropane alkaloids was reported in (144).

4.4.5. Fourier Transform Infrared Detector (FT-IR)

Like GC-MS, GC-FT-IR produces -on-line - spectra of compounds eluting from a GC column. Although FT-IR spectra are useful on their own, this relatively recently introduced detector supplements the mass spectrometric detector well. Many compounds which possess similar or identical mass spectra have different IR spectra, thus increasing the reliability of an identification in the case of known compounds or providing more clues on the identity of a still unknown compound. FT-IR was for instance useful for distinguishing between perillene [56] and a C-11 homoterpene [55] as potential structures of a kairomone produced by spider mites on bean leaves (34). Surprisingly these compounds have very similar mass spectra and retention times. An alternative way to distinguish between [55] and [56] is GC-MS with accurate mass measurements. It has also been used to screen aroma extracts from plants for acetylenic flavour precursors (145). Owing to the characteristic absorption of the acetylenic bond (2100-2250 cm $^{-1}$) FT-IR was a very selective screening method. Disadvantages of the FT-IR detector are its high cost, its low sensitivity relative to MS or FID, and the fact that the vapour-phase spectra are different from solid-phase IR spectra, thus necessitating special vapour-phase libraries. Because FT-IR measurements are non-destructive, in principle on-line GC-FT-IR-MS is possible. Reviews on the usefulness of GC-FT-IR in the essential oil field and tropical fruit analysis have been published in (146) and (147) respectively.

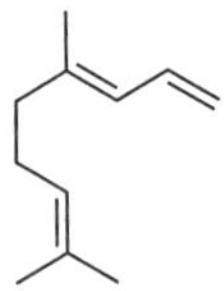

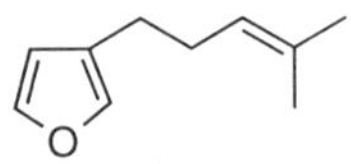

C-11 homoterpene [55] perillene [56]

4.4.6. Mass Spectrometric detector (MS)

Mass spectrometry as detection technique for GC combines universality, sensitivity and specificity. Capillary columns can be directly connected to the ion source of the mass spectrometer. Upon interacting with a beam of energetic electrons (70 eV EI), the eluting compounds form molecular and fragment ions. These ions can be mass measured by a scanning magnetic or quadrupolar field. Mass spectra are structurally very informative and also large databases are available on GC-MS datasystems. Nevertheless in some compound classes like terpenes only subtle differences may exist in the spectra of different compounds and confirmation of identity by another technique, e.g. retention on a GC column (Section 3.8), is required. For compounds which do not show a molecular ion peak in their EI spectrum, chemical ionization (CI) can be used. A reagent gas, e.g. methane, is introduced in the ion source at such a high pressure that ion-molecule reactions generate relatively stable protonated even-electron molecular ions (MH^+). A review on CI-MS in the essential oil field is given in (148). A short review on GC-MS, GC-FTIR and GC-AED (another GC detector not discussed here) was published in (149). For advantages and disadvantages of the mass spectrometric detector in general, see Table XX.

4.5. Biological Detectors

One special type of detector is the "biological" detector. This type of detector makes use of the electrophysiological or behavioural response which a certain plant extract or fraction evokes in an intact organism or part of an organism. Every bioassay-assisted fractionation (see Section 1) makes use of them to locate the compounds causing the response of interest. Most often, samples are tested off-line. Just a few examples of the hundreds of different types of off-line biological detectors are the simple brine shrimp assay for generalised toxicity (150), numerous *in vivo* and *in vitro* antitumour assays (151, 152), tests for allelopathics with the aim of finding leads for new herbicides and tests for antifeedants (153, 154) and oviposition deterrents (155) or oviposition stimulants (55) as possible new crop-protection agents.

On-line assays also exist but are less common. An example is the sniffing port to locate volatiles with a strong olfactory impact eluting from a GC column (156-158). At the end of the GC column there is a splitter which divides the carrier gas flow at a certain ratio (e.g. 1:9) between a flame ionization detector and a cap which fits a human nose. To avoid peak broadening and dehydration of the nose, sufficient humidified air (make-up gas) is added to the flow just after elution from the GC. During elution of compounds with an odour the trained test person makes notes of the type of smell and its impact. Later the two traces (FID and sniffing port) can be compared for possible direct identification with GC-MS or for guiding further preparative separation. Often strong-smelling peaks are detected in areas in the chromatogram where the FID does not show any response. This indicates that the compounds responsible for the aroma are only present in trace quantities. It also shows the extreme selectivity and sensitivity of biological detectors for certain compounds. For identification of such trace constituents large quantities of starting material will be needed.

4.6. Enzyme Immunoassays

An enzyme immunoassay is a specialised quantitation technique with both significant advantages and disadvantages (see Table XXV). The biggest disadvantage is that not every researcher can use or develop the technique because antibodies are needed. These highly-specific antibodies are produced by rabbits or mice some time after injecting them with the compounds of interest bound to a protein. Due to the specialised nature of this method a full discussion is outside the scope of this chapter. A review on immunoassays for small molecular weight products can be found in (159). Four recent examples of the development of enzyme immunoassays in the field of secondary metabolites can be found in (160-163). The development of a regenerable fibre-optic cocaine biosensor using monoclonal antibodies and capable of selectively detecting cocaine and some related compounds at 5 ng/ml in a flow cell was recently reported (164).

Table XXV. Advantages and disadvantages of enzyme immuno assays

Advantages	Disadvantages
Very selective → limited clean-up	Neither easy nor cheap to develop
High sensitivity	Cross reactivity between derivatives
High sample throughput once available	Limited accuracy

4.7. Ion-Selective Electrodes

Ion-selective electrodes for the analysis of plant constituents offer in principle some significant advantages over chromatographic methods: simplicity, fast response, compatibility with turbid samples and low costs. Problems are selectivity and sensitivity. A cocaine-selective membrane electrode was recently shown to be capable of accurately determining the cocaine content of drug mixtures in the absence or presence of morphine. Concentrations from 10^{-2} M to 10^{-6} M could be measured. The lower limit of detection was 4×10^{-7} M (165).

5. Artefact Formation and Decomposition of Genuine Compounds

One problem which has always been plaguing natural product chemists is artefact (also spelled as artifact) formation and/or decomposition. An artefact is defined as a compound which is eventually isolated or identified but which was originally either not present at all or present at a much lower concentration. Metaphorically, as soon as one picks a leaf from a tree, all kinds of metabolic (enzymatic or "chemical" in origin) processes may set in which may or may not alter the concentration of compounds. An example everyone knows well is a banana. During ripening the colour, texture, flavour and smell change noticeably and this is caused by a large number of chemical reactions taking place. It is even possible that the eventual compounds of interest were not present at all in the original plant material. Various types of artefacts may be distinguished:

(I) Well-defined products formed by chemical conversion, e.g. hydrolysis, oxidation, dehydration, isomerisation, and/or cyclisation, from plant products originally present. This can happen at any stage of the analytical chain: transportation, crushing, drying, extraction and/or purification. The products often look "natural" and it is therefore difficult to tell whether or not a compound is genuine, especially if it occurs at low concentration. The above-mentioned reactions can also be catalysed by native enzymes. Substrates and enzymes occur in different compartments and only after slicing does the conversion take place (see garlic example below). If one is tracking a promising but labile biologically-active product it may be that all activity is lost after a certain purification step. The best course of action is then to combine all the inactive fractions and to see whether the activity returns. If so, the problem is likely to be biological in origin. If not, the purification may well be the cause and it is best to go back one step, to reproduce the activity and to try an alternative, preferably milder fractionation approach.

A nice example of the strong influence of the extraction on the final product is garlic. With steam distillation diallyl disulphide [57] is the product, with alcohol and water at 25° allicin [58], the odorous principle of garlic is isolated, and with alcohol at sub-zero temperatures the genuine alliin [59] (= (+)-S-allyl-L-cysteine sulphoxide) is found. The decomposition of alliin into allicin is initiated by the enzyme alliinase (166).

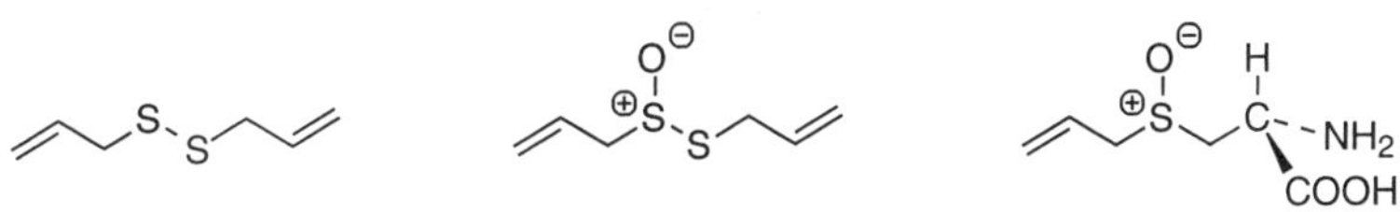

diallyl disulphide [57]   allicin [58]   alliin [59]

Another example of the formation of a product which is not originally present in the plant is the formation of chamazulene [60] during the steam distillation of chamomile flowers. During the distillation a precursor, matricin [61], is dehydrated, hydrolysed and decarboxylated to the artefact chamazulene (see Figure 23). Because of the intense blue colour of chamazulene which is present at ≈1%, chamomile oil is also deep blue. If chamomile oil were to be steam distilled under vacuum the reaction would be much slower and consequently the oil would be only light blue or colourless.

The alkaloids from *Taxus baccata*, collectively called "taxines", degrade slowly under neutral conditions (e.g. methanol), and rapidly under basic conditions, to the corresponding cinnamates with the emission of dimethylamine (Figure 24). Probably several of the neutral taxanes reported in the literature are in reality mere artefacts formed during the extraction with hot organic solvents. Another reaction which takes

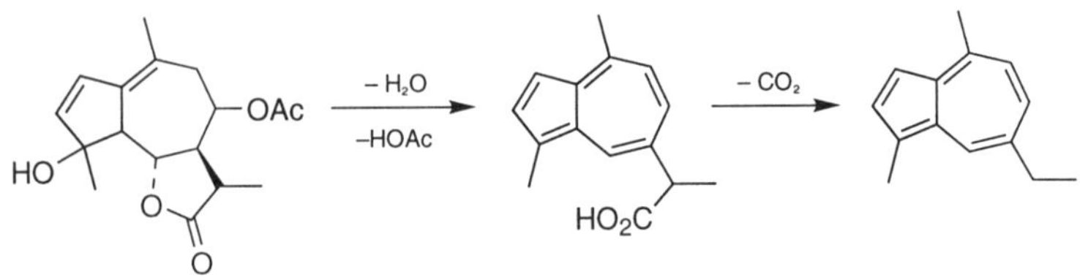

matricin [61] chamazulene [60]

Figure 23. Decomposition of matricin [61] into chamazulene [60] during steam distillation.

place very rapidly under both neutral and basic conditions is an internal trans-esterification (167). Thus taxine alkaloids are difficult to isolate and purify because basic conditions which occur at some stage during the isolation and purification of most alkaloids (see Section 1.2) should be avoided. Under weakly acidic conditions the alkaloids are reasonably stable. Under strongly acidic conditions hydrolysis of the esters can occur. Finally these alkaloids can be degraded under the influence of light in methanol, ethanol or tetrahydrofuran (reductive deamination, photocyclization and acyl migration) (168). Artefact formation of this type can also be catalysed by chromatographic techniques or chromatographic media. For a short review see (169).

(II) Well-defined products formed by the reaction of genuine compounds with materials used during the processing of plant materials. One can think of reactions with solvents or impurities in solvents or chromatographic media. Once identified these products are usually easily recognised as artefacts. A well documented example is the reaction of CH_2Cl_2 with tertiary amines. This particular solvent reacts so easily that once the question was posed whether it is a solvent or a reagent (170). Azoug *et al.* isolated after separation of the crude alkaloid mixture by flash chromatography on silica gel with a CH_2Cl_2 - CH_2Cl_2 / MeOH gradient besides affinisine [62] also chloromethylene-affinisinium [63]. This compound could be synthesized by refluxing affinisine in CH_2Cl_2 in the presence of silica and [63] was thus shown to be an artefact (171).

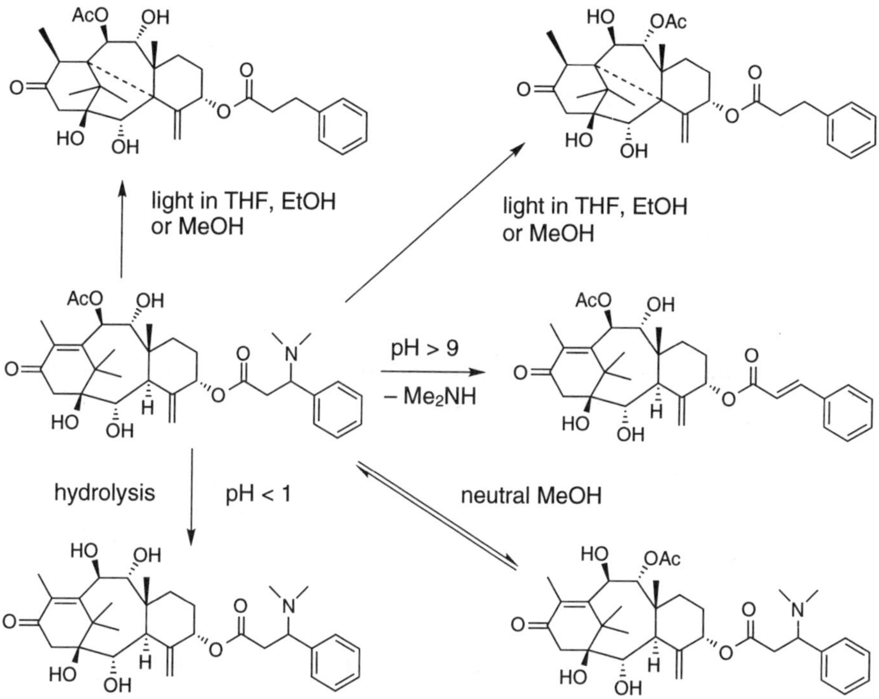

Figure 24. Decomposition of taxine under acidic, basic and neutral conditions.

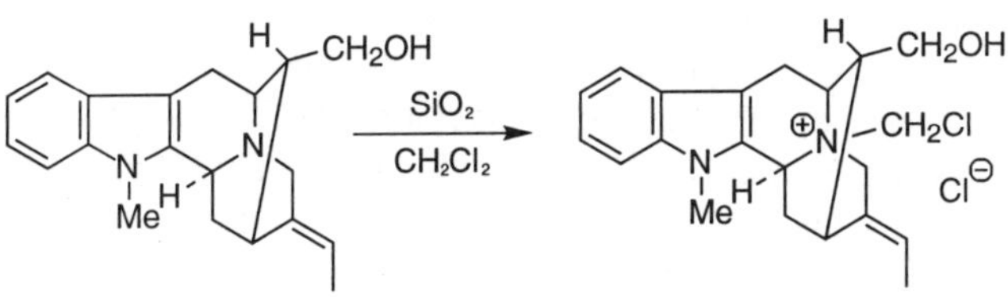

affinisine [62]

chloromethylene-affinisinium [63]

(III) More or less defined products not originally present in the material of study which were introduced during the cultivation, harvesting, storage or processing. One can think of other plant species, fungal growth, plasticizers (phthalates), silicone grease, detritus from human skin or impurities from (non-)distilled solvents. In reference (172) there is a listing of all extraneous compounds - often with mass spectra - which have been encountered as artefacts. Entries include, among others, aluminium foil contaminants, Amberlite®, antifoaming emulsion, charcoal, drying agents, filter paper contaminants, Florisil® contaminants, glassware adsorption, lubricants, silica gel, stabilizers and Tenax® artefacts.

Impurities can be a severe problem in the case of preparative isolations where large amounts of solvents are evaporated. There are always involatile impurities present and these are likewise concentrated. Even if analytical grade solvents are used this is a problem if only mg quantities, or less, of the material of interest are present: 0.001% involatile residue in 1 l solvent still means 10 mg unwanted material which could disturb both spectral measurements and biological tests. For the final separation or purification step it is most advantagous to use small amounts of freshly bidistilled solvents.

(IV) Ill-defined products formed by oxidation, thermal- or photodegradation and/or polymerisation of genuine plant compounds during processing. The products formed are often highly polar, brownish-coloured and insoluble in organic solvents. This can be a problem during preparative TLC. Decomposition to unknown, irreversibly-bound products by chromatographic media can also occur. An example is the 80% loss of the sesquiterpene bilobalide [38] when it is in contact with alumina for only 1 min (28).

The examples above make it clear that reasonable care should be exercised during the collection, drying, storage, isolation, pre-purification or final analysis of any plant part or compounds contained therein. No extraneous materials should adhere to the material, decay should be prevented, solvents should be distilled, chromatographic media should be washed with strong solvents prior to use, and aggressive chemicals, contact with oxygen, excessive light and high temperatures should be avoided. If chemotaxonomically unexpected products have been isolated the identity of the voucher specimen in the herbarium should be checked. If this is satisfactory, the isolation and/or purification should be repeated using other protocols to see if the same product is again isolated. If compounds are isolated which are suspected to be artefacts, repeating the procedure with a possible precursor is a way to check the genuine nature of the compound. For quantitative procedures (GC, HPLC), blanks and recovery experiments are absolutely essential.

6. Conclusion

In the previous paragraphs many techniques for the isolation, separation and detection of secondary metabolites from plants were discussed. Some general guidelines - such as the difference in physical and chemical properties of the analyte(s) and the matrix - for setting up a phytochemical method were mentioned in the first paragraph. Of equal importance is a good definition of the problem. Due to the large number of available techniques there will be many good but sometimes radically different methods for solving a single problem.

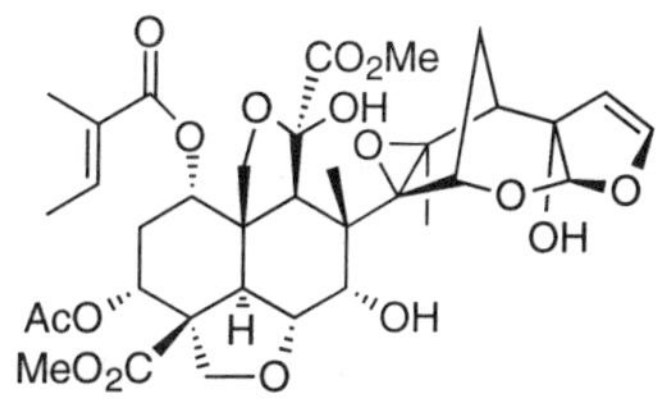

azadirachtin [64]

For the isolation of azadirachtin [64], the best natural antifeedant isolated so far, an eight-step isolation procedure has been published (119). With this method large amounts of pure azadirachtin became available for the first time. Azadirachtin occurs with many similar tetranortriterpenes in seeds of *Azadirachta indica*. The extraction is schematically depicted in Figure 27. After a pre-extraction with hexane to remove fats, the intermediately-polar [64] was extracted with alcohol. After two partitions which removed apolar and polar impurities a filtration over silica removed some other polar compounds which would have otherwise interfered with the next VLC step. Some other closely related structures which were difficult to remove by normal-phase chromatography could be removed by a selective crystallisation from tetrachloromethane. The last impurities were removed by flash chromatography on silica with a totally different solvent from that used for the VLC separation. Although the method consists of quite a number of steps it uses only two normal-phase low pressure chromatographic steps.

A more recently published method (140), also starting from 2 kg seeds, used a 5-step procedure (Figure 28). The initial two steps are very similar. In this procedure however the partition steps were omitted and instead of a filtration step the methanolic extract was directly coated onto a normal-phase stationary phase. The column was eluted with a step gradient to give, amongst others, crude azadirachtin. This fraction was rechromatographed under isocratic conditions to give pure [64].

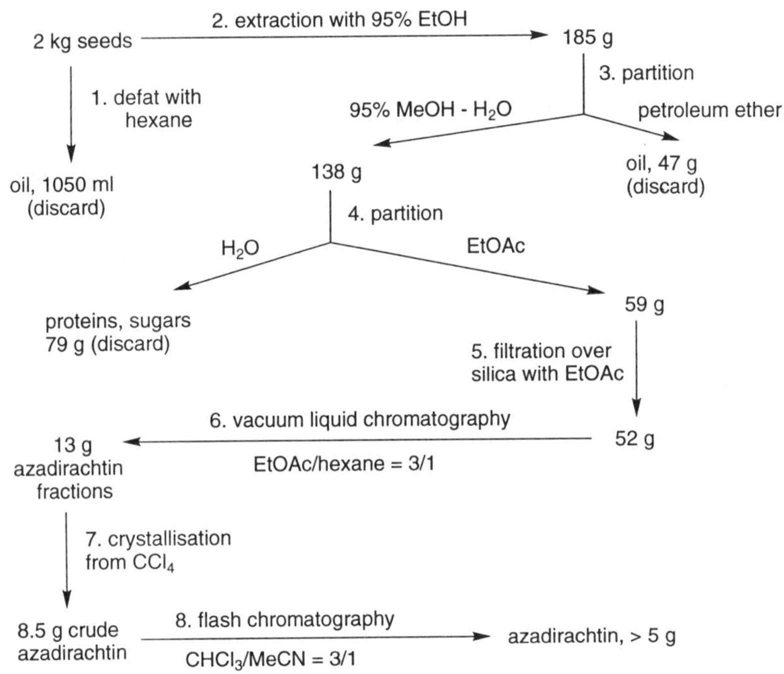

Figure 27. Schematic representation of an extraction procedure for azadirachtin [64] from seeds of *Azadirachta indica*. Taken from (119).

Without further studies it is difficult to tell which method is better. The latter procedure is certainly less labour-intensive, probably uses slightly less solvent and avoids the use of halogenated solvents, but the yield is 60% lower. On the other hand, even though both methods started with the same weight of seeds, the initial azadirachtin concentration may have been quite different. Finally the exact purity of the isolated azadirachtin is not known. Although for the purposes of the authors of these methods a purity of either 97% or 99% may be sufficient, a purity of 99% is

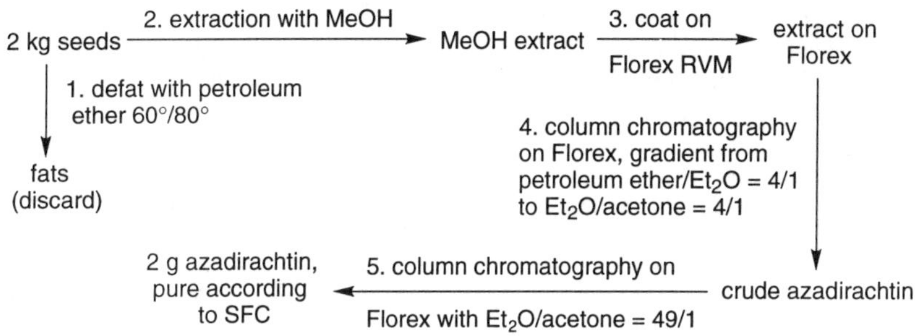

Figure 28. Schematic representation of an extraction procedure for azadirachtin [64] from seeds of *Azadirachta indica*. Taken from (140).

much, much harder to achieve and such a difference may be important for biological tests. The conclusion is that there is a choice of methods for solving a specific phytochemical problem and one is not always necessarily better or worse than the other. In the above example they are even remarkably similar. Both start with a defatting step and both use a polar organic solvent to extract the required polar material from the remaining matrix. Both methods use two low-pressure chromatographic steps on a normal-phase packing for removing most of the other interfering terpenes. One could almost say that for this particular problem, the chemical properties of the analyte and the matrix virtually dictate the approaches used. Thus the essential conclusion, as indicated at the start, is that the solution to any phytochemical problem should be the optimal interaction between the plant material itself, the aim of the investigation and the specialist skills, equipment and other resources - especially time - which can be made available. The skill and challenge of the phytochemist lies in achieving this optimal interaction.

7. List of abbreviations

AED	Atomic Emission Detector
AMD	Automated Multiple Development
APCI	Atmospheric Pressure Chem. Ionization
AU	Absorption Unit
BHT	2,6-Di-*tert*-butyl-4-methylphenol
BuOH	n-Butanol
CC	Column Chromatography
CE	Capillary Electrophoresis
CI	Chemical Ionization
CPCC	Centrifugal Partition Chromatography
CZE	Capillary Zone Electrophoresis
C8	Type of apolar phase for LC or HPLC

C18	Type of apolar phase for LC or HPLC
DAD	Diode Array Detector
DB-1	Type of apolar capillary GC column
DB-5	Type of apolar capillary GC column
DCCC	Droplet Counter Current Chrom.
DNA	Deoxyribonucleic acid
ECD	Electron Capture Detector
EI	Electron Impact
ELSD	Evaporative Light Scattering Detector
EOF	Electroosmotic Flow
EPF	Electrophoretic Flow
ESI	Electrospray Ionization
Et_2O	Diethylether
EtOAc	Ethyl acetate
EtOH	Alcohol
FAB	Fast Atom Bombardment
FI	Flame Ionization
FID	Flame Ionization Detector
FT-IR	Fourier Transform Infrared
GC	Gas Chromatography
H	Height of a theoretical plate
HOAc	Acetic Acid
HPLC	High Pressure Liquid Chromatography
HPTLC	High Performance TLC
HS	Headspace
I.d.	Internal diameter
IR	Infrared
K	Partition Coefficient
K′	Chromatographic capacity factor
L	Length
LC	Liquid Chromatography
LPLC	Low Pressure Liquid Chromatography
MECC	Micellar Electrokinetic Capillary Chromatography
MeCN	Acetonitrile
MEKC	Micellar Electrokinetic Capillary Chromatography
MeOH	Methanol
MHE	Multiple Headspace Extraction
MPLC	Medium Pressure Liquid Chromatography
MS	Mass Spectrometry
NIR	Near Infrared
NPD	Nitrogen Phosphorus Detector
OPLC	Overpressured Layer Chromatography
pTLC	Preparative TLC
PVPP	Polyvinylpolypyrrolidone
RI	Refractive Index
RLCC	Rotation Locular Counter Current Chromatography
RP	Reversed Phase
RPC	Rotation Planar Chromatography
R_s	Resolution
SDE	Steam Distillation Extraction
SDS	Sodium Dodecyl Sulphate
SFC	Supercritical Fluid Chromatography
SFE	Supercritical Fluid Extraction
SPE	Solid Phase Extraction
SPME	Solid Phase Microextraction
TAS	Tetrabutylammonium Sulphate
TCD	Thermal Conductivity Detector
THF	Tetrahydrofuran
TLC	Thin Layer Chromatography
UV	Ultraviolet
VLC	Vacuum Liquid Chromatography
XAD	Type of apolar synthetic resin

8. Acknowledgements

I wish to thank Prof. dr. L.T. Taylor (Blacksburg, VA, USA) and dr. M.A. Posthumus (Wageningen, The Netherlands) for critically reading the manuscript and useful suggestions.

9. References

1. D. M. Dolman, D. W. Knight, U. Salan and D. Toplis, *Phytochem. Anal.* **3** (1992) 26-31.
2. E. Stahl and W. Schild, *Drogenanalyse II: Inhaltstoffe und Isolierungen* (G. Fischer Verlag, Stuttgart, 1981).
3. J. B. Harborne, *Phytochemical methods - a guide to modern techniques of plant analysis* (Chapman and Hall, London, 1984).
4. K. Hostettmann, M. Hostettmann and A. Marston, *Preparative chromatography techniques* (Springer, Berlin, 1986).
5. K. Hostettmann, M. Hostettmann and A. Marston, *Preparative chromatography techniques - applications in natural product isolation*, 2nd ed. (Springer, Berlin, 1997).
6. C. F. Poole and S. K. Poole, *Chromatography today* (Elsevier, Amsterdam, 1993) 735-946.
7. E. L. Ghisalberti, in *Bioactive natural products - detection, isolation and structural determination*, eds. S. M. Colegate and R. J. Molyneux (CRC Press, Boca Raton, 1993) 9-57.
8. R. Ikan, *Natural products - a laboratory guide*, 2nd ed. (Academic Press, San Diego, 1991).
9. S. Natori, N. Ikekawa and M. Suzuki (eds.), *Advances in natural products chemistry - extraction and isolation of biologically active compounds* (Kodansha, Tokyo, 1981).
10. B. G. Osborne, T. Fearn and P. H. Hindle, *Practical NIR spectroscopy with applications in food and beverage analysis* 2nd ed. (Longman, Harlow, 1993).
11. D. Fengel and G. Wegener, *Wood: chemistry, ultrastructure, reactions* (de Gruyter, Berlin, 1989).
12. K. D. Bartle, A. A. Clifford, S. B. Hawthorne, J. J. Langenfeld, D. J. Miller and R. Robinson, J. *Supercrit. Fluids* **3** (1990) 143-149.
13. A. A. Clifford, *in Supercritical fluid extraction and its use in chromatographic sample preparation*, ed. S. A. Westwood (Blackie Academic & Professional, London, 1993) 1-38.
14. T. A. van Beek, P. Maas, B. M. King, E. Leclercq, A. G. J. Voragen and Æ. de Groot, *J. Agric. Food Chem.* **38** (1990) 1035-1038.
15. S. Budavari (ed.), *The Merck Index - An encyclopedia of chemicals, drugs, and biologicals, 11th ed.* (Merck & Co., Rahway, 1989).
16. D. Lide (ed.), *Handbook of chemistry and physics, 76th ed.* (CRC, Boca Raton, 1995).
17. *Nederlandse Farmacopee, achtste uitgave, deel II,* (Staatsuitgeverij, 's-Gravenhage, 1978).
18. F. Höfler, D. Jensen, J. Ezzel and B. Richter, GIT *Chromatographie* **15** (1995) 68-71.

19. B. E. Richter, B. A. Jones, J. L. Ezzell, N. L. Porter, N. Avdalovic and C. Pohl, *Anal. Chem.* **68** (1996) 1033-1039.
20. C. E. Bevan and P. S. Marshall, *Nat. Prod. Rep.* **11** (1994) 451-466.
21. R. M. Smith, LC-GC **9** (1996) 8-15.
22. W. K. Modey, D. A. Mulholland and M. W. Raynor, *Phytochem. Anal.* **7** (1996) 1-15.
23. U. Braumann, H. Hèndel, K. Albert, R. Ecker and M. Spraul, *Anal. Chem.* **67** (1995) 930-935.
24. L. T. Taylor, Anal. Chem. **67** (1995) 364A-370A.
25. S. A. Westwood (ed.), *Supercritical fluid extraction and its use in chromatographic sample preparation* (Blackie Academic & Professional, Glasgow, 1993).
26. L. T. Taylor, *Supercritical fluid extraction* (Techniques in analytical chemistry Vol. 2) (John Wiley, New York, 1996).
27. G. P. Blanch, E. Ibáñez, M. Herraiz and G. Reglero, *Anal. Chem.* **66** (1994) 888-892.
28. T. A. van Beek and L. T. Taylor, *Phytochem. Anal.* **7** (1996) 185-191.
29. E. Guenther, *The essential oils, Vol. 1* (D. Van Nostrand Company, Toronto, 1948) 85-226.
30. A. Koedam, in *Capillary gas chromatography in essential oil analysis*, eds. P. Sandra and C. Bicchi (Huethig, Heidelberg, 1987) 13-27.
31. R. A. J. Kaiser, in *Bioactive volatile compounds from plants* (ACS Symp. Ser. Vol. 525), eds. R. Teranishi, R. G. Buttery and H. Sugisawa (Am. Chem. Soc., Washington, 1993) 240-268.
32. G. W. Robertson, D. W. Griffiths, W. M. Smith and R. D. Butcher, *Phytochem. Anal.* **4** (1993) 152-157.
33. V. Karl, A. Dietrich and A. Mosandl, *Phytochem. Anal.* **4** (1993) 158-164.
34. M. Dicke, T. A. van Beek, M. A. Posthumus, N. Ben Dom, H. van Bokhoven and Æ. de Groot, *J. Chem. Ecol.* **16** (1990) 381-396.
35. K. Tatsuka, S. Suekane, Y. Sakai and H. Sumitani, *J. Agric. Food Chem.* **38** (1990) 2176-2180.
36. C. Bicchi and D. Joulain, *Flav. Fragrance J.* **5** (1990) 131-145.
37. R. E. Shirey, C. L. Woolley and R. F. Mindrup, *Int. Labmate* **20** (3) (95/96) 50-53.
38. D. Ulrich, S. Eunert, E. Hoberg and A. Rapp, *Deutsche Lebensm.-Rndsch.* **91** (1995) 349-351.
39. M. A. Jeannot and F. F. Cantwell, *Anal. Chem.* **68** (1996) 2236-2240.
40. H. Kuwajima, M. Morita, K. Takaishi, K. Inoue, T. Fujita, Z.-D. He and C.-R. Yang, *Phytochemistry* **31** (1992) 1277-1280.
41. C. F. Poole and S. K. Poole, *Chromatography today* (Elsevier, Amsterdam, 1993).
42. K. Robards, P. R. Haddad and P. E. Jackson, *Principles and practice of modern chromatographic methods* (Academic Press, London, 1994).
43. A. Braithwaite and F. J. Smith, *Chromatographic methods* (Chapman and Hall, London, 1990).
44. B. S. Furniss, A. J. Hannaford, P. W. G. Smith and A. R. Tatchell, *Vogel's textbook of practical organic chemistry, 5th ed.* (Longman, Harlow, 1989).

45. G. M. Tschan, G. M. KÜnig, A. D. Wright and O. Sticher, *Phytochemistry* **41** (1996) 643-646.
46. J.-Y. Zhou, Q.-C. Fang and Y.-W. Lee, *Phytochem. Anal.* **1** (1990) 74-76.
47. I. Slacanin, A. Marston, K. Hostettmann, D. Guédon and P. Abbe, *Phytochem. Anal.* **2** (1991) 137-142.
48. A. Marston, I. Slacanin and K. Hostettmann, *Phytochem. Anal.* **1** (1990) 3-17.
49. K. Hostettmann, M. Hamburger, M. Hostettmann and A. Marston, in *Modern phytochemical methods* (Recent advances in phytochemistry Vol. 25) eds. N. H. Fischer, M. B. Isman and H. A. Stafford (Plenum, New York, 1991) 1-32.
50. A. Marston and K. Hostettmann, *Nat. Prod. Rep.* **8** (1991) 391-413.
51. A. P. Foucault, *Anal. Chem.* **63** (1991) 569A-579A.
52. S. J. Bloor, J. P. Benner, D. Irwin and P. Boother, *Phytochemistry* **41** (1996) 801-802.
53. C. Van Sumere, P. Faché, K. Vande Casteele, L. De Cooman, E. Everaert, R. De Loose and W. Hutsebaut, *Phytochem. Anal.* **4** (1993) 279-292.
54. J. M. Betz and S. W. Page, *Proceedings of Bonn Bacans Symposium* (July 17-22, 1990) 103-104.
55. J. J. A. van Loon, A. Blaakmeer, F. C. Griepink, T. A. van Beek, L. M. Schoonhoven and Æ. de Groot, *Chemoecology* **3** (1992) 39-44.
56. M. Sato, T. Nakano, M. Takeuchi, N. Kanno, E. Nagahisa and Y. Sato, *Phytochemistry* **42** (1996) 1595-1597.
57. G. Dobson, W. W. Christie and B. Nikolova-Damyanova, J. Chromatogr. **671B** (1995) 197-222.
58. T. A. van Beek and D. Subrtova, *Phytochem. Anal.* **6** (1995) 1-19.
59. G. Subramanian (ed.), *A practical approach to chiral separations by liquid chromatography* (VCH, Weinheim, 1994).
60. S. Ahuja (ed.), *Chiral separations by liquid chromatography* (ACS Symp. Ser. Vol. 471), (Am. Chem. Soc., Washington, 1991).
61. F. A. Tomás-Barberán, M. Amparo-Blázquez, C. Garcia-Viguera, F. Ferreres and F. Tomás-Lorente, *Phytochem. Anal.* **3** (1992) 178-181.
62. J. E. Dellar, M. D. Cole and P. G. Waterman, *Phytochemistry* **41** (1996) 735-738.
63. Y. Kawasaki, Y. Goda and K. Yoshihira, *Chem. Pharm. Bull.* **40** (1992) 1504-1509.
64. E. Dagne, B.-E. Van Wyk, M. Mueller and W. Steglich, *Phytochemistry* **41** (1996) 795-799.
65. S. W. Pelletier, H. P. Chokshi and H. K. Desai, *J. Nat. Prod.* **49** (1986) 892-900.
66. J. C. Coll and B. F. Bowden, *J. Nat. Prod.* **49** (1986) 934-936.
67. N. Handjieva, H. Saadi, S. Popov and I. Baranovska, *Phytochem. Anal.* **2** (1991) 130-133.
68. B. S. Joshi, K. M. Moore, S. W. Pelletier and M. S. Puar, *Phytochem. Anal.* **2** (1991) 20-25.
69. R. Capasso, A. Evidente, A. Cutignano, M. Vurro, M. C. Zonno and A. Bottalico, *Phytochemistry* **41** (1996) 1035-1039.

70. G. T. Maatooq, D. K. Stumpf, J. J. Hoffmann, L. K. Hutter and B. N. Timmermann, *Phytochemistry* **41** (1996) 519-524.
71. A. Carotenuto, V. De Feo, E. Fattorusso, V. Lanzotti, S. Magnio and C. Cicala, *Phytochemistry* **41** (1996) 531-536.
72. H. Engelhardt, *High performance liquid chromatography* (Springer, Berlin, Germany, 1979).
73. W. J. Lough and I. W. Wainer (ed.), *High performance liquid chromatography - fundamental principles and practice* (Blackie Academic & Professional, London, 1996).
74. J. W. Dolan and L. R. Snyder, *Troubleshooting LC systems* (Humana Press, Clifton, 1989).
75. M. Verzele and C. Dewaele, *Preparative high performance liquid chromatography* (De Muyter, Deinze, Belgium, 1986).
76. J. L. Glajch, J. J. Kirkland, K. M. Squire and J. M. Minor, *J. Chromatogr.* **199** (1980) 57-59.
77. A. Kato, M. Moriyasu, Y. Nishiyama, M. Ichimaru, F. D. Juma and J. O. Ogeto, *Phytochem. Anal.* **4** (1993) 72-75.
78. G. D. Manners and J. A. Pfister, *Phytochem. Anal.* **4** (1993) 14-18.
79. H. D. Friedel and R. Matusch, *J. Chromatogr.* **407** (1987) 343-348.
80. T. A. van Beek, N. v. Dam, Æ. de Groot, T. A. M. Geelen and L. H. W. van der Plas, *Phytochem. Anal.* **5** (1994) 19-23.
81. K. C. Van Horne (ed.), *Sorbent extraction technology* (Analytichem Int., Harbor City, USA, 1985).
82. H. Wagner, S. Bladt and E. M. Zgainski, *Drogenanalyse* (Springer, Berlin, 1983).
83. H. Wagner and S. Bladt, *Plant drug analysis, 2nd ed.*, (Springer, Berlin, 1996).
84. T. A. van Beek, R. Verpoorte, A. Baerheim Svendsen, A. C. Santos and L. P. Olay, *J. Nat. Prod.* **46** (1983) 226-231.
85. A. S. Gupta and S. Dev, *J. Chromatogr.* **12** (1963) 189-195.
86. N. Grinberg (ed.), *Modern thin-layer chromatography* (Chromatographic Science Series Vol. 52), (M. Dekker, New York, 1990).
87. V. S. Govindarajan, *CRC Crit. Rev. Food Sci. Nutr.* **17** (1982) 189-258.
88. S. Nyiredy, C. A. J. Erdelmeier, K. Dallenbach-Toelke, K. Nyiredy-Mikita and O. Sticher, *J. Nat. Prod.* **49** (1986) 885-891.
89. C. A. J. Erdelmeier and G. M. König, *Phytochem. Anal.* **2** (1991) 3-14.
90. T. A. van Beek and G. P. Lelyveld, *Phytochem. Anal.* **4** (1993) 109-114.
91. D. S. Letham, S. Singh and D. A. Willcocks, *Phytochem. Anal.* **3** (1992) 218-222.
92. P. J. Houghton and I. M. Osibogun, *Phytochem. Anal.* **1** (1990) 22-25.
93. D. R. Baker, *Capillary electrophoresis* (Techniques in analytical chemistry Vol. 1), (John Wiley, New York, 1995).
94. R. Kuhn and S. Hoffstetter-Kuhn, *Capillary electrophoresis: principles and practice* (Springer, Berlin, 1993).
95. S. F. Y. Li, *Capillary electrophoresis, principles, practice and applications* (Journal of Chromatography Library, Vol. 52), (Elsevier, Amsterdam, 1993).

96. F. A. Tomás-Barberán, *Phytochem. Anal.* **6** (1995) 177-192.
97. E. Kenndler, C. Schwer, B. Fritsche and M. Pöhm, *J. Chromatogr.* **514** (1990) 383-388.
98. T. K. McGhie and K. R. Markham, *Phytochem. Anal.* **5** (1994) 121-126.
99. S. Terabe, K. Otsuka, K. Ichikawa, A. Tsuchiya and T. Ando, *Anal. Chem.* **56** (1984) 111-113.
100. C.-T. Chen and S.-J. Sheu, *J. Chromatogr.* **710 A** (1995) 323-329.
101. R. Weinberger and I. S. Lurie, *Anal. Chem.* **63** (1991) 823-827.
102. J. P. Foley and J. A. Crow, in *Modern phytochemical methods* (Recent advances in phytochemistry, Vol. 25) eds. N. H. Fischer, M. B. Isman and H. A. Stafford (Plenum, New York, 1991) 113-147.
103. R. M. Smith (ed.), *Supercritical fluid chromatography* (RSC chromatography monographs), (Royal Society of Chemistry, London, 1988).
104. B. Wenclawiak, *Analysis with supercritical fluids: extraction and chromatography* (Springer, Berlin, 1992).
105. T. A. Berger, *Packed column SFC* (RSC chromatography monographs), (Royal Society of Chemistry, London, 1995).
106. D. L. Mount, G. D. Todd and V. Navaratnam, *J. Chromatogr.* **666 B** (1995) 183-187.
107. M. Hadj-Mahammed, Y. Badjah-Hadj-Ahmed and B. Y. Meklati, *Phytochem. Anal.* **4** (1993) 275-278.
108. C. Bicchi, P. Rubiolo, M. Fresia, F. David and P. Sandra, *Phytochem. Anal.* **7** (1996) 37-41.
109. G. L. Alonso, M. R. Salinas, F. J. Estebau-Infantes and M. A. Sánchez-Fernández, *J. Agric. Food Chem.* **44** (1996) 185-188.
110. M. R. Koupai-Abyazani, C. S. Creaser and G. R. Stephenson, *Phytochem. Anal.* **3** (1992) 80-84.
111. K. Blau and J. Halket (eds.), *Handbook of derivatives for chromatography, 2nd ed.* (Wiley & Sons, Chichester, 1993).
112. A. Bernreuther and P. Schreier, *Phytochem. Anal.* **2** (1991) 167-170.
113. A. Zlatkis and V. Pretorius (eds.), *Preparative gas chromatography* (Interscience, New York, 1971).
114. T. A. van Beek and G. P. Lelyveld, *Phytochem. Anal.* **2** (1991) 26-34.
115. T. A. van Beek, R. Kleis, G. P. Lelyveld and Æ. de Groot, *Chromatographia* (1989) 126-128.
116. P. Weyerstahl, H.-C. Wahlburg, H. Marschall and A. Rustaiyan, *Liebigs Ann. Chem.* (1993) 1117-1123.
117. D. Cortes, G. P. Wannigama, J. Saez and A. Cavé, *Phytochemistry* **25** (1986) 2693-2695.
118. K. Okabe, K. Yamada, S. Yamamura and S. Takada, *J. Chem. Soc.* (1967) 2201-2206.
119. D. R. Schroeder and K. Nakanishi, *J. Nat. Prod.* **50** (1987) 241-244.
120. P. M. Bramley, *Phytochem. Anal.* **3** (1992) 97-104.

121. T. A. van Beek, H. A. Scheeren, T. Rantio, W. C. Melger and G. P. Lelyveld, *J. Chromatogr.* **543** (1991) 375-387.
122. A. J. Shah, D. A. Younic, M. W. Adlard and C. S. Evans, *Phytochem. Anal.* **3** (1992) 20-25.
123. A. Stolyhwo, H. Colin and G. Guiochon, *J. Chromatogr.* **265** (1983) 1-18.
124. M. H. J. Bergqvist and N. U. Olsson, *Phytochem. Anal.* **3** (1992) 215-217.
125. F. F. Camponovo, J.-L. Wolfender, M. P. Maillard, O. Potterat and K. Hostettmann, *Phytochem. Anal.* **6** (1995) 141-148.
126. M. Dreux, M. Lafosse and L. Morin-Allory, *LC-GC* **9** (1996) 148-156.
127. J.-G. Chen, S. J. Woltman and S. G. Weber, in *Advances in chromatography,* eds. P. R. Brown and L. Grushka (Marcel Dekker, New York, 1996) 273-313.
128. N. Acton, D. L. Klayman and I. J. Rollman, *Planta Med.* **51** (1985) 445-446.
129. M. J. Donaldson, H. Brody, M. W. Adlard and C. Bucke, *Phytochem. Anal.* **1** (1990) 18-21.
130. A. Saija, M. P. Germano, D. Marzullo, C. Imperatore, S. Campo and R. D. Pasquale, *Phytochem. Anal.* **3** (1992) 254-257.
131. C. Bicchi and P. Rubiolo, *Phytochem. Anal.* **4** (1993) 171-177.
132. T. Moritz, Phytochem. Anal. **3** (1992) 32-37.
133. W. M. A. Niessen and J. van der Greef, *Liquid chromatography-mass spectrometry - principles and applications* (Chromatogr. Sci. Ser. Vol. 58), (Marcel Dekker, New York, 1992).
134. P. Cremin, D. M. X. Donnelly, J. L. Wolfender and K. Hostettmann, *J. Chromatogr.* **710 A** (1995) 273-285.
135. J.-L. Wolfender, M. Maillard, A. Marston and K. Hostettmann, *Phytochem. Anal.* **3** (1992) 193-214.
136. J.-L. Wolfender, M. Maillard and K. Hostettmann, *Phytochem. Anal.* **5** (1994) 153-182.
137. R. Verpoorte and W. M. A. Niessen, *Phytochem. Anal.* **5** (1994) 217-232.
138. J. C. Lindon, J. K. Nicholson and I. D. Wilson, in *Advances in chromatography*, eds. P. R. Brown and L. Grushka (Marcel Dekker, New York, 1996) 315-382.
139. K. Albert, G. Schlotterbeck, U. Braumann, H. Hèndel, M. Spraul and G. Krack, *Angew. Chem.* **107** (1995) 1102-1104.
140. S. Johnson, E. D. Morgan, I. D. Wilson, M. Spraul and M. Hofmann, *J. Chem. Soc. Perkin Trans.* I (1994) 1499-1502.
141. T. A. van Beek, R. Verpoorte and A. B. Svendsen, *J. Chromatogr.* **298** (1984) 289-307.
142. M. Monforte-González, T. Ayora-Talavera, I. E. Maldonado-Mendoza and V. M. Loyola-Vargas, *Phytochem. Anal.* **3** (1992) 117-121.
143. T. A. van Beek, E. Bombardelli and P. Morazzoni, *Fitoterapia* **67** (1996) in press.
144. A. Martinsen, K. Hiltunen and A. Huhtikangas, *Phytochem. Anal.* **3** (1992) 69-72.
145. P. Winterhalter, G. Full, M. Herderich and P. Schreier, *Phytochem. Anal.* **2** (1991) 93-96.

146. S. L. Smith, in *Capillary gas chromatography in essential oil analysis*, eds. P. Sandra and C. Bicchi (A. Huethig, Heidelberg, 1987) 367-385.
147. P. Schreier, H. Idstein and W. Herres, in *Analysis of volatiles - methods and applications*, ed. P. Schreier (W. de Gruyter, Berlin, 1984) 293-306.
148. A. P. Bruins, in *Capillary gas chromatography in essential oil analysis*, eds. P. Sandra and C. Bicchi (Huethig Verlag, Heidelberg, 1987) 329-357.
149. F. David and P. Sandra, *Phytochem. Anal.* **3** (1992) 145-152.
150. A. De Giulio, S. De Rosa, G. Di Vincenzo and G. Strazzullo, *Phytochem. Anal.* **3** (1992) 122-124.
151. J. E. Anderson, C. M. Goetz, J. L. McLaughlin and M. Suffness, *Phytochem. Anal.* **2** (1991) 107-111.
152. M. R. Boyd and K. D. Paull, Drug Development Res. **34** (1995) 91-109.
153. T. A. van Beek and Æ. de Groot, *Recl. Trav. Chim. Pays Bas* **105** (1986) 513-527.
154. M. van Helden, H. P. N. F. van Heest, T. A. van Beek and W. F. Tjallingii, *J. Chem. Ecol.* **21** (1995) 761-774.
155. A. Blaakmeer, A. Stork, A. van Veldhuizen, T. A. van Beek, Æ. de Groot, J. J. A. van Loon and L. M. Schoonhoven, *J. Nat. Prod.* **57** (1994) 90-99.
156. T. E. Acree, J. Barnard and D. G. Cunningham, in *Analysis of volatiles - methods and applications*, P. Schreier (W. de Gruyter, Berlin, 1984) 251-267.
157. F. Drawert and N. Christoph, in *Analysis of volatiles - methods and applications*, ed. P. Schreier (W. de Gruyter, Berlin, 1984) 269-291.
158. O. Nishimura, *J. Agric. Food Chem.* **43** (1995) 2941-2945.
159. R. J. Robins, in *Immunology in plant sciences* (Modern methods of plant sciences - new series, Vol. 4), eds. H. F. Linskens and J. F. Jackson (Springer, Berlin, 1986) 86-141.
160. M.-A. Fliniaux, F. Manceau and A. Jacquin-Dubreuil, *Phytochem. Anal.* **3** (1992) 223-226.
161. P. G. Grothaus, G. S. Bignami, S. O'Malley, K. E. Harada, J. B. Byrnes, D. F. Waller, T. J. G. Raybould, M. T. McGuire and B. Alvarado, *J. Nat. Prod.* **58** (1995) 1003-1014.
162. F. Bourgaud, V. Bouque, S. Marchal, G. Innocenti, F. Dall'Acqua and A. Guckert, *Phytochem. Anal.* **6** (1995) 306-312.
163. J. F. S. Ferreira and J. Janick, *Phytochemistry* **41** (1996) 97-104.
164. P. J. Devine, N. A. Anis, J. Wright, S. Kim, A. T. Eldefrawi and M. E. Eldefrawi, *Anal. Biochem.* **227** (1995) 216-224.
165. K. Watanabe, K. Okada, H. Oda, K. Furuno, Y. Gomita and T. Katsu, *Anal. Chim. Acta* **316** (1995) 371-375.
166. E. Block, *Sci. Amer.* **252** (3) (1985) 94-99.
167. L. H. D. Jenniskens, E. L. M. v. Rozendaal, T. A. v. Beek, P. H. G. Wiegerinck and H. W. Scheeren, *J. Nat. Prod.* **59** (1996) 117-123.
168. G. Appendino, G. Cravotto, P. Gariboldi, B. Gabetta and E. Bombardelli, *Gazz. Chim. Ital.* **124** (1994) 1-4.

169. J. Garnero and R. Tabacchi, in *Capillary gas chromatography in essential oil analysis*, eds. P. Sandra and C. Bicchi (Huethig Verlag, Heidelberg, 1987) 359-366.
170. E. Besselièvre, N. Langlois and P. Potier, *Bull. Soc. Chim. France* (1972) 1477-1478.
171. M. Azoug, A. Loukaci, B. Richard, J.-M. Nuzillard, C. Moreti, M. Zèches-Hanrot and L. L. Men-Olivier, *Phytochemistry* **39** (1995) 1223-1228.
172. B. S. Middletich, *Analytical artifacts - GC, MS, HPLC, TLC and PC* (Journal of Chromatography Library, Vol. 44), (Elsevier, Amsterdam, 1989).

Chapter 4

STRUCTURE ELUCIDATION OF PLANT SECONDARY PRODUCTS

GEORGES MASSIOT, CATHERINE LAVAUD AND JEAN-MARC NUZILLARD

1. Introduction

Structural elucidation is an exercise based on chemistry, physics, intuition and good literature knowledge. This field, belonging entirely to organic chemistry, has learnt a great deal from natural products, of both plant and animal origin. Challenges continue to arise with the discovery of new and more complex molecules in nature. The structural elucidation of strychnine in 1951 (1) and of cholesterol by Windaus and Rosenheim in 1955 (2) are landmarks in the history of natural products chemistry and of chemistry *per se*. These structures have been established through the work of many laboratories worldwide during several decades, and must have used kilograms of pure substances. In 1997, identification of such structures has become a matter of days and milligrams! The field of structural elucidation has changed, the tools are new, the objects are more sophisticated and the purpose of this chapter is to give an overview of the techniques in use and some insight into future developments.

The determination of a structure involves many factors: atom number and nature, bonds, configuration (relative, absolute), and conformation. Each of these items could have easily been the object of a separate subchapter. Instead, we have chosen to address the problems in terms of techniques and present a brief summary of each of the contributions of the main tools. For the sake of brevity, and in order to limit the discussion to strategic considerations, this chapter only deals with the techniques which offer a global view of molecules (X-ray, MS, NMR) and with computer-assisted structural elucidation. Techniques which yield information on parts of molecules (IR, UV, CD...) have not been considered here. It is, however, obvious that they should be used from the very beginning of an investigation.

2. Diffraction Studies

When an X-ray beam hits a crystal, the light is reflected (diffracted) in a multitude of spatial directions. The phase, intensity and direction of these reflections is linked to the repartition of the electronic density in the crystal. X-ray crystallography is the science which translates diffraction patterns into structures. It is considered as the ultimate tool in structural elucidation and is now a well established field, despite the fact that major research in this area has shifted from smaller molecules to macromolecules of biological interest. Progress in this area has benefited from advanced systems of data acquisition (four-circle automated diffractometers, image plates), increased calculation capacities and X-ray beams of higher quality.

The main problem with the use of X-ray crystallography is the preparation of single crystals. In the fifties and sixties, natural products chemists knew how to crystallize compounds and had little or no access to diffraction. In the nineties, diffractometers are numerous but the art of crystallisation is no longer mastered by a new generation of impatient chemists. It must be said however that once a crystal is placed on the instrument, the structure is not automatically obtained at the end of a series of well-precedented operations (although this is mostly true 80% of the time). Unfortunately, in some cases, the compound crystallizes in an uncommon group of symmetry, or what is thought to be a 30-atom problem becomes a 120 atom problem because of particular spatial arrangements of independent units in the crystal lattice; or, sometimes, the nice crystal with bright faces has grown from twinned systems. All these particular instances require the attention of professional and skilled crystallographers and it would be unreasonable to think that the technique has become "push-button" and could be placed into the hands of pure bench chemists.

The structure of strychnine was solved by X-ray in 1951 (1) and reinvestigated many times since (3). Figure 1 shows the evolution of the presentation of data coming from X-ray analysis during this period.

Crystallography is now immediately used when just a tiny amount of highly valuable compound is available. This was the case of brassinolide (Figure 2) isolated in milligram quantity from nearly half a ton of rape pollen (4). It is worth noting that in this case, the question of the configuration of the four contiguous asymmetric centres on the sidechain would have been particularly difficult to settle. Another example of the use of crystallography when the determination of configurations and/or of substitutions is difficult is provided in a recent article on *seco*terpenoids from *Ruptiliocarpon caracolito* where three different structures from the same plant were established by this technique (5). The example shown in Figure 2, spirocaracolitone A, contains 13 asymmetric centres.

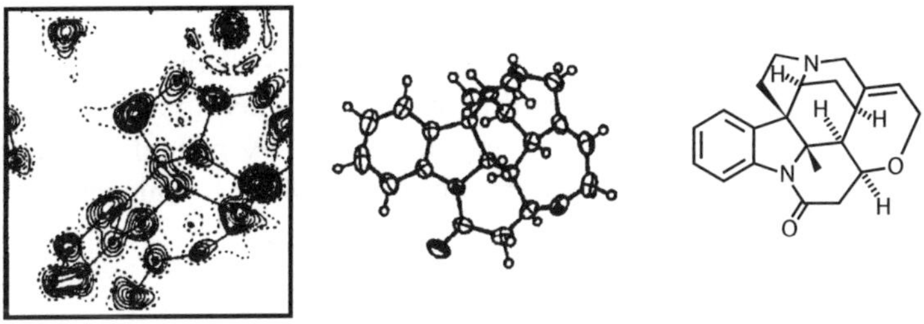

Figure 1. From left to right : strychnine according to X-ray in 1951, strychnine 1985 (ORTEP plot) and the structure of strychnine.

Technically, the major difficulty remaining in the resolution of structures using X-ray diffraction is related to the phase problem. In the early days of X-ray crystallography, this was solved by the "heavy atom" technique, i.e. by taking advantage of the contribution of a single heavy atom to the diffraction pattern of the whole assembly of atoms. This approach also has the advantage of providing the absolute configuration of the molecule using anomalous dispersion effects. The introduction of a heavy atom (I, Br) is the best way to determine the absolute configuration of a molecule in a novel series of compounds. This is best done by preparing salts of amines (with CH_3I or HBr) or esters of "heavy" acids (*para*bromobenzoates).

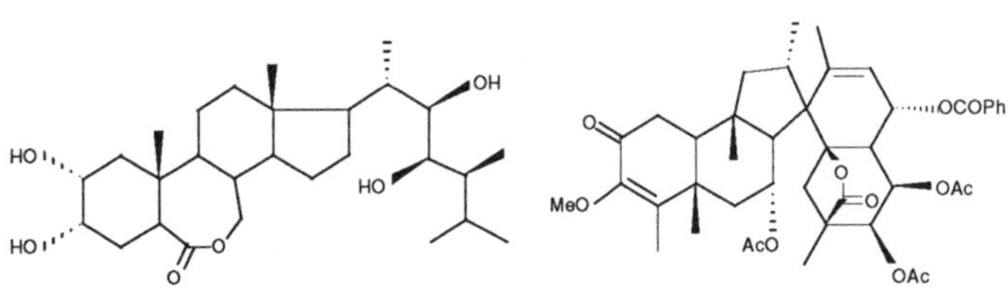

Figure 2. The structures of brassinolide (left) and of spirocaracolitone A (right).

Another solution to the problem of absolute configuration determination consists of making a derivative of a substance of known absolute configuration; interestingly, this was recently done with a lactate of strychnine and neutron diffraction (6).

Progress in computing capacities now allows structure-solving in the absence of heavy atoms; this is the so-called direct technique (limited to a finite number of "light" atoms quite compatible with the most complex natural products). The structures of the spirocaracolitones were solved using these techniques.

3. Mass Spectrometry

Trajectories of ions in magnetic and electrostatic fields depend on the ratio between their mass and charge. This is the basic principle of mass spectrometry.

Mass spectrometry is probably the most sensitive technique in use in the field of natural products, yielding information on molecular weight and fragments on submicrogram quantities of material. The study of mass spectral fragmentation of natural products has been a domain of considerable activity and the subject of several textbooks (7). In the not-so-distant past, it was the main tool available to identify small assemblies of atoms, elements of a molecular puzzle. This particular application of mass spectrometry is now mostly limited to what may be referred to as "sequencing", i.e. the identification of fragments linked by a single bond to the rest of the molecule (amide bonds in peptides, anomeric bonds in oligosaccharides...). Figure 3 provides an example of the sequencing of a complex saponin chain of sugars by mass spectrometry (8).

Fragmentation study remains important, however, when it is difficult to obtain a clear count of atoms through other techniques (NMR for example). This is particularly true with lipids, and with compounds containing long chains of saturated carbon atoms. For example, mass spectrometry is widely used in the structural elucidation of acetogenins, which are molecules with 35-37 carbon atoms in a single almost unramified chain, substituted by 6-8 oxygen atoms. Figure 4 shows the typical fragmentation observed in the acetogenin, squamocin (9). All occur in rings of α to heteroatoms, which are thus easily located on the chain.

Research in mass spectrometry focuses on three points: sensitivity, resolution and desorption of ions from non-volatile molecules or of molecules of high molecular weight (10). In terms of sensitivity, sample quantities in the picomolar range yield

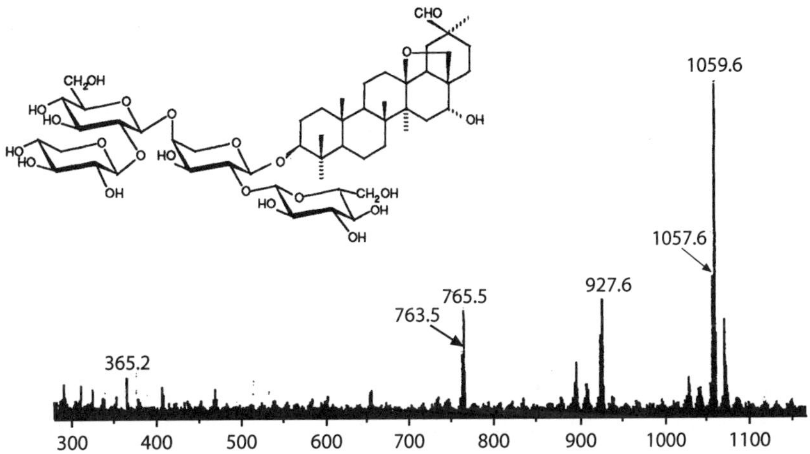

Figure 3. Negative FAB mass spectrum of saxifragolin B isolated from *Myrsine pellucida*. Fragments at *m/z* 765 and 927 correspond to the successive losses of a pentose and of a hexose.

molecular ions and spectra. The sample itself is no longer a limitation; rather, it is the sampling conditions which may present problems. The high sensitivity of mass spectrometry is not used at its best under direct introduction conditions but after coupling with a separator device GC or LC. It is not necessary to introduce here the usefulness of high resolution MS; it suffices to say that in most cases, exact mass measurements have replaced elemental analysis in the determination of molecular composition.

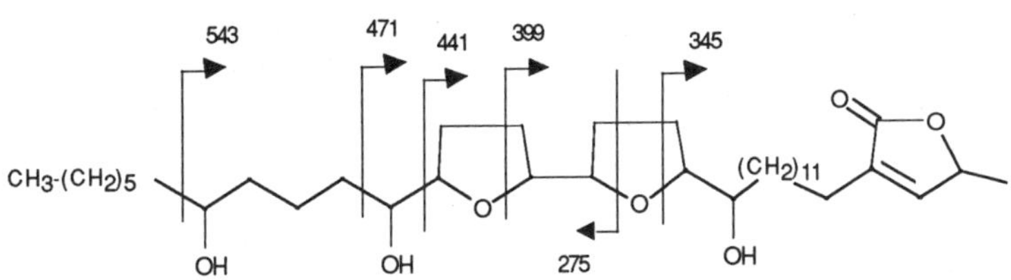

Figure 4. Mass spectral fragmentations of squamocin (9).

Despite new developments in ionization techniques, the chemical methods of ionization in the gaseous phase: electron impact and chemical ionization, are still widely used. Interestingly, while mass spectrometry has long considered molecules as "flat" species, stereochemistry is now studied with encouraging results. An example has recently been provided with the study of the diastereoisomeric indole alkaloids dregamine and tabernaemontanine (Figure 5) (11) whose mass spectra are distinguishable by an ion at m/z 252 (of unknown structure!).

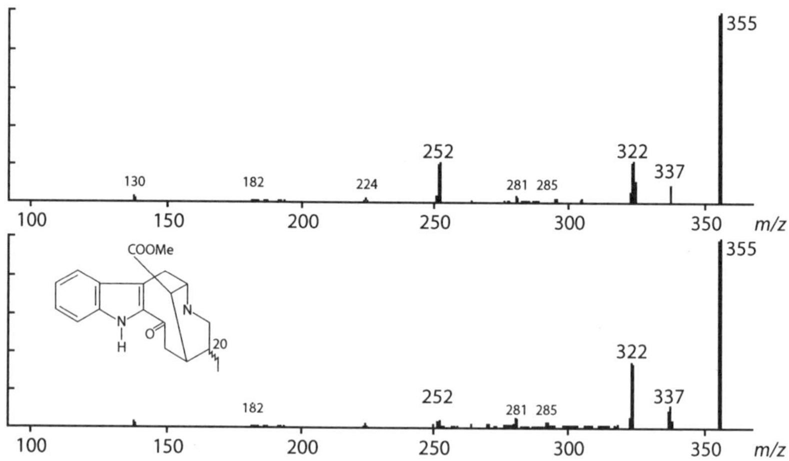

Figure 5. Mass spectral fragmentations of dregamine (20bH) and tabernaemontanine (20aH) (11).

A landmark case in the mass spectrometry of natural products was the acquisition of a molecular ion at m/z 2704 for palytoxin published in 1980 $(M+Na)^+$ (12). Palytoxin remains the "heaviest" natural product with non-repetitive units and its molecular ion was obtained with a new soft ionization technique "Californium Plasma Desorption" and time of flight MS. Difficulties in handling Californium sources may have tempered developments of this type of instrument but interesting results have been achieved with glycoconjugates of triterpenes and steroids, native or derived.

FAB (Fast Atom Bombardment) is probably the most widespread means of soft ionization of complex and fragile molecules. It is the best technique for the investigation of saponins, notoriously non-volatile molecules, yielding molecular ions and fragments. For example, negative ion FAB-MS of saponin yield a quasimolecular ion at m/z 1059 featuring $[M-H]^-$ and fragments at 765 and 927 correspond to the successive losses of a pentose and of a hexose (Figure 3).

4. Nuclear Magnetic Resonance

The interaction of a radiofrequency field with certain nuclei placed in a static magnetic field induces an energy absorption which is at the basis of nuclear magnetic resonance. These nuclei are those which have a non-zero spin value and natural product NMR is mostly concerned with proton, with the 13-isotope of carbon and to a lesser extent with ^{15}N. The most abundant ^{12}C and ^{16}O nuclei are silent in terms of NMR.

Nuclear magnetic resonance is a major tool in the structural elucidation of natural products. The main reason for its success is probably due to the fact that the technique is readily accessible from both an instrumental and a data interpretation standpoint. The introduction of new physical concepts and a constant flow of technical improvements have considerably modified the field during the past two decades. Even though chemical shifts (δ) and spin-spin coupling measurements (J) have been, and still are, the basic parameters for NMR structural analysis, their acquisition is no longer based only on first and higher order analysis of multiplets. The purpose of this chapter is to give a brief summary of the techniques now in use, and to contribute to the definition of a strategy for the better use of NMR, with a prospective view on expected developments.

At a time when nominal magnetic fields did not allow the observation of proton NMR at frequencies above 100 MHz, the practice of NMR was an art which consisted of deciphering complex and overlapping multiline, multispin systems. This was a difficult task, often complicated by higher order effects (which meant that line splitting did not always equate with coupling constants!). NMR books contained the ingredients for assignments : examples of spin system analysis (with mathematical formulae included), and tables of chemical shifts for protons, carbons and a few other nuclei in different environments (13). The advent of superconducting magnets with fields now allowing proton observation at up to 800 MHz has brought a first revolution, and the almost complete disappearance of higher order systems. A corollary of the race to higher field is a much welcomed dive in the quantities of material necessary to record experiments: milligrams in the routine use of ^{1}H and ^{13}C NMR, micrograms when specialized equipment is available. The second revolution was brought about by the introduction of concepts which predict the behaviour of spin systems after sophisticated sequences of impulsions and delays (14). Accompanying software and hardware improvements have led to the development of what is now called 2D (3D, ...) NMR and which is improperly thought of as a specialized field of NMR.

4.1. One-dimensional ^{1}H and ^{13}C NMR

1D ^{1}H NMR rapidly yields information on the nature (and purity) of a compound. A limited set of patterns, which ought to be recognized, allows the detection of protons in their immediate environment (the interested reader is invited to refer to one of the many available textbooks on the basic contributions of NMR to structural analysis)(15). One outstanding comment should be made here: only in rare cases of low complexity does ^{1}H NMR give a complete picture of a molecule. ^{13}C NMR is completely different and the broad range of shifts (200 ppm *vs* 10 ppm for proton) allows detection of all individual atoms in molecules of medium size with few degenerate elements. Furthermore, simple experiments exist which record individual subspectra for CH_3, CH_2, CH or quaternary carbons, thus increasing the resolving power of the technique (DEPT or J-modulated spectra; see Figure 10 for an example of a J-modulated ^{13}C spectrum).

1D ^{1}H and ^{13}C NMR experiments are thus used to introduce an unknown substance into a primary family of compounds: terpene, alkaloid, lipid, aromatic, peptide, glycoside, etc.

4.2. Two-dimensional NMR

The two main categories of 2D NMR experiments produce δ-δ and δ-J correlation maps. Historically, correlations between chemical shifts and coupling constants were first introduced mainly for sake of computer memory size. If one spreads ^{13}C chemical shifts in one dimension and $^{1}J_{CH}$ in the other, one obtains a map in which multiplicities (number of attached protons) and coupling constants are directly read. Owing to the development of other powerful methods of ^{13}C subspectra edition, this δ-J experiment is now seldom used except as a means of measurement of ^{1}J (useful in determination of configurations or hybridization states). Figure 6 shows part of a homonuclear J-resolve experiment and slices through the map illustrating the resolving power of the technique. In cases of highly crowded systems, the ^{1}H homonuclear δ-J experiment is used to untangle overlapping systems and to obtain "computed" pure δ proton spectra.

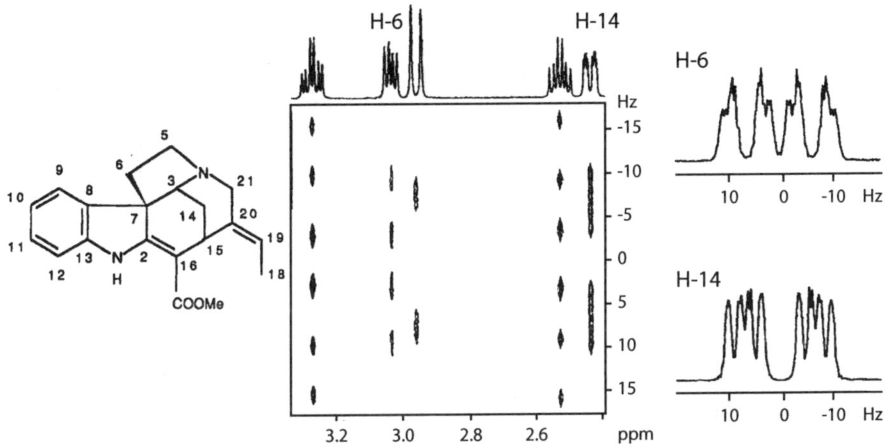

Figure 6. Part of a ${}^{1}H$ δ-J 2D NMR experiment (J-resolve). The molecule is akuammicine; $CDCl_3$ - 500 MHz. Expansion on the right are slices through the map.

The COSY spectrum gives δ-δ correlations through J couplings; this is the most popular 2D NMR experiment, easy to run, almost devoid of calibration problems, fast to record and not too complex to master. In the homonuclear case, it is usually displayed in a symmetrical fashion (with respect to a diagonal) and, roughly speaking, correlations on the diagonal correspond to chemical shifts while off-diagonal peaks correspond to couplings. Figure 7 illustrates the usefulness of such an experiment on a simple alkaloid. When one signal is assigned (according to its multiplicity or chemical shift for example), all the J-connected signals are assigned one after another. The COSY experiment replaces a series of decouplings and its only limitation is severe multiplet overlap.

There exist many variants of the COSY experiment aimed at providing correlation figures with better shape and resolution (double quantum filtered COSY for example). They are especially useful when overlap is such that coupling constants cannot be measured in the 1D spectrum and must be extracted from the 2D map (which, due to computing space and storage, are of lower resolution than the 1D spectrum).

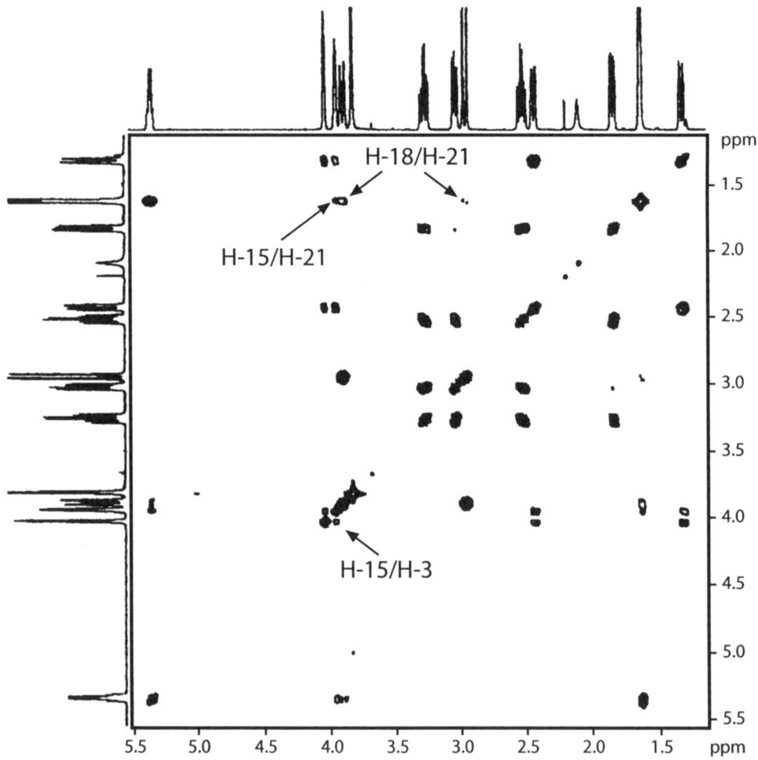

Figure 7. Part of a COSY map of akuammicine ; $CDCl_3$ - 500 MHz. Correlations corresponding to long-range couplings are indicated with arrows.

A particularly useful and simple modification of the COSY sequence puts emphasis on very small couplings : the delayed COSY. In this experiment, one can see correlations between protons coupled with a small J (< 2 Hz) and which produce broadening rather than splitting. This is the case with the so-called long range (four bond J and more) and allylic couplings. This experiment is unique in determining configurations of groups separated by quaternary carbon atoms (W couplings) in rigid systems or locating elements such as alkyl- or O-alkyl groups on aromatic rings.

It is not difficult to imagine that the superimposition of several 6 or 7 spin systems such as the ones produced in the ^{1}H NMR spectra of sugars rapidly leads to a map of high complexities. The relayed COSY experiment allows one to progress in the step-by-step proton assignment. If A couples to B and B to C, the COSY experiment gives cross peaks between A and B and between B and C; the relayed COSY shows an A to C correlation through B as a relay. Figure 8 illustrates the principle of the

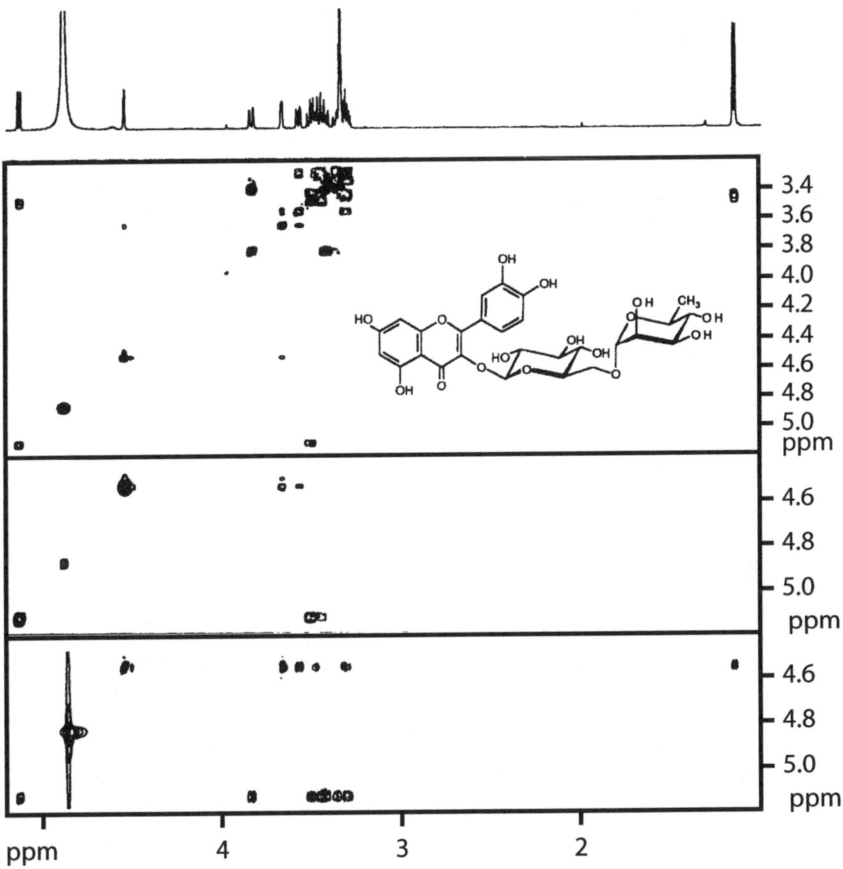

Figure 8. From top to bottom : COSY, relayed COSY and HOHAHA (parts of) on a flavonoid disaccharide, rutin. At 5.1 ppm is the anomeric proton of glucose; H-2 is at 3.45 ppm (COSY) and H-3 is at 3.4 ppm (relayed COSY). H-1 of rhamnose is at 4.5 ppm, H-2 at 3.6 ppm, H-3 at 3.5 ppm; the rest of this spin system down to the terminal methyl (1.1 ppm) is observable in the HOHAHA experiment.

experiment on a concrete example. The motor of the relay which permits the B to C rebound is a J coupling ; if one of the couplings along a path is zero, the relay is interrupted. The concept behind the experiment has been extended to sequential multirelayed COSY experiments. For example, a COSY experiment will show a 1,2 correlation between H-1 and H-2 of sugar, the relayed COSY will show a 1,3 correlation (if J 2-3 is not zero), the doubly relayed will show a 1,4 correlation and so on.

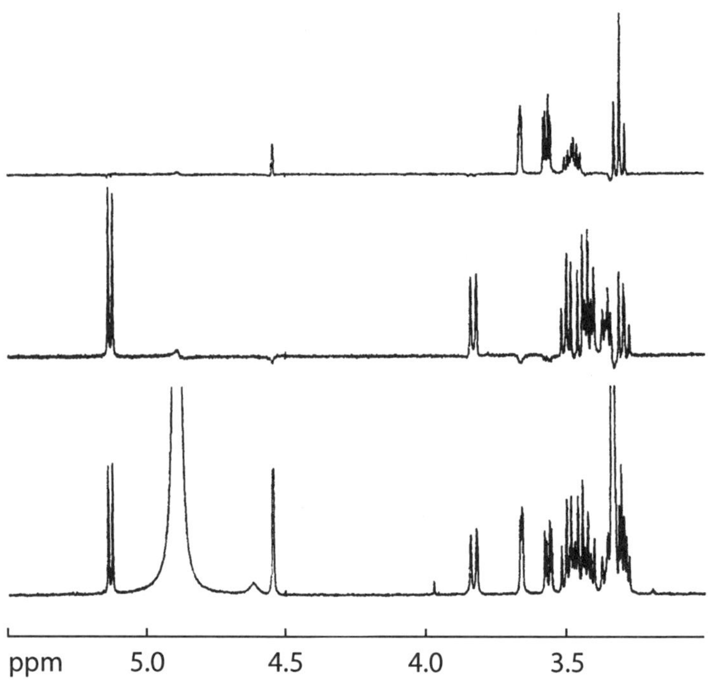

Figure 9. 1D HOHAHA subspectra on rutin (sugars are glucose and rhamnose).

Another concept, isotropic mixing, allows the observation of total correlation maps: TOCSY (Total Correlation Spectroscopy) or HOHAHA (Homonuclear Hartmann Hahn). The maps show all the correlations that one can dream of in a multirelayed operation, provided that there is coupling and that the mixing time is long enough. It is especially useful in situations of strong overlap where it allows one to distinguish and count networks of unrelated spin systems.

Heteronuclear δ-δ correlation and especially ^{1}H-^{13}C correlations now are of widespread use in structural elucidation. They may be run in two different ways : ^{13}C signal recording with ^{1}H excitation during the evolutionary phase of the experiment, or ^{1}H signal recording with ^{13}C excitation. The second mode (also called reverse mode) is more sensitive, and was experimentally introduced later for technical reasons (for the sake of lack of duality between transmitter and decoupler pulses). The two domains of ^{1}H to ^{13}C couplings correspond to two major families of experiments. The direct ^{1}J couplings have a magnitude of 120-200 Hz, very different from the ^{n}J ($n \geq 2$) which range from 0 to 15 Hz. It is therefore possible to devise experiments able to separate their interactions.

The ^{1}J coupling experiment is called X-H COSY in the normal mode or HMQC in the reverse mode (Heteronuclear Multiple Quantum Correlation). It yields information on direct couplings, allowing assignment of one nucleus when that of the other is known. While this sequence may seem trivial, its consequences are not so, especially as far as structural elucidation or assignments are concerned. A ^{1}H signal at δ 5.2 ppm may belong to an olefin or to an anomeric proton; in one case, its attached carbon resonates at 120-140 ppm, while in the other, it appears at 90-110 ppm. A C-H correlation easily distinguishes between these situations. Conversely, a methyl at 51 ppm is an O-methyl ester ; its proton signal may be expected anywhere between 3 and 4 ppm, in the range of O- or N-methyl signals for example, and may thus be difficult to assign. In this case the assignment of N- *vs* O-methyl, ester or ether is straightforward using a CH correlation map. Another application of this experiment is the determination of C multiplicities by counting the number of attached protons. A different and related experiment HSQC (Heteronuclear Single Quantum Correlation) gives the same information with increased sensitivity and improved resolution ; it may be J-modulated and phased according to multiplicities: CH_2 correlations appear on one side of the surface, with CH and CH_3 on the other face. The presentation of negative and positive contours with different colours allows one to immediately distinguish signals of different parities.

This family of experiments does not allow observation of non-protonated carbon atoms. They do however appear on long-range correlation maps provided that they are coupled to a proton within a 3- or 4- bond range. The experiments which show long-range CH correlations are named COLOC (in the normal mode) or HMBC

(Heteronuclear Multiple Bond Correlation) in the reverse mode. They are unique in their ability to link elements with no directly-seen couplings. For example, if a methyl is situated on a quaternary carbon, it will show correlations with this quaternary carbon and with the three other atoms (if carbon) situated one bond away. A unique feature of this experiment is that it allows the connection of elements through a silent segment (a set of two non-protonated atoms): two contiguous

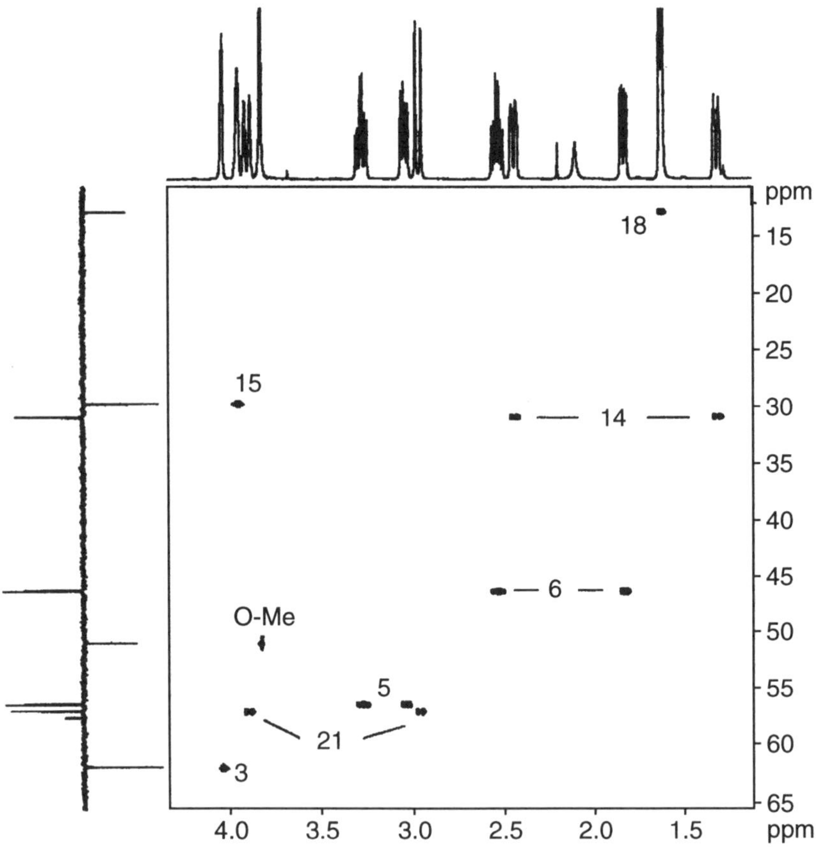

Figure 10. Part of a HMQC 2D δ^1H-$\delta^{13}C$ correlation map. Note that simple chemical shift considerations allow assignment of ^1H and ^{13}C. (Akuammicine, $CDCl_3$, 500 MHz). The ^{13}C spectrum on the left is a J-modulated spectrum (CH_3 and CH negative, C and CH_2 positive).

quaternary carbon atoms, heteroatom + quaternary carbon, ester links etc. This experiment is certainly the most important one in order to obtain information on the carbon skeleton of molecules (see below for computer-assisted uses of this experiment). It is now also run to detect ^{15}N to ^{1}H correlations (16), thus allowing full characterization of the connectivities within the N and C skeleton of molecules (^{17}O unfortunately does not yet lend itself routinely to such experiments).

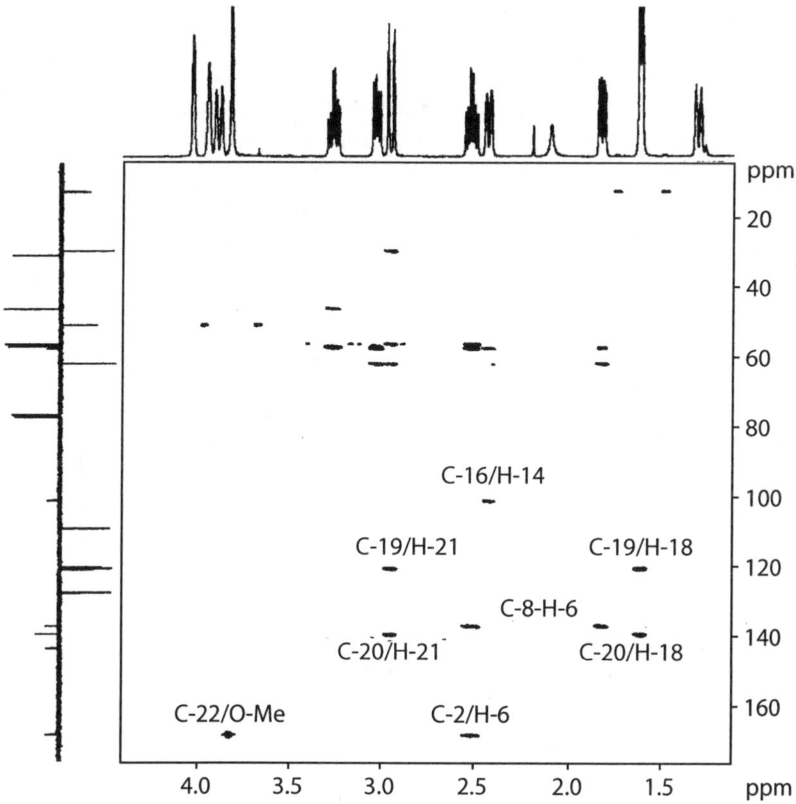

Figure 11. Part of the HMBC map of akuammicine showing correlations between protons linked to sp_3 carbons and sp_2 carbons.

4.3. The Ultimate NMR Experiment : INADEQUATE

The INADEQUATE experiment provides a map of connectivities between carbon atoms. Since the natural abundance of ^{13}C averages 1%, pairs of interacting ^{13}C atoms are met in one out of ten thousand molecules! The first three letters of the acronym stand for Incredible Natural Abundance (...DoublE QUAntum TransfEr) and this tells all the difficulty of the experiment. It is designed to eliminate the signals belonging to monolabeled species and to keep only those coming from molecules with two coupling ^{13}C nuclei (generally adjacent since the sensitivity of the response depends on the coupling constant). INADEQUATE maps show all the pairs of adjacent carbons and the determination of the carbon framework of a molecule is obtained in a straightforward manner from the experiment by linking all the pairs. Owing to the low natural abundance of ^{13}C pairs in natural products, the experiment requires large quantities of material *(ca* 200 mg) and is rarely run. However, the recently introduced liquid-nitrogen-cooled probe head and preamplifier technology, and improvements in signal treatment, may yet give a second birth to the technique. This has recently been illustrated by the structural elucidation of bistramide A, a marine natural

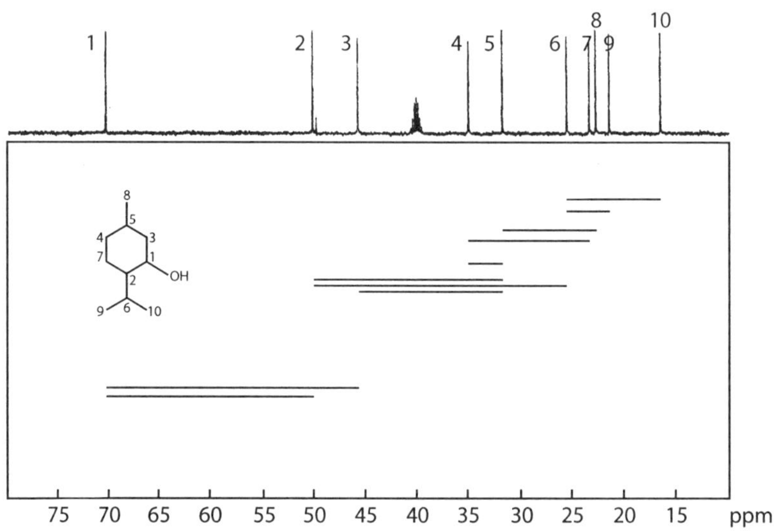

Figure 12. INADEQUATE map. Menthol, 500 MHz, DMSO. On the vertical line corresponding to C-1 (70 ppm), one finds two correlations with carbons at 50 and 46 ppm which are C-2 and C-3. From these atoms, one reaches C-5 and C-6 and C-7 and so on.

product of MW 705, where an INADEQUATE experiment was run with only 20 mg of material (17). The proton-detected version of this experiment also brings the necessary improvement in sensitivity, but on protonated carbon atoms only.

4.4. NMR and Stereochemistry

One of the main successes of NMR is the determination of relative configurations in complex structures. 1D proton NMR provided the first tools for these determinations with the discovery of the angular dependence of vicinal coupling magnitude (the famous Karplus Laws). Today, this still remains the safest way to determine configurations provided that adequate modelization is used. In rigid three- or six-membered rings, the law holds perfectly but care must be taken when dealing with five-membered rings or open-chain compounds. Computer and molecular mechanics may be used in conjunction with NMR to settle the most delicate cases. Long-distance couplings, which only pertain to certain spatial arrangements, also provide information on stereochemistry in the cases where three-bond coupling does not operate.

Carbon NMR chemical shifts are also subject to geometrical dependence, and may be used to assign relative configurations. For example, the γ effect rule is still widely used. This rule states that when a substituent is close to a γ-situated CH, then the particular C in question is shielded. However, to apply this rule, it is best to know the spectra of the isomer pair or to have reference compounds available (18).

The nuclear Overhauser effect is the change in the integrated intensity of a signal when a nearby other spin is saturated (19). It shows magnetic field and distance dependence, and as such gives invaluable information on molecular geometry. One of the first examples of the use of the Overhauser effect (nOe) in structural elucidation was obtained by Nakanishi and coworkers in their investigation of ginkgolides from *Ginkgo biloba* (20). Irradiation of the tertiary butyl group signals in the series of ginkgolides led to selective increases of the signals belonging to protons in the vicinity of this group. It is worth noting here that these configurations could not have been obtained by coupling constant measurements since several protonated asymmetric centers were separated by quaternary carbon atoms. The measurement of nOe through spectra integration has now been replaced by the integration of the difference between spectra recorded with and without decoupling (nOe difference experiment). When a qualitative approach to nOe is sufficient, two-dimensional techniques are used : NOESY or ROESY.

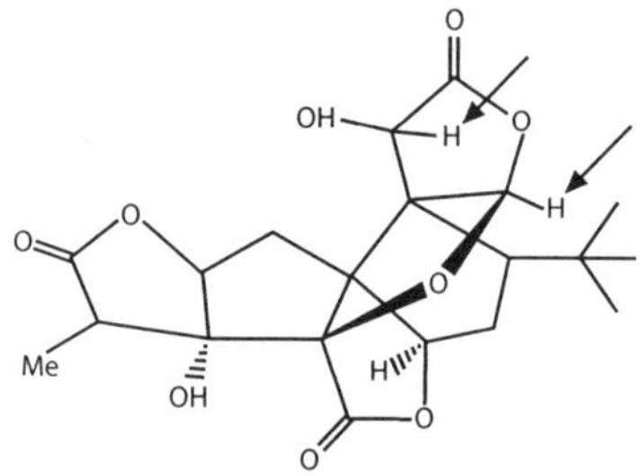

Ginkgolide A

Figure 13. The structure of Ginkgolide A. NOE were observed after irradiation of the t-butyl group (arrows).

4.5. A General Strategy for the Use of NMR

Structural elucidation with a strong dependence on NMR may involve three steps: obtaining a working hypothesis, followed by a flat formula and then the final structure. The most difficult to obtain is probably the first working hypothesis which must take into account the crude formula (obtained by high resolution mass spectrometry and combustion analysis) and which must incorporate the most striking features of the 1D proton and carbon NMR spectra. At this stage, it is important to be able to make the best possible count of proton and carbon atoms, including those which might be exchangeable. Chemical shifts will characterize the hybridization state of the carbon atoms and the "edited" carbon spectra will allow their arrangement into methyls, methylenes etc. A combination of homo and heteronuclear shift correlation maps (COSY, HMBC, HMQC, HOHAHA) will then allow fragment identification. The game then consists of linking those fragments, without forgetting the heteroatoms. Only a limited number of hypotheses should survive this step and selection of the correct one may be done by careful model inspection and introduction of stereochemistry. The nOe experiment will be introduced at this moment and may provide the definitive conclusion. At the end of the exercise all the data should be confronted again with the proposed formula until everything fits into the picture.

4.6. NMR Improvements to be Expected

A general problem with NMR is sensitivity, and experiments are frequently devised which can cost up to a full day of instrument time. Gradient-enhanced NMR spectroscopy is the most recent practical and general innovation which aims at producing cleaner and faster experiments (21). The use of static B0 field gradient pulses is as old as the NMR technique itself, and was initially proposed as the localization tool for spin tomography. The technical introduction of actively-shielded gradient coils allowed the development of the gradient-enhanced NMR spectroscopy.

Most non-gradient pulse sequences rely on phase-cycling, z-filters, or radiofrequency gradient fields in order to remove unwanted parts of the NMR signal that would otherwise lead to spectral artefacts. Phase-cycling requires the repetition of the experiment changing only the pulse and the receiver phases. A standard unphased proton COSY experiment is a collection of a high number (256, 512 or more) of free induction decay signals (FID), each resulting from the addition of at least four FIDs designed to eliminate F-1 quadrature artefacts and axial peaks. The overall recording time is reduced by a factor of four when the selection of the useful peaks is performed by introduction of gradient pulses between RF pulses. This also efficiently solves the problems of solvent peak suppression of pure phase selective excitation.

At the level of the natural product chemist, the most impressive impact of gradient availability is the quality improvement achieved in HMQC and HMBC experiments. Residual signals from protons bound to ^{12}C lead to artefact lines referred to as "T_1 noise" which are cleanly removed by the selective excitation of ^{1}H-^{13}C spin systems. This improvement is so spectacular that otherwise-invisible ^{4}J correlations become detectable, thus leading to risks of data misinterpretation.

5. Expert Systems for Structural Elucidation

5.1. General Principles and Background

Chemists facing a structure elucidation problem develop solving strategies that involve data collection and analysis steps. As early as the 1960's, computer scientists were tempted to mimic the analytical chemists' way of deducing molecular structures from physical and chemical experimental data. It is worth noting here that modern analytical techniques have grown in parallel with computer hardware and software engineering.

Clearly, the various analytical techniques do not deliver the same kind of information. Two of them, MS and NMR, are considered by organic chemists to provide a richer content and were therefore investigated very early as primary sources of computer-assisted analytical systems. The DENDRAL project was a federation of the major initial attempts towards automatic structure generation (22).

Their strategy involved a three step process :

First, the experimental data (say, the spectra) are reduced, meaning that raw data are freed from informationless portions. The result is generally a list of spectral peaks characterized by their position and intensity.

In a second step, these spectral signatures are converted into molecular fragments. Unfortunately, there is no technique for which the signature-structure relationship is unambiguous and generally, this relationship is retrieved from spectroscopic databases, built by the compilation of the world's chemical literature. As a result, spectra are interpreted as a list of alternative fragment sets.

The third stage combines the elements of the list by choosing the alternatives which will lead to a valid molecular structure. In this step the computer program delivers readable drawings for the solution structures, a much more complicated affair than one realizes.

Many programs inspired by this general philosophy are still developed today, and have achieved rather impressive results. One can wonder why these programs have not reached chemical laboratories, at least at the extent of molecular modeling. What are the weaknesses of the existing systems and how can they be improved?

NMR plays a particular role in structural elucidation, as it reveals the existence of molecular fragments or “NMR chromophores” which can be connected by means of scalar and dipolar couplings. It it thus possible not only to assert that a given molecule bears, for example, acetate groups, but also to know, as well, the chemical shifts of the nuclei in their neighbourhood. Connectivity or proximity are used as a primary source of structural information. In other words, steps two and three of the general scheme of structure elucidation become partially mixed. The widespread use of 2D NMR, which now complements 1D NMR, has led program developers to include proximity relationships in their structure generators, even though chemical shift databases still play a central role. Indeed, molecular structure analysis by means of the gross molecular formula and of 1D and 2D NMR spectra can be effectively carried out using only very simple chemical shift arguments. With this goal in mind, we have written a structure generation program (LSD), in which connectivity is the main source of information, requiring no need for databases. The result of the straightforward analysis of the 1D spectra is added by the user as a supplementary constraint to the problem.

5.2. The LSD Program: Principles

The LSD program uses Logic for Structure Determination. Its initial version was written in Prolog (*Pro*gramming in *Log*ics), a language in which facts (data) and rules (the way of manipulating them) are coded according to the methodology of first-order logics (23). Encoding in C language was then carried out for speed improvement and transportability.

The treatment of a structural problem using LSD requires five NMR spectra : ^{1}H and ^{13}C (in J-modulated or DEPT mode, preferably) 1D spectra, ^{1}H-^{1}H COSY, HMBC and HMQC (J-modulated, if possible) 2D spectra. This combination meets the constraints imposed by sensitivity, reliability and richness of content criteria. Limitation solely to the ^{1}H NMR spectroscopy does not provide enough useful and unambiguous data for an automated process. Conversely, the above mentioned ^{13}C-^{13}C INADEQUATE experiment provides most of the structure but its limited sensitivity makes it difficult in practice.

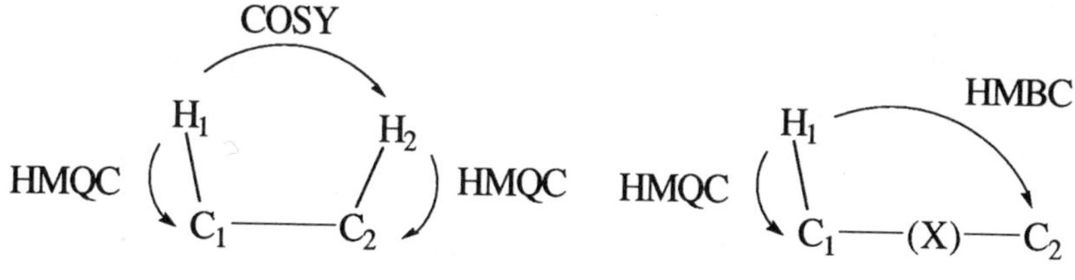

Figure 14. Proximity relationship obtained by combination of HMQC and COSY and of HMBC and HMQC experiments. Atom X is an atom of the list or no atom at all.

Combination of HMQC and COSY data implicitly provides molecular fragments (Figure 14). The COSY map essentially shows correlations arising from geminal (two-bond) and vicinal (three-bond) couplings. The geminal couplings are easily detected by the HMQC experiment, except in the case where the chemical shifts are identical. The remaining correlations in the COSY are thus three-bond correlations and therefore, when one observes a three-bond coupling between two protons H1 and H2, it may be deduced that their attached carbon atoms C1 and C2 are connected. This result is, in some respects, the one expected from an INADEQUATE experiment but the use of this simple deduction is limited by spectral superimpositions generally observed in the COSY spectrum of a complex molecule.

Fortunately, more information is gathered by the combination of an HMQC and an HMBC, which, like a COSY, mainly shows correlations mediated by two- and three-bond heteronuclear couplings. When proton H1 correlates in the HMQC spectrum with C1 (one-bond) and in the HMBC spectrum with C2, then C1 and C2 are either directly bound together or both bound to the same atom X (Figure 14). Combining HMQC and HMBC spectra allows one to explore circles around each protonated carbon, whose radius measured in bonds is either 1 or 2. The primary goal of the LSD program is to determine this radius and the nature of the intermediate atom X that eventually separates C1 and C2.

Even though a trained chemist may succeed in this task, the exhaustiveness of the search cannot be proved and may extend to proportions requiring the help of a computer. Moreover, the chemist generally introduces his own knowledge of chemical shifts, often for the best, but sometimes for the worst.

5.3. The LSD Program. An Example : Gibberellic Acid

The input file for the LSD program, the way of dealing with problems, and the results it provides will be exemplified through the structural elucidation of a $C_{19}H_{22}O_6$ molecule: gibberellic acid (24).

Carbon atoms are first numbered in decreasing order of their chemical shifts (numbers on Figure 15) : the J-modulated ^{13}C spectrum and the HSQC spectrum reveal the presence of seven quaternary carbons, six methines, five methylenes and one methyl group accounting for a $C_{19}H_{19}$ partial formula. The three protons which are not bound to carbon atoms belong to hydroxy groups. The carbon resonances at 177 and 173 ppm indicate that the molecule contains two sp2 oxygen atoms bound to sp2 carbon atoms. Four other sp2 carbon atoms (numbers 3 to 6) are also identified.

The hybridization state and multiplicity (number of bound hydrogen atoms) of each atom is a requirement for LSD, allowing computation of the number of non-hydrogen neighbours at the beginning of the resolution process.

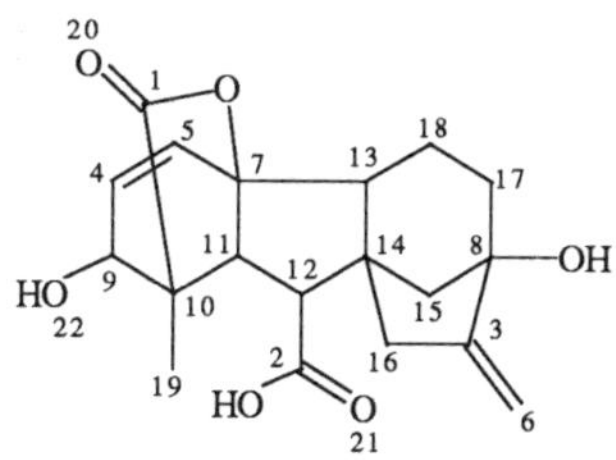

Figure 15. The structure of gibberellic acid. Numbering follows carbon chemical shift order.

The HMBC experiment gives 45 correlations, six of which yield redundant information, thus leading to the same carbon-carbon proximity relationships. Three COSY cross peaks are also taken into account, as well as the C-1/O-20 and the C-2/O-21 carbonyl bonds which are deduced from ^{13}C chemical shift data. This information considerably reduces the size of the search space and it is also added that the neighbour of C-19 is necessarily a quaternary carbon atom since atom C-19, a methyl (14.7 ppm,) gives a sharp singlet peak in the 1H NMR spectrum. Such data is referred to as "atom property" and is encoded by means of atom list operators.

The resolution of a structural problem by LSD then proceeds in four steps :

1) The input file is analyzed and checked against trivial inconsistencies.

2) HMBC correlations are exploited and bonds are subsequently established between atoms. A HMBC correlation opens three levels of indetermination : the first one is the choice of a responsible carbon atom when the correlation concerns a group of atoms with similar chemical shifts, the second one is the choice of the path length (2 or 3 bonds) between the carbon and the hydrogen atom, and the third one concerns the intermediate atom X (Figure 14) when the path length is 3. When the introduction of a correlation contradicts the growing structure, it is deduced that a previous choice is inappropriate and the most recent choice is reconsidered by a mechanism called backtracking. The program stops when all choices for all correlations are explored.

3) Pairing step: when all correlations have been successfully assigned, atoms

generally remain having free positions for bond formation. These free positions are systematically paired, but this leads to the proposal of chemically equivalent structures differring by the numbering of atoms for which no bonding or correlation information is available. Such atoms are thus grouped into equivalence classes during step 1 and only a single member of a class is available at a time for bond formation. Classes are updated when their members are used, and this mechanism stays active during step 2 as well, when intermediate atom X is a member of a class. Backtracking takes place if pairing cannot be completed (due to atom properties for example). The piece of structure established during step 2 is reconsidered when no complete pairing is possible.

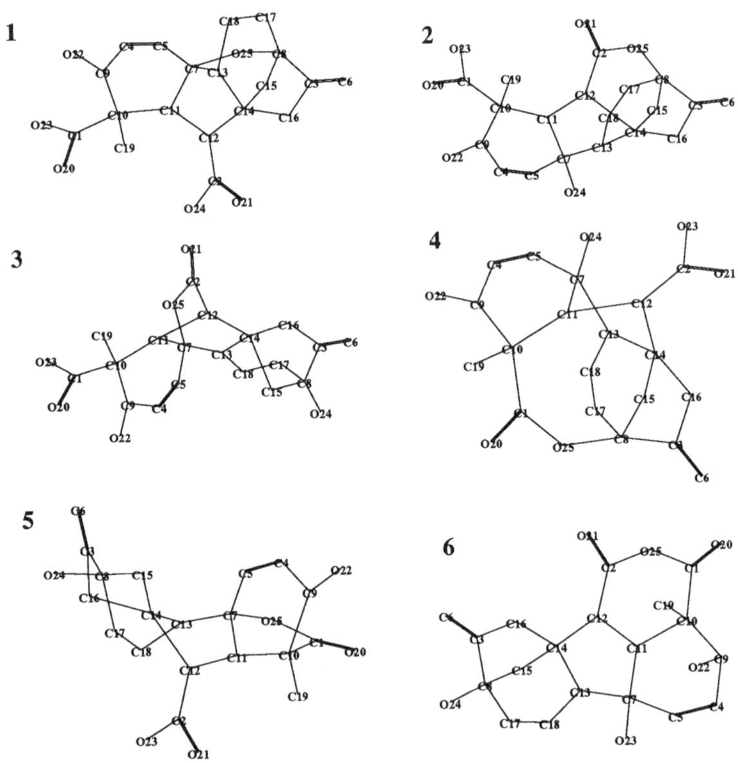

Figure 16. Structures proposed by LSD for gibberellic acid.

4) The final step groups checking procedures.

Six structures are proposed by the LSD program for gibberellic acid (Figure 16). The LSD program is also able to produce partial solutions in which all of the correlations are integrated, but without pairing. LSD however warrants that at least one pairing possibility exists that can lead to a fully acceptable solution structure. The production of partial structures helps to classify by families when the problem is far from being determined. The partial solution obtained here corresponds to the fact that quaternary carbon atoms C-1, C-2, C-7, and C-8 (asterisked on Figure 17) and oxygen atoms O-23, O-24 (hydroxyls) and O-25 do not present any correlations that lead to their binding to other atoms. Solution 5 in Figure 16 is the commonly accepted structure for gibberellic acid.

The LSD program yields solutions as text files that can be decoded by another program. Possible outputs are lists of bonds, 2D coordinates, 3D coordinates under Macromodel format, or SMILES chains that are a compact alphanumeric coding of molecular structures. The 2D coordinates can be translated into a Postcript file, a possibility used to draw Figures 16 and 17.

Many other substances were submitted to the LSD protocol, known and unknown substances in the terpene and indole alkaloids fields and for which solutions were successfully proposed by the program (25).

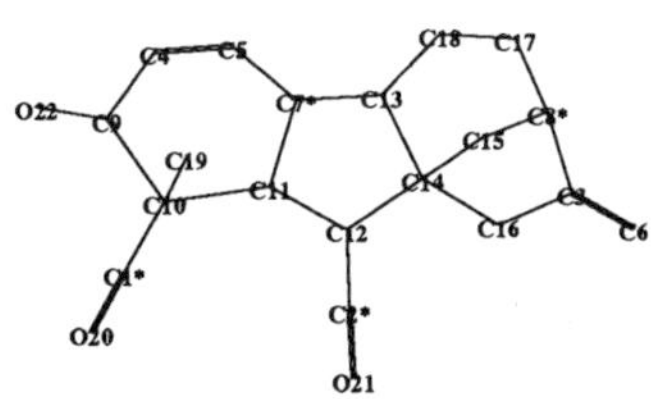

Figure 17. Partial structure proposed by LSD for gibberellic acid.

6. Concluding Remarks.

Natural product chemistry now benefits from tools allowing precise structural elucidation of most secondary metabolites provided that a minimum amount of compound in a minimal state of purity is available. Methods now exist to solve structures, but a general and fully automated method to solve all structures does not yet exist. Structural analysis still strongly depends on man, on literature knowledge and on a basic training in all the main techniques described here and elsewhere (IR, UV, CD...). As a final word, the link with synthetic chemistry (full or partial) should not be forgotten. The structure of palytoxin, the most complex natural product with non-repetitive units was settled by a total synthesis (26)!

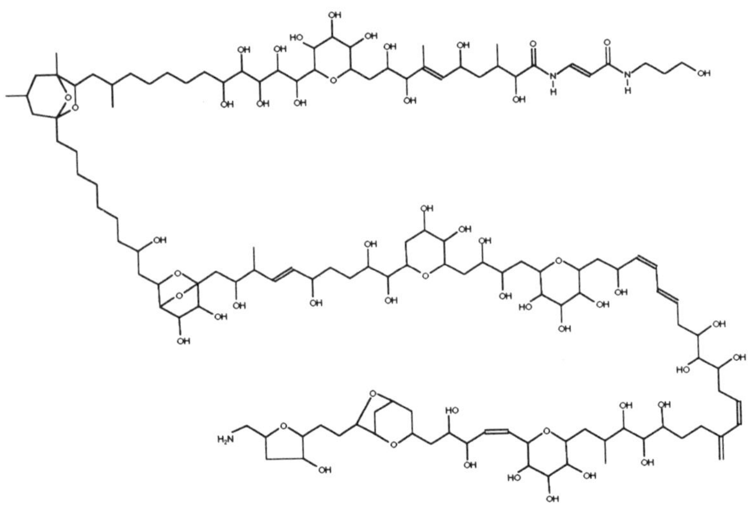

Figure 18. The structure of palytoxin.

7. References

1. J.H. Robertson and C.A. Beevers, *Acta Cryst.* **4** (1951) 270; C. Bokhoven, J. C. Schoone and J. M. Bijvoet, *Acta Cryst.* **4** (1951) 275.
2. L.F. Fieser and M. Fieser, *Steroids* (Reinhold Publishing Corporation, New York, 1959).
3. S.B. Glover, R.O. Gould and M.D. Walkinshaw, *Acta Cryst.* C41 (1985) 990 ; A. Mostad, Acta Chem. Scand. **B39** (1985) 705.
4. M.D. Grove, G.F. Spencer, W.K. Rohwedder, N. Mandava, J.F. Worley, J.D. Warthen, G.L. Steffens, J.L. Flippen-Anderson and J.C. Cook Jr, *Nature* **281** (1979) 216.
5. S.L. MacKinnon, C. Bensimon, J.T. Arnason, P.E. Zanchez-Vindas and T. Durst, *J. Org. Chem.* **62** (1997) 840.
6. H.S.H. Yuan, R.C. Stevens, R. Bau, H.S. Mosher and T.F. Koetzle, *Proc. Natl. Acad. Sci. USA* **91** (1994) 12872.
7. H. Budzikiewicz, C. Djerassi and D.H. Williams, *Structure Elucidation of Natural Products by Mass Spectrometry* (Holden-Day, San Francisco, London, Amsterdam, 1964).
8. C. Lavaud, G. Massiot, J.B. Barrera, C. Moretti and L. Le Men-Olivier, *Phytochemistry* **37** (1994) 1671.
9. O. Laprevote and B.C. Das, *Tetrahedron* **50** (1994) 8479.
10. E. De Hoffmann, J. Charette and V. Stroobant, *Mass Spectrometry, Principles and Applications* (J. Wiley, New York and Masson, Paris, 1996).
11. O. Laprevote, A.M. Bui, B.C. Das, B. Charles and J.-C. Tabet, *Org. Mass Spectrom.* **26** (1991) 621.
12. M.D. Mc Farlane, D. Uemura, K. Ueda and Y. Hirata, *J. Amer. Chem. Soc.* **102** (1980) 875.
13. L.M. Jackman and S. Sternhell, *Applications of Nuclear Magnetic Resonance Spectroscopy in Organic Chemistry* (Pergamon, Oxford, 1969).
14. R.R. Ernst, G. Bodenhausen and A. Wokaun, *Principles of Nuclear Magnetic Resonance in One and Two Dimensions* (Clarendon Press, Oxford, 1987).
15. T. Kusumi, T. Iwashita and H. Naoki in *One-dimensional and Two-dimensional NMR Spectra by Modern Pulse Technique* (K. Nakanishi ed., Kodansha, Tokyo, 1990).
16. R.C. Crouch and G.E. Martin, *J. Heterocyclic Chem.* **32** (1995) 1665.
17. M.P. Foster, C.L. Mayne, R. Dunkel, R.J. Pugmire, D.M. Grant, J.-M. Kornprobst, J.-F. Verbist, J.-F. Biard and C. M. Ireland, *J. Amer. Chem. Soc.* **114** (1992) 1110.
18. H.-O. Kalinowski, S. Berger and S. Braun, *Carbon-13 NMR Spectroscopy* (John Wiley and Sons, New York, 1988).
19. D. Neuhaus, M. Williamson, *The Nuclear Overhauser Effect in Structural and Conformational Analysis* (VCH, New York, 1989).

20. M.C. Woods, I. Miura, Y. Nakadeira, A. Terahara, M. Maruyama and K. Nakanishi, *Tetrahedron Lett.* **(1967)** 321.
21. P. Barker and R. Freeman, *J. Magn. Res.* **64** (1985) 334.
22. D.H. Smith, N.A.B. Gray, I.G. Nourse and C.W. Crandell, *Anal. Chim. Acta* **133** (1981) 471.
23. J.-M. Nuzillard and G. Massiot, *Tetrahedron* **47** (1991) 3655.
24. G. Stork, A. Newmann, *J. Amer. Chem. Soc.* **81** (1959) 5518.
25. J.-M. Nuzillard and G. Massiot, *Anal. Chim. Acta* **242** (1991) 37. ; J.-M. Nuzillard and G. Massiot, *Phytochemical Analysis* **3** (1992) 153 ; S.V. Ley, K. Doherty, G. Massiot and J.-M. Nuzillard, *Tetrahedron* **50** (1994) 12267.
26. R.W. Amstrong, J.M. Beau, S.H. Cheon, W.J. Christ, H. Fujioka, W.H. Ham, L. D.Hawkins, H. Jin, S.H. Kang, Y. Kishi, M.J. Martinelli, W.W. McWhorter, M. Mizuno, M. Nakata, A.E. Stutz, F.X. Talamas, M. Taniguchi, J.A. Tino, K. Ueda, J.I. Uenishi, J.B. White and M. Yonada, *J. Amer. Chem. Soc.* **111** (1989) 7525, 7530.

Chapter 5

DRUG DISCOVERY AND DEVELOPMENT

MONIQUE S.J. SIMMONDS AND RENÉE J. GRAYER

1. Introduction

Plants provide man with many of the essential materials required for day-to-day living, including food, oxygen, clothing, timber for housing, as well as providing compounds such as oils, resins, rubbers, gums, dyes, pesticides and drugs that are economically important and improve the quality of life. However, in recent years there has been a move away from assessing the drug potential of plant-derived compounds. Thus many of the more recent drugs have been developed as a result of advances in synthetic chemistry. Today there is a widespread belief within the pharmaceutical industry that computational chemistry will result in the synthesis of the next generation of drugs. This raises the question as to whether plant-derived compounds still have a role in drug discovery programmes. This chapter will address this question by reviewing the principles and practice of drug discovery and illustrate the argument with some examples of plant-derived drugs.

Plants can be considered as chemical factories that have been evolving their biosynthetic programmes over the last 400,000 million years. During this evolution plants have synthesised an array of compounds whose structural diversity is often beyond the dreams of even the most imaginative organic chemists. The ability of chemists to realise the full economical potential of the chemistry of the flora has often been limited by political, economic and moral constraints. Plants appear to have evolved the chemical pathways to produce compounds capable of curing many diseases, yet despite their uses we indulge in the careless destruction of them. It is well recognised that the problems of pollution, global warming and de-forestation pose a threat to the welfare of plants in many parts of the world and that these factors have already decreased the diversity of our global flora. It is important that plant-based drug discovery programmes help highlight the need to conserve our flora and do not directly result in the loss of threatened plants by uncontrolled exploitation.The majority of the biologically-active compounds isolated from plants are secondary metabolites (1) - thought for a long time to be waste-products resulting from primary metabolic processes, but now known in the majority of cases to be by- or end-products of complex biosynthetic processes (Chapters 1 and 2). Secondary metabolites may have no essential function in plant metabolism but there is evidence that many have an important ecological function, for example in defence against herbivores, bacteria and fungal infections. Others, for example flavonoids, may play an important part in protecting the plant from environmental damage due to UV radiation (1).

Many of the compounds used by the plants for a particular purpose, for example protection against fungi, can be used by man for a similar purpose in a different context. Conversely, the use made by man of plant-derived constituents may not relate directly to the ecological use that the plants make of the compounds. For example, the non-protein amino acid L-Dopa is found in many legumes (e.g. *Vicia faba)* where it functions as an antifeedant and protects the plant from herbivory (2), whereas in medicine it is used as a drug in the treatment of Parkinson's Disease (3).

Why should pharmacologists and chemists want to study the potential pharmaceutical value of plant-derived compounds? The most compelling reason is that they have provided, and are still providing, leads for drugs design, despite the move away from the plant-related screening that took place in the 1960's (4).

A few drug discovery programmes based on plants re-started in the late 1980's when pharmacologists, physicians and chemists started to re-investigate the active ingredients in plants used for medicinal purposes. The more recent advances in chromatographic and spectrographic technical equipment have decreased the time taken to identify and determine the molecular structures of the active compounds in these plants. This has resulted in an increase in the number of natural products being isolated and identified every year, although few will be developed as drugs. However, because of the novel structures of some of these active compounds, they might serve as templates for drugs.

Drug discovery is a long and challenging process that involves many different disciplines. Once an active compound has been structurally characterised, effort is expended on optimising its activity as part of a "rational drug design programme". Such a programme requires knowledge about the structure of the drug and its receptor. Many pharmaceutical companies spend a great deal of their efforts on this latter approach to drug discovery because the modification of an existing active molecule is financially less risky than developing a new compound with a new mode of action. This is because before a new compound is released as a drug it would have to undergo a series of clinical trials to provide knowledge about its toxicity as well as about its absorption, distribution in, and elimination from the body. It is also necessary to know whether the new compound is likely to cause allergic responses. Thus, from start to finish a drug discovery programme is both lengthy and expensive.

So how does one start a plant-based drug discovery programme? One could select a human disease, and test all available plants in assays designed to identify extracts/compounds that could inhibit enzymes associated with that disease, a method adopted with success by the American National Cancer Institute (NCI) in their anti-cancer and anti-viral assays (5). Alternatively one could select plants known to be used in the treatment of a specific disease, or plants with known "toxic" activity, and screen extracts from these plants in a range of assays: an approach used by the American company Shaman Pharmaceuticals with success. Perhaps the best approach is to combine both methods.

2. Collecting Plants

Whatever type of screening programme a company or institute adopts, a strategy has to be formulated as to how plant material will be collected.

It has been estimated that 20-30 % of the world's flora of 250,000-500,000 species has been subjected to phytochemical and pharmacological investigations (6). However, it is at present impossible to give an accurate figure of the number of plants that have been subjected to phytochemical analysis. It is even more difficult to estimate the number of plants that have been screened for any form of biological activity. Firstly, most companies do not publish the results of their screens. Secondly, only a small proportion of academic studies get published. At present many phytochemical and medicinal databases are based on the information contained in published papers. Yet very few academic journals publish papers that just contain information about the re-discovery of a known compound from another species or "negative results" (lack of activity of a compound in an assay). Thus much potentially useful information about the distribution and specificity of biological activity of plant-derived compounds does not get added to those databases used most frequently in drug discovery programmes. Unless this situation alters there is a real possibility that many institutions will waste time and effort in isolating known compounds from previously-studied plants.

A key issue in a plant-based drug discovery programme is the selection of plants. A variety of criteria have been used which include: random collections, collections based on ethnobotanical or chemotaxonomic knowledge, or collections based on geographical surveys. Whatever criteria are used, the logistical problems in collecting and collating the information about plants are great. This approach involves the employment of botanists, ethnobotanists and natural product chemists; a range of expertise that is not required if a synthetic approach to drug discovery is adopted.

3. Different Approaches to Plant Collection

3.1. Random Collection

It is very difficult to make a truly random collection of plants. This would involve collecting plants as encountered, without using any specific criteria. Most collecting methods involve some form of selection. For example, a country or habitat may be selected, and within that habitat a type of plant (e.g. tree or shrub). One method used by many pharmaceutical companies involves taking plants at random from a specific habitat and making extracts of these plants, which are then exposed to an array of assays.

In recent years there has been an increased awareness of the need to conserve rainforests and other areas of high plant diversity that are being threatened by environmental changes. These habitats are the focus of many bioprospecting programmes as they contain a high proportion of plants that have not yet been phytochemically studied. Plants for these programmes are collected randomly because there is often a lack of documented ethnobotanical information about the uses of plants within these areas.

Random collecting can provide samples that meet the numbers for high-throughput screens against an array of *in vitro* receptor-based assays. In plant-rich habitats from which only a small proportion of plants have been screened for biological activity, the random approach to plant collecting would most likely result in the isolation of novel biologically-active compounds. One of the pioneering organisations in plant screening was the National Cancer Institute (NCI) in the United States. They adopted a random approach to plant collecting for their initial anti-cancer screening programme, which took place from the 1950's to 1982, although they did select some plants for screening as a result of having ethnobotanical information about their uses (7). During this period two drugs, Taxol® and camptothecin were developed as anti-cancer drugs. It was estimated that there was a 1:8,000 chance of finding a marketable anti-cancer plant-derived drug using random screens. This ratio compares well with the ratio of 1:10,000 associated with drug discovery based on the random screening of synthetic compounds (7).

3.2. Rational Selection of Plants

3.2.1. Use of ethnobotanical information

Although the random screening programme adopted by NCI has been relatively successful, other organisations have had less success in screening randomly collected plant material. These organisations are either moving out of plant-screening or adopting alternative rational approaches to plant selection.

For example, some organisations now use ethnobotanical information about the medicinal uses of plants to select plants for their drug discovery programmes (8). This method has been shown to be a valuable way to select plants which yield biologically active compounds. Farnsworth *et al.* (9) estimated that about three-quarters of the biologically active plant-derived compounds used in medicine have been discovered through ethnobotanical routes, although the use made of the compounds isolated from the plant is not always that associated with the ethnopharmacological uses reported for the plant. For example, the Madagascar periwinkle (*Catharanthus roseus*) was initially studied for its anti-diabetic activities, but during the phytochemical studies of this plant, compounds with potent anti-cancer activity were isolated (vincristine and vinblastine) and it is these compounds that have been developed as drugs (see below).

There are three stages to an ethnobotanical-based drug discovery programme (10). Firstly, the basic information about the use of plants by a community has to be collated. Secondly, this information should be quantitatively analysed to identify whether factors like availability or potent activity determine the use of a plant. Thirdly, there follows the experimental phase, which involves collecting the plants, followed by the isolation and evaluation of the biologically- active molecules.

The gathering, collating and then quantitative and qualitative analysis of ethnobotanical information is complex (11). It is often too costly for smaller pharmaceutical companies or university departments. For example, the three-staged ethnobotanical approach to bio-prospecting would involve the cost of employing botanists, ethnobotanists and field workers for a period of at least one-to-two years gathering, collating information and then selecting plants. Natural product chemists and biologists would then need to be employed to extract and evaluate the activity of the plants.

Studies by Phillips and Gentry (12,13) in Peru have shown the importance of assessing how information is gathered. They showed that the age of the informants and their role in society can influence the type and quality of information people can offer about the uses of their local plants. Young people were able to identify edible plants, whereas those in the 30-50 age group had knowledge about the commercial potential of plants. The knowledge about medicinal plants was greatest in those aged over 50 years.

Van den Berghe *et al.* (14) showed that plants selected using ethnobotanical information have provided more active leads than those gathered randomly. A wide range of plants could contain compounds that show activity in a specific therapeutic area. This could be due to the plants sharing the same or very similar active compounds or a range of different compounds. Balick (15) examined some of the data obtained from the NCI screens and was able to show that if tannins and polysaccharides are removed from many active plant-fractions the numbers of leads from random or ethno-directed collections are equal. This suggests that these compounds are widely distributed among plants and might explain the activity of many medicinal herbs.

This example might lead one to question the value of ethnobotanical information in drug discovery programmes. However, other studies have shown that 74% of the currently-used drugs based on plant compounds were discovered from ethnobotanical leads (4,9,16). Despite this proven track record and increasing interest in the use of ethnobotanical information in the selection of plants, very few companies use biological assays which are relevant to the ethnobotanical uses of the plant to guide the fractionation of extracts.

Thus despite the wealth of ethnobotanical information that exists, it is surprising how few plants have been fully studied and the active components identified. For example, a survey of natural product and pharmaceutical databases suggests that [illegible] plants belonging to the family Bignoniaceae have medicinal properties, but less [illegible]% have been subjected to extensive phytochemical studies.

Companies that use ethnobotanical information to select their plants have adopted low- through-put screens, although these screens can involve a diversity of receptors. For example, an American-based company, Shaman Pharmaceuticals, has been using ethnobotanical information to screen the activity of plants in tropical forests. The company started in 1990 and after two years had two products in clinical trails, Provir®, an oral product for the treatment of respiratory viral infections and Virend® for the treatment of herpes (17). Their strategy is to obtain and collate primary ethnobotanical information about the flora in a specific geographical area. They then search the ethnobotanical and chemical databases for information about the uses and chemistry of plants within this area. Plants are subsequently selected for screening, using a set of specific therapeutic criteria. For example, in one study they selected plants that were either used for the treatment of disorders of the central nervous system, or had anti-fungal or anti-viral activity. Viral *in vitro* assays showed that of the 134 plants used traditionally for anti-viral activity, 67% had this activity; of the 61 plants selected for their anti-fungal activity, 59% had anti-viral activity and of the 40 plants selected for their use in CNS disorders, 62.5% showed anti-viral activity (17). Thus the chances of finding plants with anti-viral activity were independent of the criteria used for their selection. However, there was no control group of plants studied to indicate whether the fact that they had medicinal properties increased the probability of being active compared to a randomly collected group of plants. This study indicates that it is important to test plants with a known medicinal activity in a range of screens, as plants selected for activity in one therapeutic area can express activity in other therapeutic areas.

This study also shows that the use of criteria such as anti-fungal and anti-viral properties are good indicators of plants that contain biologically-active molecules. However, although a high proportion of the aqueous extracts from the plants showed potent anti-viral activity they were not selective in their activity. As occurred in the NCI screens described earlier, the aqueous extracts of most of these anti-viral plants contained known non-specific-binding tannins and polysaccharides (17). Thus the compounds were unsuitable for drug development. The majority of pharmaceutical companies now spend a great deal of effort in removing phenolics and polysaccharides from plant extracts in order to decrease the proportion of "false positives".

Shaman Pharmaceuticals are presently selecting plants used in the treatment of non-insulin- dependent diabetes mellitus (17). They integrate information gained from the traditional uses of plants with the current phytochemical information about them. They then select a small range of plants which are then exposed to state-of-the-art assays. These assays are being used to isolate and purify active principles which might lead to the development of new anti-diabetes drugs.

3.2.2. Use of chemotaxonomic information

Chemotaxonomy involves the use of biomolecules as characters to study the phylogenetic relationships among plants. The diversity of chemicals in a particular plant species is usually associated with its position within the plant kingdom. Knowledge about the genus to which a species of plant belongs and, in turn, the family to which that genus belongs will often help predict the spectrum of compounds that could be found in a species. However, phytochemical studies have shown that the ability to synthesise the same type of compound has occurred in more than one family; often in unrelated families. In these cases the compounds are frequently biosynthesised via different biosynthetic pathways.

Taxonomic information could be used to drive a drug-discovery programme; once an active compound has been isolated from one species of plant there is the potential of finding more active analogues in related species. However, in companies, an active compound is usually given straight away to organic chemists to synthesise derivatives for testing. What are the relative costs of these two approaches? Modern chromatographic techniques can be used to quickly identify and isolate compounds and it might be cost-effective to make more use of chemotaxonomic surveys before starting a synthetic programme.

An example of a chemotaxonomic study that resulted in the identification of biologically-active molecules is that undertaken into the distribution of polyhydroxyalkaloids by researchers at the Royal Botanic Gardens, Kew, England (18). An array of these compounds (Figure 1) have been isolated from different genera within the legumes but also from other families, such as Araceae, Moraceae and Myrtaceae. Ethnobotanical literature on the plants containing these compounds indicated reports of toxicity to cattle or resistance to damage by insects. Some of this activity could be explained by the fact that the compounds are potent glucosidase inhibitors (19). Research has shown that they decrease insect feeding by blocking the insects' responses to phagostimulants such as sugars (20) and they kill plant nematodes (21). The compounds also have medicinal properties as they interfere with glycoprotein processing, which could explain their anti-viral and anti-cancer activity (18, 22, 23).

The use of botanical collections for chemotaxonomic investigations in drug discovery

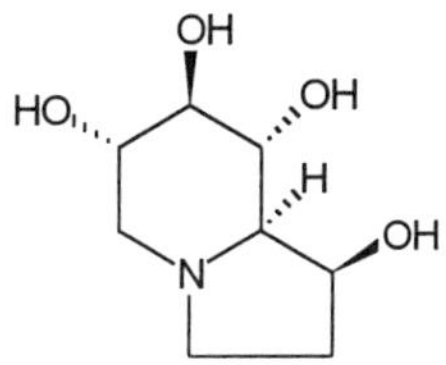

Castanospermine

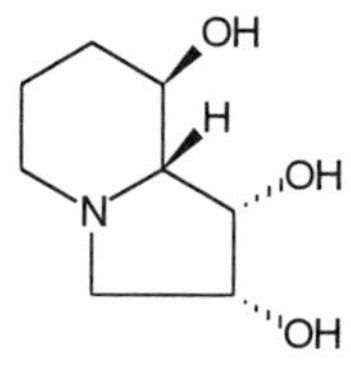

Swainsonine

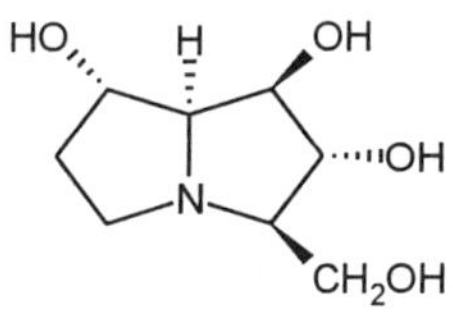

Alexine

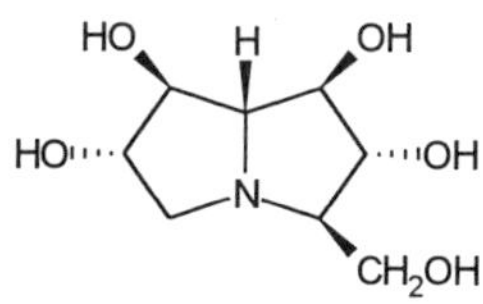

Casuarine

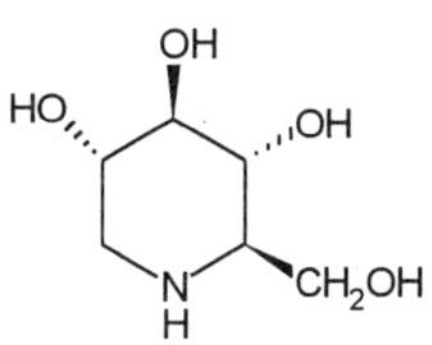

Deoxynojirimycin

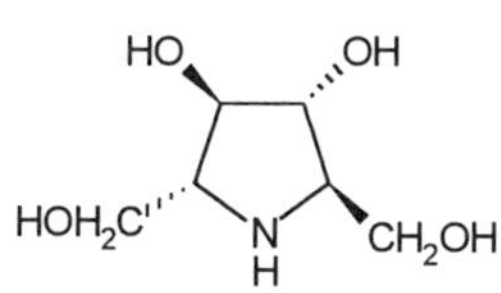

DMDP

Figure 1. Different polyhydroxy alkaloids isolated from plants. Castanospermine (24) from *Castanospermum australe*, swainsonine (25) from *Swainsonia canescens*, alexine (26) from *Alexa leiopetala*, casuarine (27) from *Casuarina equisetifolia* and 2*R*,5*R*-Dihydroxymethyl-3*R*,4*R*-dihydroxypyrrolidine (DMDP) (28) from *Derris elliptica*.

programmes has not yet been fully explored. Firstly, plants rich in a specific group of compounds, e.g. alkaloids or terpenoids, could be selected and subjected to an array of screens. Secondly, companies with lead compounds could make more use of plant collections to screen plants that might contain similar, more active compounds. Chemotaxonomic surveys can also be of use in the identification of alternative sources of active compounds. At present the majority of companies involved in bioprospecting plants for drug discovery do not use chemotaxonomic information. Instead they mostly employ random screening, although more are adopting a strategy that uses ethnobotanical information to target specific plant genera.

4. Methods of Extraction

The majority of natural-product-based screening programmes involve extracting the plant material in methanol or ethanol, then removing the phenolics before testing. The phenolics are responsible for many "false positives", so it is understandable why they need to be removed. However, if ethnobotanical information is used in the selection of plants, then it is important that the screening system has some flexibility to allow the testing of an extract based on the extraction method used by indigenous people. For instance, extraction methods could involve fermentation without which the extract would not be active.

5. Screening

The success of a drug discovery programme will depend not only on good-quality extracts, but also on the availability of a range of assays relevant to the selected diseases. Since 1988 the NCI has encouraged institutes world-wide to submit their compounds for screening against an array of receptor-based cancer and HIV assays.

When NCI started to screen natural compounds for anti-cancer activity, it used a leukaemia model (L1210 lymphocytic leukaemia in mice), a sarcoma assay (sarcoma 180 in rats) and a range of carcinoma screens (29,30). Some of the assays were changed because they were too sensitive to phytosterols, saponins and tannins, resulting in too many false positives. In the period 1960 to 1986 over 35,000 species of plants (108, 330 extracts) were screened in their anti-cancer assays. About 3.8% of these plants showed activity, which represented 1,410 genera and 2,935 species. Receptor-based assays were used to guide the fractionation of the extracts and about

2,000 compounds were isolated. Eleven of these compounds showed highly potent activity and were approved for testing in Phase I clinical trials. These included bruceantin, maytansine, colubrinol, indicine-*N*-oxide, tripdiolide, Taxol®, homoharringtonine, ellipticine and bouvadrin (28,29,30).

Some general conclusions can be drawn from the results of these assays. Representatives from nearly every class of plant-derived compound have been shown to exhibit anti-cancer activity and, within a plant, often two or more compounds can be responsible for the observed anti-cancer activity. The assays used to evaluate the activity of these plants change with time, to reflect the increased knowledge about the receptor-based mechanisms involved in the different forms of cancer. Many of the assays have been automated to increase throughput and decrease the variability often associated with different operators undertaking the assays. NCI has recently increased its capacity to screen over 40,000 samples per year (30,31). This is low compared to the 50,000 a day capacity of Glaxo Wellcome (32).

An example of an integrated approach to plant screening is illustrated by the programme operated by an international group which includes two American institutes, the University of Illinois and the Research Triangle Institute, and a British company, Glaxo Wellcome. This group tests plant extracts across an array of normal and drug-resistant cell lines to evaluate their ability to disrupt different stages of mitotic activity and bind to hormone receptors. In the later tests they monitor inhibition of endocrine hormones, inhibition of human collagenase as well as agonism and antagonism properties of compounds against oestrogen- and androgen-receptor binding (29). Assays also assess cytotoxicity and include known plant-derived compounds as positive controls. For example, some of the positive controls include vincristine and vinblastine which are known to affect tubulin depolymerisation, Taxol® which affects tubulin stabilisation, camptothecin which inhibits topoisomerase I and adriamycin which inhibits topoisomerase II.

Overall, although cytotoxicity in a drug is not wanted, a cytotoxicity assay is an important part of a screening programme. Cytotoxicity usually equates to anti-tumour activity, although cytotoxicity is not a sufficient determinant for anti-cancer activity. The concept of selective cytotoxicity suggests the presence of a cell-specific receptor and activity against such receptors can be used to differentiate one tumour-type from another.

Many companies are trying to design very selective drugs. To identify compounds that express selectivity requires highly-specific receptor-based assays. Such screens are not always suitable for the testing of complex plant extracts because of "false positives" (see above) and interactions that can occur among compounds in extracts. Therefore an array of robust *in vitro* assays and, to avoid chasing known compounds, a good library of UV profiles or mass spectra of known compounds needs to be used when fractionating plant extracts for potent novel compounds. It is now possible for highly-sophisticated modern robot- automated biological and chemical screening technology to screen plants for novel compounds with selective activity (29).

The plant-based drug discovery projects being run by different organisations vary in complexity from the large GlaxoWellcome and NCI programmes to more modest screening programmes. For example, The Royal Botanic Gardens, Kew, screens plant extracts and isolates compounds against a range of pharmaceutical and insecticidal assays. Plants for these screens are either taken at random from the gardens, selected as part of chemotaxonomic studies into the distribution of potentially-active compounds or selected for their activity based on ethnobotanical information. The majority of plants studied in this programme are from the 35,000 species growing at Kew. To date, the most active pharmaceutical and agrochemical novel compounds, such as the polyhydroxyalkaloids, have been isolated as a result of combining knowledge gained from ethnobotanical and chemotaxonomic investigations (18,21).

6. Drugs From Plants

The use of plant-based medicines either as drugs or herbal remedies varies greatly amoung countries. Recent surveys estimate that over 80% of the population in parts of the developing world still rely on plant-derived medicines for their primary health care (33). Thus plants are still the main immediate source of medicine available to the majority of people in the world. Yet globally more money is being directed to the production of synthetic drugs for the benefit of the minority than to improving the production and conservation of medicinal plants for the majority. In some areas of the world the prevalence of western medicines has supressed the use of medicinal plants. Although people within these communities have benefited from the use of western drugs, the availability of drugs is often restricted to people with an adequate income to purchase them. This has resulted in a decrease in the availability of both western and traditional medicine for the poorer members of the community. The poor do not have the finance to purchase the western drugs and the traditional healers or "medicine men" who used to treat the people have moved from these communities. In other countries, such as China and to a lesser extent India, westernised medicine goes hand-in-hand with traditional medicine. In Europe, countries such as Germany and France have a long established history of the use of plant-derived medicines, whereas in Britain this is not so, but the use of plant-based medicines is increasing.

The western European phytochemical market in prescriptions was $2.2 billion (70% Germany) in 1989 and growing. Figures show that the trade in plants used within Europe for non-conventional medicines is increasing by 15-20% a year, with an import value of $3.6 billion in 1995 (34). However, the exact value of the medicinal/phytochemical drug market is difficult to establish because although some of these products are sold via prescription, others are sold over the counter by a range of different retailers.

So far we have dealt with the issues relating to drug discovery but not the actual plants or compounds used. There are many publications that highlight the active components in medicinal plants, for example, specific information about the array of active compounds in plants can be obtained from the numerous National Pharmacopoeiae, Dictionaries or databases (35-39). These can be consulted to get a further insight into the plants and chemicals used in medicine.

This section of the Chapter will serve to illustrate a few of the plants that have been studied and the diversity of compounds that have been isolated. Although many medicinal plants have been well studied, the compounds associated with their uses have not always been confirmed. It will be interesting to see whether the next generation of drugs will result from further studies of these plants or via computational chemistry.

Aconitum spp. Family Ranunculaceae

Extracts from the roots of these plants, often known as monkshood, were used by the Japanese as arrow poisons and are used in Chinese and Japanese medicine for the treatment of rheumatism and neuralgia. The roots contain two types of diterpenoid alkaloids: the toxic aconitine type, which includes aconitine, mesaconitine, hypaconitine, jesaconitine and neopelline and the less toxic atisine type, which includes anisine, atidine, hysognavine, kobusine and ignarine (Figure 2). The plants also contain the cardiac-activating alkaloids higenamine (40) and coryneine (41) as well as the glucan aconitan A (42). The latter compound stimulates glucose utilisation in the liver and peripheral tissue and can be used as a hypoglycaemic agent (42). Extracts from these plants, as well as purified compounds, are still used as drugs in different parts of the world. Care has to be taken in collecting the right plant material at the right time as the alkaloid quality and quantity varies greatly among plant varieties and is influenced by environmental conditions.

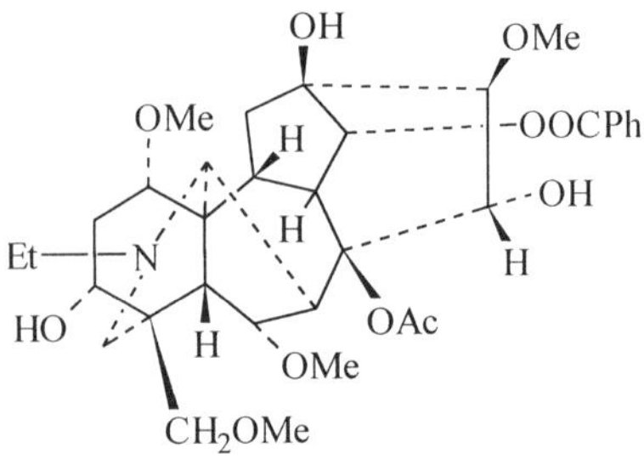

Aconitine

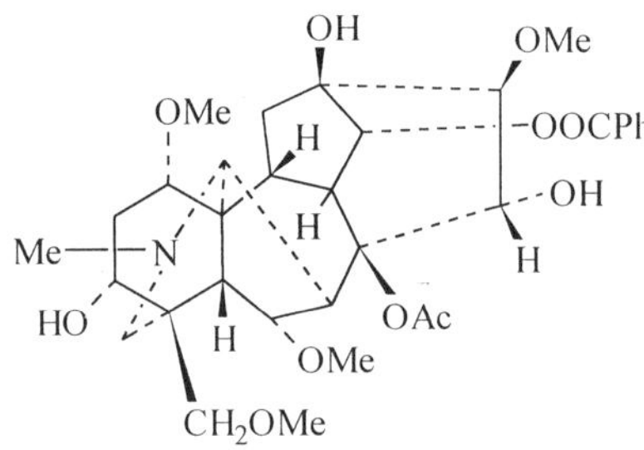

Mesaconitine

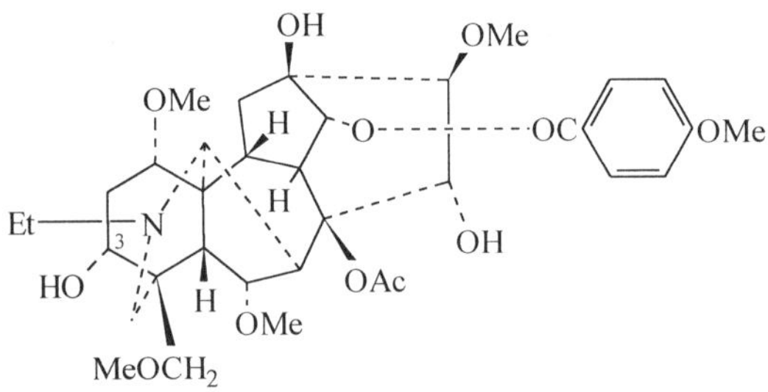

Jesaconitine

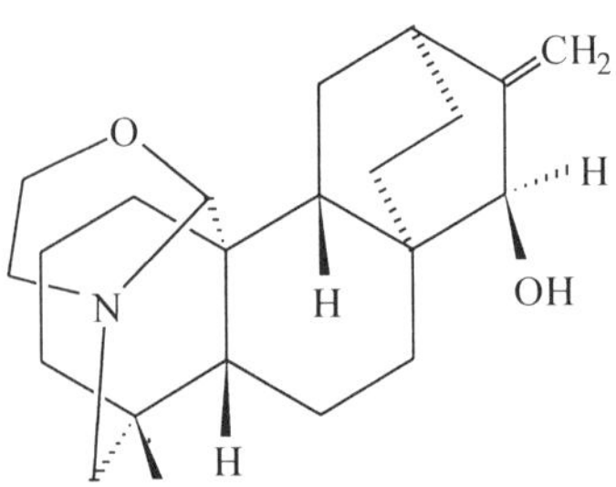

Atisine

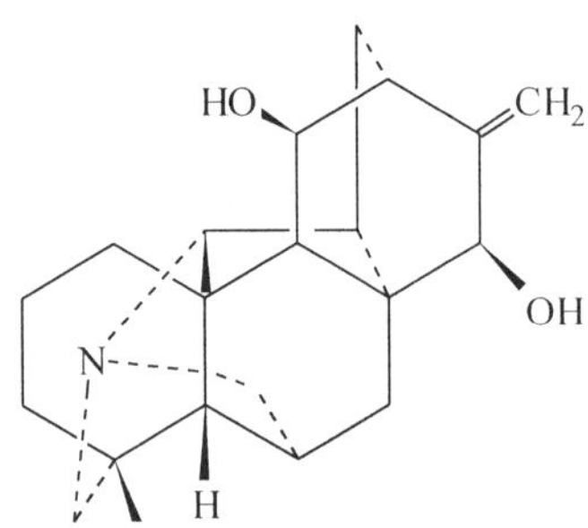

Kobusine

Figure 2. Aconitine, mesaconitine, jesaconitine, atisine and kobusine are some of the pharmacologically-active compounds in *Aconitum* spp.

Allium spp. Family Alliaceae

Many species of *Allium* have medicinal properties, for example, garlic (*Allium sativum*), onion (*A. cepa*), chives (*A. schoenoprasum*) and leek (*A. porrum*) (43, 44). These species are used to treat an array of different ailments: abdominal pains, cholera, dermatitis, hypertension, heart conditions, rheumatism, stomach disorders, haemorrhoids, parasitic diseases and skin rashes. The medicinal properties of the plants are usually attributed to sulphur-containing compounds, each species differing in the range and quantity of these constituents. Alliin, as well as a breakdown product, ajoene, are thought to be the most important biologically-active molecules in garlic (Figure 3). For example, the ability of garlic to decrease the concentration of serum lipids, serum fibrinogen and increase fibrinolytic activity has been attributed to ajoene (45). The activity of this compound could be due to its affecting the physico-chemical properties of cells by interacting with sulphydryl groups on cell membranes (46). Other compounds, such as the mono-terpenoids linalool, citral and geraniol as well as saponins are thought to contribute to the medicinal activity of these species. The flavonoids which occur in low concentrations are thought to add to the antioxidant properties of *Allium* spp. (47). Several studies have shown an inverse correlation between antioxidant flavonoid intake and the risk of coronary

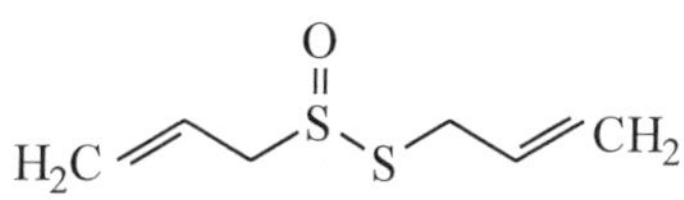

Allicin

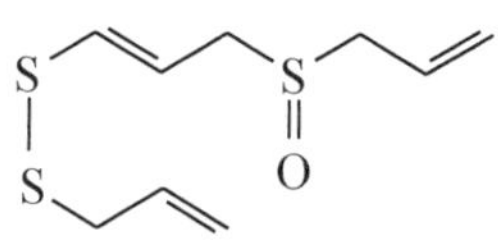

E - Ajoene

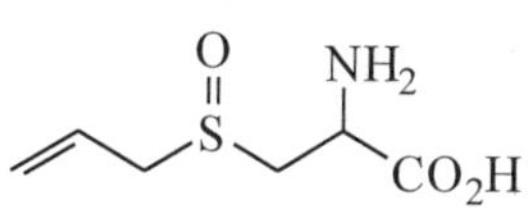

Alliin

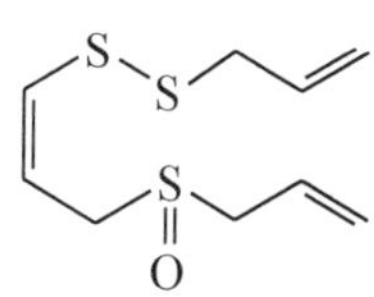

Z - Ajoene

Figure 3. Some of the pharmacologically-active sulphur-containing compounds isolated from *Allium* spp.

heart disease. Thus increasing the proportion of *Allium* in diets could decrease heart diseases (48). Despite the interest in these species, knowledge about which compounds are active and the concentrations required to produce the varied pharmacological effects remains scant. Clinical studies have shown that extracts of garlic can decrease serum levels of cholesterol and triglycerides (45).

In recent years there has been an increase in the trade of *Allium* products for food flavouring as well as for over-the-counter drugs for medicinal purposes. All these products are extracted from plants.

Ammi majus L. Family Apiaceae.
This plant, known as Bishop's weed or the toothpick plant, is widely distributed in North Africa and the Mediterranean area and has been used since biblical times for the treatment of respiratory diseases. In 1879 one of the major active components, khellin, was isolated and its structure was determined in 1938. This compound led to the synthesis of chromolyn, a drug marketed as a bronchodilator and known for its anti-allergenic properties (49). The fruits of this plant are one of the richest sources of the linear furanocoumarins called psoralens which are used in the treatment of many dermatological diseases and are included in products such as Meladinine® (Memphis Chemical Co.), Oxsoralen® (Elder, Gerot) and Ammifurin® (Medexport).

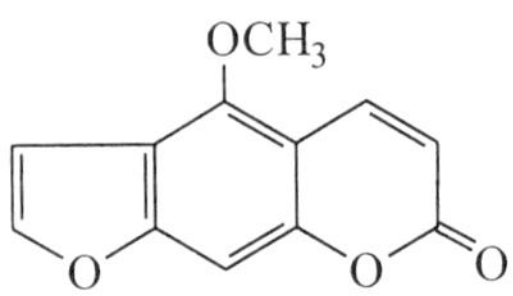

Bergapten

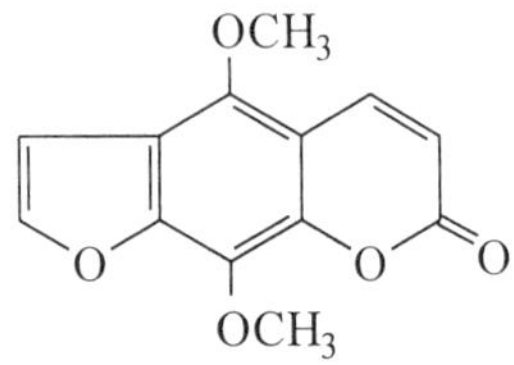

Isopimpinellin

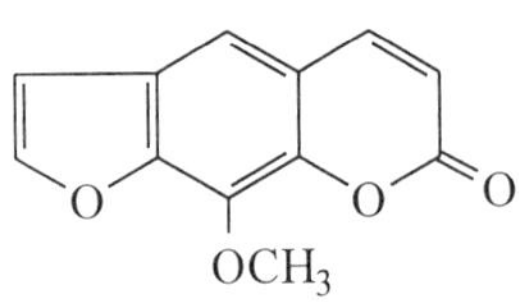

Xanthotoxin

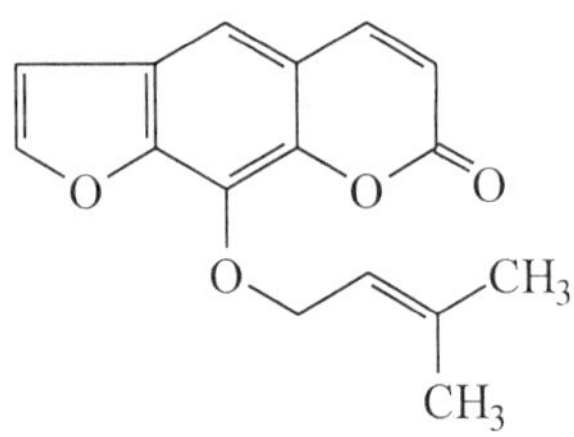

Imperatorin

Figure 4. Some of the psoralens isolated from *Ammi* spp.

The psoralens are activated by light and they are usually used in the presence of long-wavelength UV light, (UV-A, λ 320-400nm) in a dermal therapy referred to as PUVA (Psoralens + UV-A) (50). The active compounds include bergapten, xanthotoxin, imperatorin and isopimpinellin (51) (Figure 4). These compounds also act as calcium antagonists and this activity might increase their therapeutic uses. The diversity of the psoralens in *Ammi* species varies among geographical areas. Plants of Egyptian origin are regarded as the most valuable.

Angelica spp. Family Apiaceae.
The roots of *Angelica officinalis* Moench contain coumarins which could be associated with the anti-spasmodic, expectorant and anti-inflammatory properties of the plant. *A. sinensis* is an important plant in traditional Chinese medicine, and is cultivated in the Gan Su province of China where it is known as Dang Gui. It is widely used to "regulate the blood" and in the treatment of menstruation and respiratory conditions (52). Some of the active compounds have been characterised (Figure 5). For example, ligustilide is involved in the anti-asthmatic activity of extracts. The plant also contains isosafrol, bergapten, xanthotoxin, vanillic acid and scopoletin, compounds with known pharmacological activity. Because of its wide use the demand for *Angelica* has increased. Its distribution is limited and there is not enough plant material to meet these demands. Because of these limitations it is being used mostly for patients with severe diseases who have not responded to other forms of treatment, for example, those shown to be resistant to cancer treatments or resistant to drugs used in the treatment of hepatitis.

Artemisia annua L. Family Compositae
This is a plant used in traditional Chinese medicine and known within China as Quinh Hao. In Europe the plant is known as sweet Annie or sweet wormwood. The first mention of the use of the plant is in the Prescriptions for 52 Kinds of Diseases dated 168 BC when it was prescribed for the treatment of haemorrhoids. It is recorded that in AD 340 (Handbook of Prescriptions for Emergency Treatments) the Chinese herbalist Ge Hong claimed that one handful of *A. annua* in a sheng (1 litre) of water could be used to treat fevers (53). Scientists in China started to investigate the antimalarial activity of the plant and in 1972 they isolated the active anti-malarial compound which they named qinghaosu or artemisinin (Figure 6).

It is a sesquiterpene lactone and its structure was finally elucidated in 1980. The compound was shown to have potential activity against the primaquine-resistant strains of *Plasmodium falciparum*. Clinical trials were undertaken in China and by 1979 workers had successfully treated 2099 cases of malaria (53).

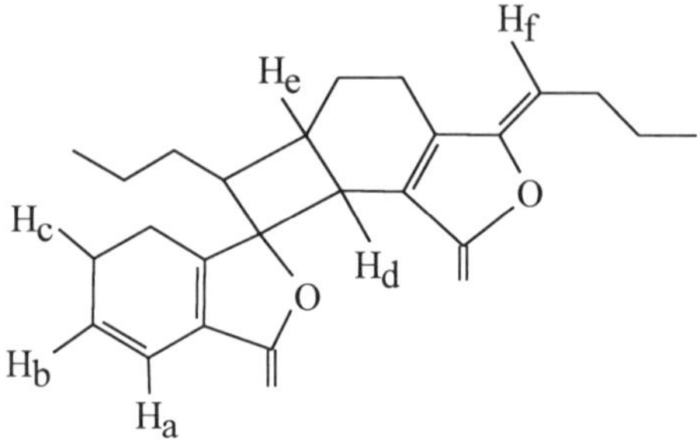

Angelicide

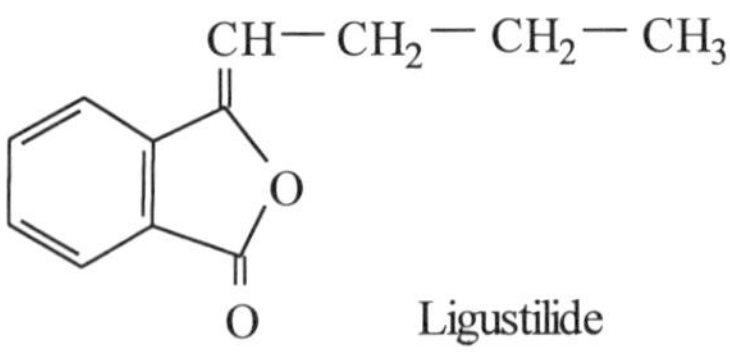

Ligustilide

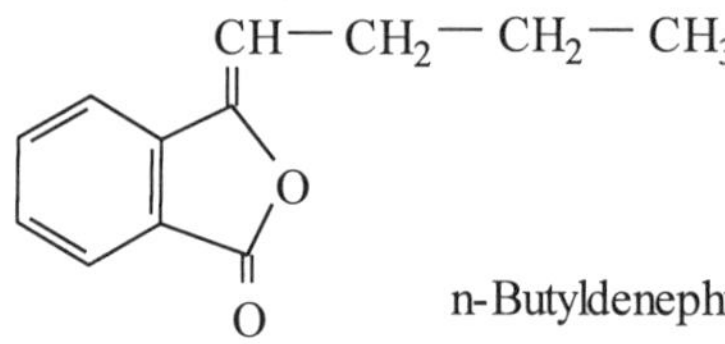

n-Butyldenephthalide

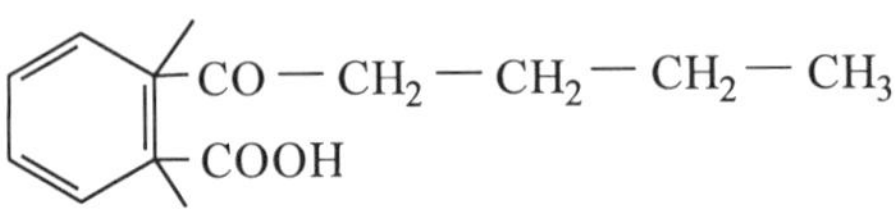

n-valerophenone-o-
carboxylic acid

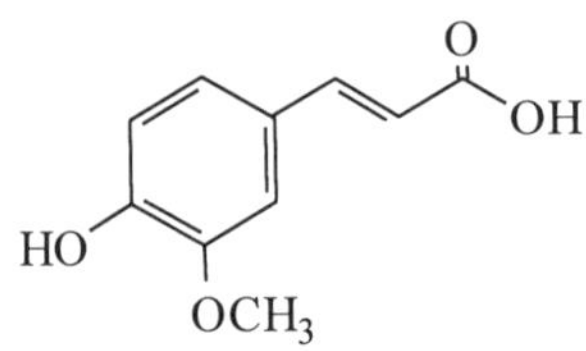

Ferulic acid

Figure 5. Pharmacologically-active compounds isolated from *Angelica spp.*

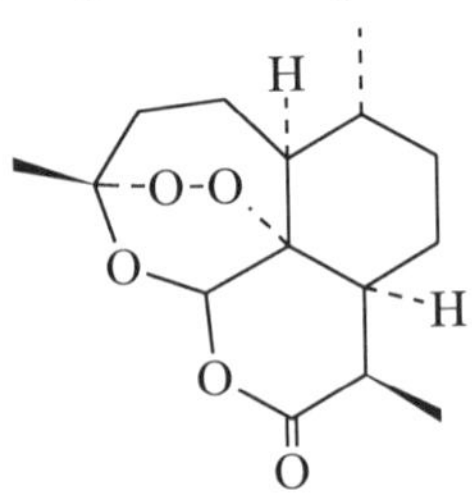

Figure 6. Artemisinin, the anti-plasmodial compound isolated from *Artemisia annua*.

Several laboratories in different parts of the world are now working on the mode of action of artemisinin and synthetic programmes are being carried out to optimise its activity (54). Artemisinin causes changes in cell membranes leading to the formation of autophagic vacuoles with the loss of cytoplasm and death of the parasitoid (55). Cytochrome oxidase in the membranes could be the target. To date the most important derivatives of artemisinin are the sodium salts of artelinic acid; these compounds are water soluble and are currently being evaluated for use as commercial drugs. However, extracts from the plant often have better levels of activity than artemisinin or these synthetic derivatives. This could be due to the presence of other compounds in the extracts such as the methoxylated flavones, artemetin, chrysoplenetin, chrysoplenol D and casticin (56). The flavones appear to strengthen the antimalarial activity of artemisinin.

At present artemisinin is not available on a global scale and many research groups are investigating how to enhance its production in cell lines of *A. annua* or whole plants (57). Chemical studies on 30 of the other 300 species of *Artemisia* have shown that *A. annua* contains the highest concentrations of artemisinin (54), although the yields are only about 0.5% dry weight. In contrast, artemisinic acid occurs in higher concentrations and this acid can be modified to produce artemisinin (56). The differences in concentrations of these compounds in other species of *Artemisia* could explain their poor anti-malarial activity. Other species contain compounds which have different medicinal properties. For example, *A. douglasiana* Besser contains dehydroleucodin, a sesquiterpene lactone with anti-ulcerous properties.

Atropa belladonna L. Family Solanaceae

Preparations of the leaves of this plant, commonly known as deadly nightshade, have been used since the Middle Ages to produce hallucinogenic effects and cause dilations of the pupils. The compounds responsible for these activities are tropane alkaloids (Figure 7). These tropane alkaloids, especially atropine and scopolamine, are thought to be responsible for the use of the plant in the treatment of toothache and for inducing sleep. They are also effective as antimuscarinic agents because of their ability to antagonize the muscarine action of acetycholine (58). Tropane alkaloids have been isolated from a range of Solanaceae species belonging to the genera *Atropa*, *Anisodus*, *Datura*, *Duboisia, Hyoscyamus,* and *Scopolia*. Atropine has been used for the effective treatment of people exposed to cholinesterase-inhibiting nerve gases (58). Atropine (hyoscyamine) is used as a premedication for anaesthesia in gastrointestinal disorders and scopolamine (hyoscine) is used to prevent motion sickness. Many synthetic drugs have been derived from these alkaloids but few have advantages over the natural compounds, which are still being isolated from plants.

Despite the biotechnological resources that have been targeted at the propagation of plants rich in tropane alkaloids, the best yields of alkaloids still come from field-grown plants (59).

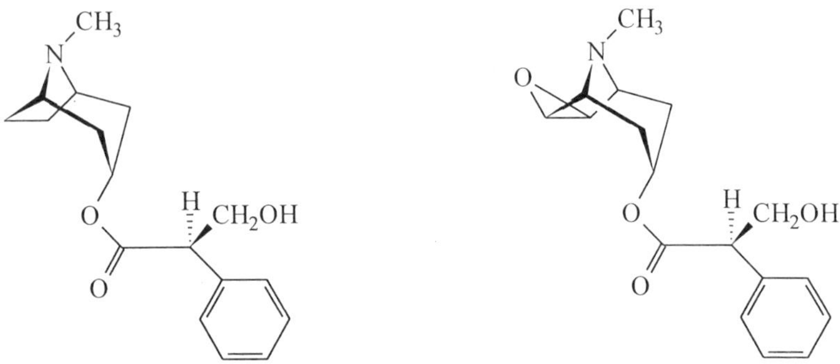

Figure 7. Atropine (left) and scopolamine (right), two of the tropane alkaloids isolated from *Atropa belladonna.*

Camptotheca acuminata Decne Family Nyssaceae

The indole alkaloid camptothecin (CPT) was isolated from the Chinese ornamental tree *Camptotheca acuminata* in 1958 and its structure elucidated in 1966. Clinical trials of CPT in China showed that it was active against a range of cancers, including cancer of the liver and bladder. However, a trial undertaken in the United States in the 1970's was terminated because the compound caused severe adverse toxic effects (31). Recent work has concentrated on the synthesis of derivatives of camptothecin that inhibit topoisomerase I (60). Currently, CPT analogues are being evaluated in clinical trials and a Japanese derivative, CPT-11, is showing activity against refractory leukaemia and another derivative, topotecan, is showing activity against liver and lung cancer (60, Figure 8). Other workers have shown that camptothecin and 10-hydroxycamptothecin, a related indole alkaloid present in *C. acuminata,* have potential use in the treatment of breast cancer; 10-hydroxycamptothecin was more effective and less toxic than camptothecin 61. These water-insoluble compounds have also shown to be effective in the control of human colon cancer when fed orally to tumour-bearing mice (62).

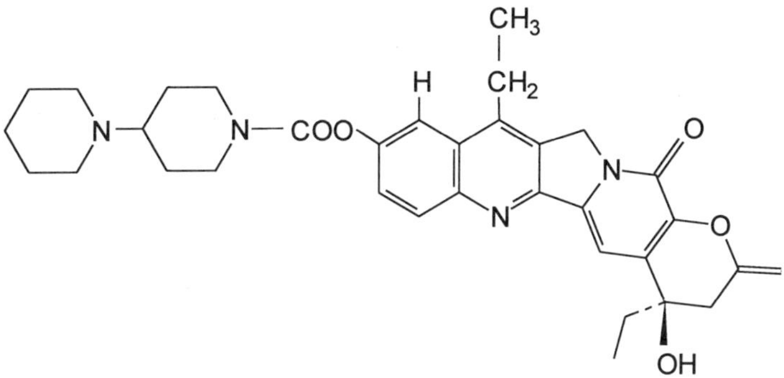

Figure 8. CPT-11, a derivative of the indole alkaloid, camptothecin, isolated from *Camptotheca acuminata.*

Camptotheca acuminata is native to China; however, in order to provide enough of the active indole alkaloids, plantations have been established in America (63). The compounds are usually extracted from the bark and seeds, but young leaves have been shown to contain high levels and could provide a non-destructive source of the compound (64).

Cannabis sativa L. Family Cannabinaceae

This plant, known as hemp or hashish, has an ancient history as a narcotic in India and Africa. In 2000 BC, people who had eaten hemp were reported to see the devil, whereas in AD 200 the Greek physician Galen reported that hemp could be added to cakes to produce hilarity and enjoyment at dinner parties (65). The plant has many uses, it has a high biomass content and the stem-fibre can be used in weaving and in the production of ropes. The seeds have a high oil content (20-40%) and this oil has been used to fuel lights since biblical times.

Tetrahydrocannabinols (THC) are the compounds associated with the psychotropic effects of smoking the plant (Figure 9). The production of cannabis is high but the estimated amount grown for use as drugs is thought to have peaked in 1980 at about 340 tons. Despite the interest in cannabis there is still no definitive explanation of its range of pharmacological activities. However, significant advances have been made in the last six years (66).

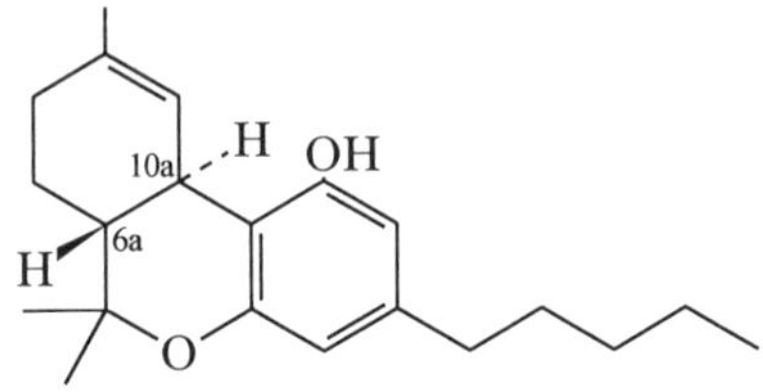

Figure 9. Tetrahydrocannabinol (THC), a pharmaceutically-active compound in *Cannabis sativa.*

Molecular probes have been used to clone the cDNA sequences encoding the human peripheral and central nervous system cannabinoid receptor. In addition, anandamide, a ligand thought to bind cannabinoids to receptors, has been isolated (66). These research tools will enable scientists to study how cannabinoids alter the levels of adrenalin, noradrenalin and serotonin in body tissue. Currently cannabis-based drugs are available in parts of the world for treatment of pain, asthma (dilation of the bronchioles), epilepsy (reduces convulsions), anorexia (stimulates appetite) and sickness resulting from chemotherapy. Only the latter use has received serious consideration in America. In 1982, the American company Eli Lilly produced an analogue of THC called Nabilone® which has been widely used to prevent sickness asssociated with cancer chemotherapy. It is possible that other products based on THC analogues will be produced as drugs in the near future (65).

Carthamus tinctorius L. Family Asteraceae

The oil from this plant, safflower, is rich in unsaturated fatty acids, especially linoleic acid, which can be used effectively as part of a balanced diet to control arterial and heart diseases. *C. tinctorius* is also rich in tocopherols (vitamin E) which could be responsible for the plant's potent anti-oxidant activity. Plants are often used as a preferred source of tocopherols as the natural constituents are more active than the synthetic racemic (*dl*) α-tocopherols (67).

Catharanthus roseus L. G. Don Family Apocynaceae

This plant, the Madagascar periwinkle, previously known as *Vinca rosea,* contains two alkaloids, vinblastine and vincristine (Figure 10), that have potent activity in the treatment of different forms of cancer. The plant was known to be used for the treatment of diabetes in the Philippines, South Africa, India and Australia. Workers in Canada investigating this anti-diabetic activity used rabbits to test whether the extracts would control glucose levels. In their tests the extracts did not affect glucose levels but some of the rabbits died from severe bacterial infections, due to a decrease in their white blood cells and a decrease in their ability to initiate an immune response (65). These researchers reported their finding in 1955 and were approached by Eli Lilly. This collaboration resulted in the isolation of the two anti-cancer compounds, which have helped save thousands of lives. Vinblastine was isolated in 1958 and vincristine in 1961. Vinblastine is used for the treatment of Hodgkin's disease, whereas vincristine is primarily used for the treatment of paediatric leukaemia. These compounds act by the depolymerization of tubulin. In 1970, the five-year survival rate for patients with Hodgkin's disease was 5% but it is now 98%

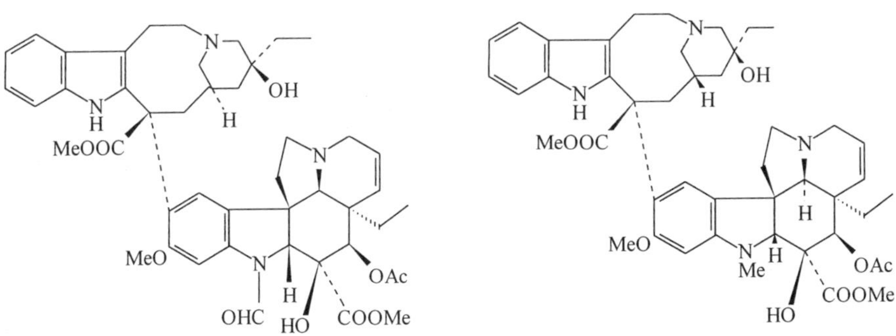

Figure 10. Vincristine (left) and vinblastine (right), the two anti-cancer drugs isolated from *Catharanthus roseus.*

with the use of drugs including vincristine. Similarly, in 1970, 95% of children with acute lymphoblastic leukaemia died within five years but with vinblastine about 60% survive. Since the 1970's vinblastine has been used as part of a combination chemotherapy for the treatment of testicular teratoma and about 90% of patients now survive for more than five years. These compounds can be produced synthetically but they are still being extracted (68). The yield of the two alkaloids is low (0.001%) and Eli Lilly requires about 8000kg flowers per year. Despite the costs of growing the plants and extracting the compounds the company makes an annual profit from these drugs of more than $100 million (65).

Cinchona spp. Family Rubiaceae
Quinine, isolated from the bark of *Cinchona* spp., has been used for centuries for the treatment of malaria (69, Figure 11). However, with the production of synthetic quinine derivatives such as chloroquine, the use of quinine decreased. Resistance of the malarial parasite, *Plasmodium,* to chloroquine has increased in the last 10 years, although the parasite is not resistant to quinine. Therefore there has been a revival of interest in quinine and in the synthesis of more analogues. Despite the many synthetic programmes on quinine no chemist has yet been able to synthesise a compound that is more active than the natural compound against a broad range of parasites (70).

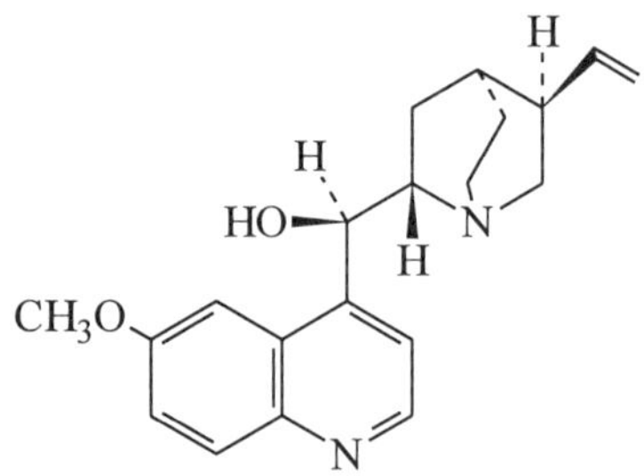

Figure 11. Quinine, the anti-plasmodial drug isolated from *Cinchona* spp.

Coleus forskohlii (Poir.) Family Lamiaceae
During the 1960's the Central Drug Research Institute in Lucknow, India, undertook an extensive screen of Indian medicinal plants (42). One of the plants studied was *Coleus forskohlii* which has been used as a medicinal plant for the treatment of colic, indigestion, haemorrhoids and heart disease. As a result of these screens forskolin, a labdane diterpenoid, was isolated from the roots and it is used as an antihypertensive drug. Forskolin activates the enzyme adenylate cylase, resulting in an increase in the biosynthesis of the second messenger, cyclic adenosine monophosphate (cAMP) (42).

Coptis spp. Family Ranunculaceae
Species in this genus contain alkaloids such as berberine and palmatine which have antibacterial and anti-inflammatory activity. These compounds, especially berberine (Figure 12), have a large market as drugs in Asia for their antihypertensive action (71). They are still extracted from plants but can be produced in cell culture (72).

Digitalis purpurea L. Family Scrophulariaceae
The first mention of the medicinal properties of foxgloves was by Fuchs and Trgus

in their sixteenth century Herbal (1542). However, it did not become widely used until the Scottish physician, Dr W. Withering, in 1785 gave an account of the uses of the plant. The main ingredients are the cardioactive cardenolide glycosides which occur in the leaves of the plant. The different glycosides vary in their pharmacokinetic properties but they all exert a similar physiological effect. Today foxgloves are still a major source of these glycosides and the plants are grown in Europe as well as in the United States of America, Argentina, Africa and Asia (74, Figure 13).

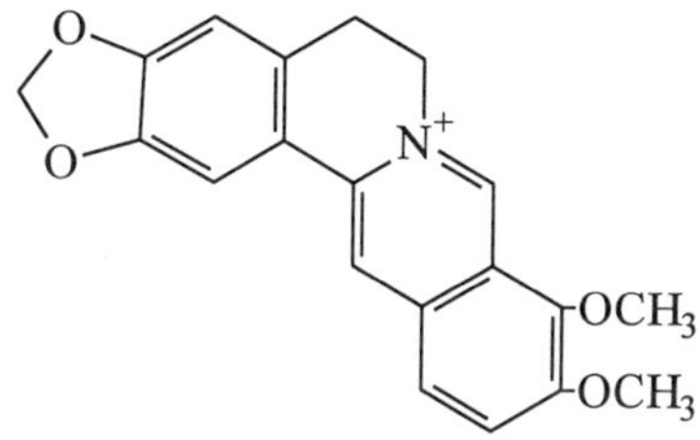

Figure 12. Berberine, one of the important pharmacologically active compounds in *Coptis* spp.

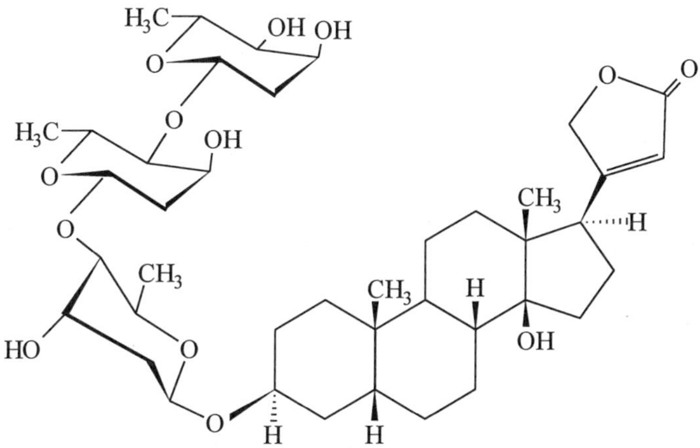

Figure 13. Digitoxin, a cardioactive cardenolide glycoside isolated from *Digitalis purpurea.*

Ginkgo biloba L. Family Ginkgoaceae

The Maidenhair Tree has a long history of medicinal use with records dating back to 2800 BC (76). It has been described as a living fossil as it is the only remaining species in the order Ginkgoales, an order that goes back to the Jurassic period. The tree is long-lived, with specimens thought to be about 1000 years old.

The popularity of this plant in herbal remedies has increased over the last decade. The plant is used in the treatment of diabetes, memory loss and asthma (77). The medicinal activity of the plant was initially attributed to flavonoids and then to the ginkgolides. Now both groups of compounds are thought to contribute to the medicinal properties of the plant. Ginkgolides were first isolated from the root bark and leaves in 1932 but their structures were not determined until 1960 (78) (Figure 14). There is evidence that ginkgolides are able to competitively inhibit the binding of specific platelet-aggregating factors to cell membranes. This type of activity could explain the use of these plants in the treatment of inflammations and for respiratory and panic disorders (79).

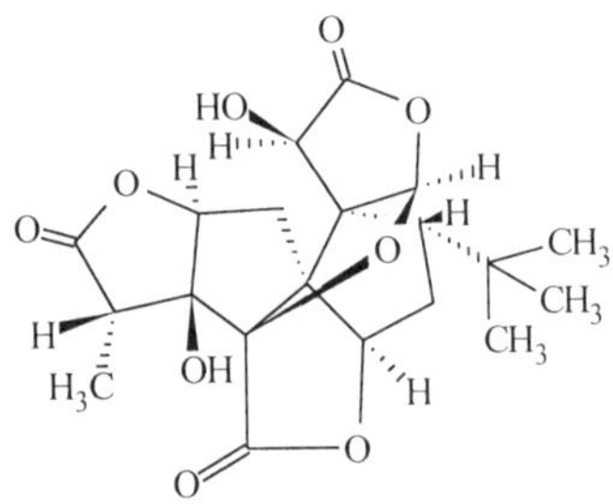

Figure 14. Ginkgolide A isolated from *Ginkgo biloba.*

Matricaria recutita L. (formerly *Chamomilla)* Family Compositae
This plant, known as German chamomile, contains a range of medicinal compounds. The flower head contains an array of pharmaceutically active compounds, such as the terpenes α-bisabolol, bisabolol oxides A and B (Figure 15) and the sesquiterpene lactone matricin, flavonoids including apigenin, luteolin and quercetin and coumarins such as umbelliferone and herniarin. Chamomile drugs are based on extracts from the flower heads and in Germany there are currently about 290 chamomile-containing formulations registered for use as drugs. An estimated 6000-7000 tons per year are harvested and a further 3000 tons imported (80). In 1995, 1 litre of chamomile oil retailed at $8,000 to 10,500 on the world market (34). The drug is used for travel sickness, inflammations of the skin and bacterial diseases of the skin, gums and respiratory tracts (81,82).

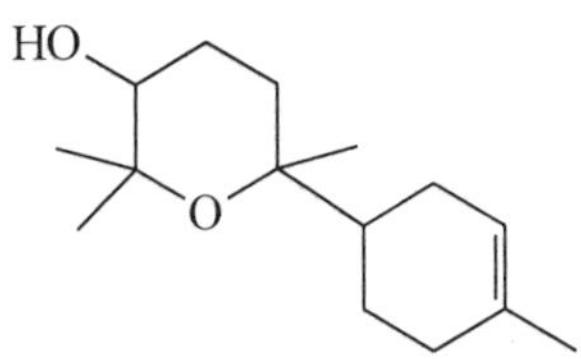

(-) - alpha - Bisabolol

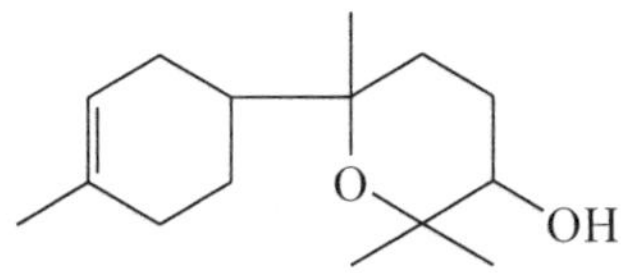

(-) - alpha - Bisabolol-oxide A

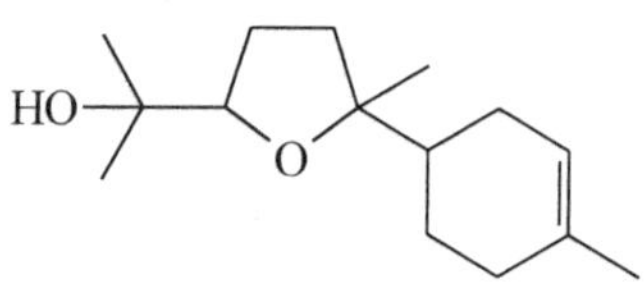

(-) alpha - Bisabolol-oxide B

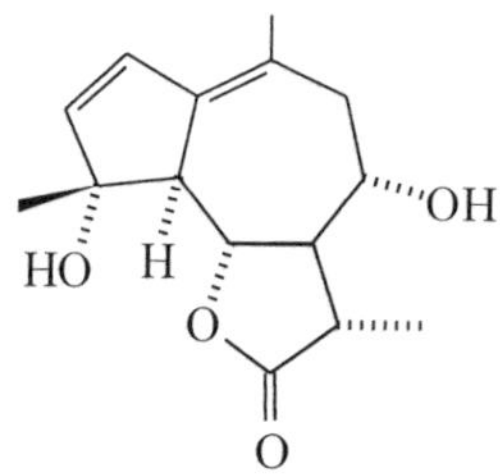

Matricin

Figure 15. Some of the medicinal compounds isolated from *Matricaria recutita*.

Recent studies have shown that apigenin and apigenin glycosides could contribute to the anti-inflammatory and anti-spasmodic activity of drugs from the plant. This is of interest, as free apigenin does not occur in living chamomile flower heads but occurs in the drugs as a result of post-harvest processing, which converts the apigenin glycosides into the aglycone, apigenin (83). Many of the uses of this drug can be explained by the compounds in the plant. For example, antiphlogistic, antimicrobial and ulcer-protective activity can be atributed to α-bisabolol and matricin (82,84). Apigenin could explain the antiphlogistic activity if the drug is applied topically (82,85,86). Chamomile-based drugs are as active as commercially-available synthetic antiphlogistics such as salicylamide and indomethacin (80). Attempts to produce the active compounds in cell culture have failed and their synthesis is expensive, so hydroalcoholic extracts from the plant represent the optimal form of chamomile. High-quality plant material is being grown in Argentina, Bulgaria, Egypt and Hungary, and the demand for the drug is increasing.

Ochrosia spp. Family Apocynaceae

The anti-tumour alkaloid, ellipticine, was isolated from *Ochrosia elliptica* Labill in 1959 (Figure 16). Since then the compound has been isolated from a range of species of *Ochrosia*, including *O. borbonica* J.F. Gmel. from New Caledonia and *O. sandwicensis* A. Gray from Hawaii (87). Ellipticine and some synthetic derivatives were shown to have potent activity against a range of cancers. However, clinical trials on these compounds were stopped due to toxic effects on the patients. Further derivatives were sythesised and tested. One of these derivatives, Celiptium® (N-methyl-9-hydroxy-ellipticinum acetate), is now being used for the treatment of breast cancer (88).

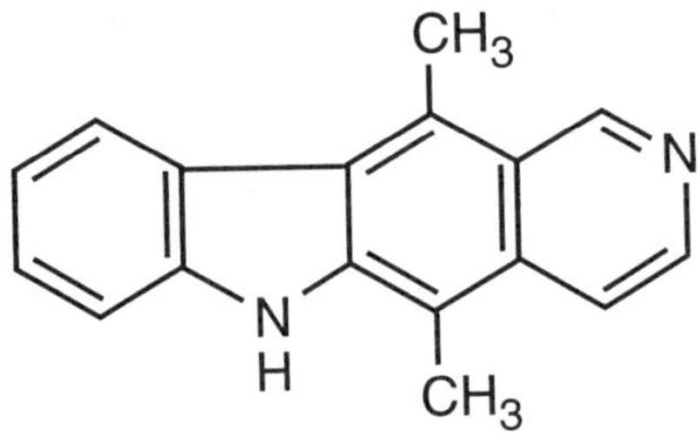

Figure 16. The anti-cancer compound ellipticine isolated from *Ochrosia elliptica.*

Papaver spp. Family Papaveraceae

Papaver somniferum L. was used by the Greeks and Romans as a sedative and to relieve pain (89). The plant contains the pharmaceutically-active alkaloids codeine and morphine. *P. somniferum* is grown commercially in many parts of the world but mainly in India, Iran, Turkey and Japan. Plants are harvested for their morphine content which is commercially converted to codeine. These opiates have served as model templates for the synthesis of a range of drugs. Recent researchers have cloned what is thought to be the primary receptor for morphine-type drugs, the Cl-opioid receptor (90). It is hoped that this receptor can be used to elucidate the mechanisms associated with the multiple actions of morphine.

Podophyllum peltatum L. Family Berberidaceae

This plant known as the American mandrake has a long history in the treatment of cancer. The plant became popular in the 1940's for the treatment of warts. The resin from the plant called podophyllin contains lignans which were shown to inhibit the growth of cancer cells. In the early 1950s the structure of one of these lignans, podophyllotoxin (see also Chapter 11), was determined and a range of analogues were produced by scientists at Sandoz, including a derivative called VP 16-213

(Etoposide) (Figure 17). This derivative was approved for use in the treatment of cancer by the United States FDA in 1983 (91). This compound is now sold in Europe and the United States (trade name Vepeside®) as well as Japan (trade name Lastet®) for the treatment of lung and testicular cancer, as well as nonlymphocyctic leukaemias and non-Hodgkin's lymphomas. Etoposide also has therapeutic value in treatment of AIDS-associated sarcoma (91). The drug is administered by injection or orally as a capsule. Etoposide is one of the most successful medicines derived from plant material. Another derivative, Teniposide, is active against acute lymphocytic leukaemia in children and non-Hodgkin's lymphoma brain tumours in adults (91).

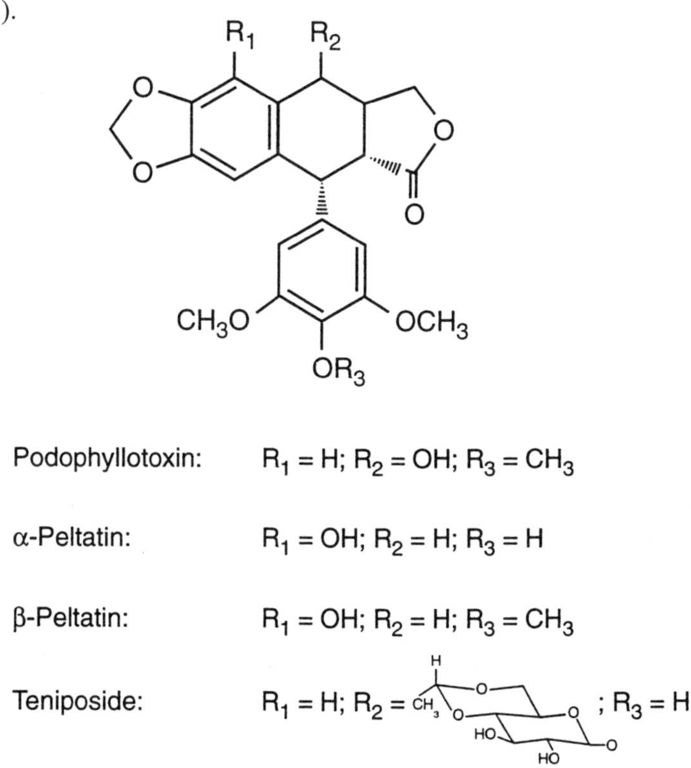

Figure 17. Lignans isolated from *Podophyllum peltatum* and etoposide, a semisynthetic derivative.

Taxus brevifolia Nutt Family Taxaceae

The anti-cancer drug Taxol® (paclitaxel) was isolated from the bark of yew trees as a result of an anti-cancer screen undertaken by the NCI in the 1960's (92; Figure 18). Extracts from *Taxus brevifolia* were first shown to have anti-tumour activity in 1962 but it was not until 1971 that the structure of the diterpenoid was determined. Further work on the biological activity of the compound was hampered because of poor supplies (it occurs in the bark at a low concentration of about 0.02% and it was estimated that it would require 3 trees for every gram of Taxol®) and because it is almost insoluble in water. Needles from a range of species of *Taxus* have been shown to provide a renewable source of taxanes (93). The level of taxanes in the needles varies, the highest concentration occuring in one-month-old flush growth (94).

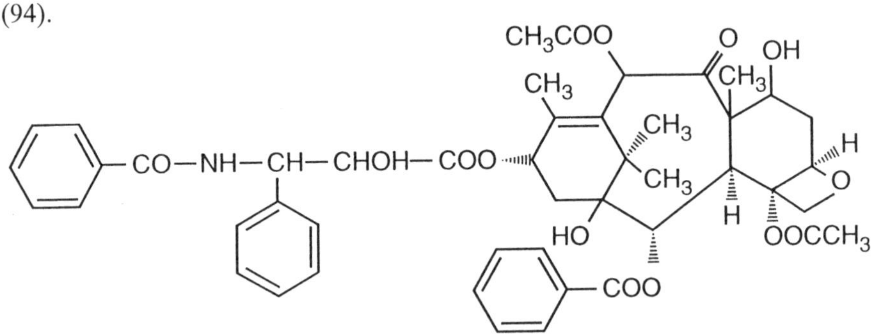

Figure 18. Taxol®, the anti-cancer compound isolated from *Taxus* spp.

In the early 1970's NCI tested Taxol® against a range of new assays and it showed significant levels of activity against various solid human tumours and a B16 mouse melanoma (31). Further studies were carried out and in 1979, Susan Horwitz, a molecular pharmacologist, discovered that Taxol® binds to the cellular protein tubulin and promotes the production of microtubules, which results in the modification of cell division and cell death. This novel mode of action stimulated further research on the molecule and Phase 1 trials were initiated in 1983. However, the trials were nearly stopped due to some patients suffering severe allegic reactions. These reactions were not to Taxol® but to the adjuvant, cremophor, with which it had to be combined. These responses were minimised by changing the adjuvant. Taxol® entered Phase 2 trials in 1985 and in December 1992, Bristol-Myers /Squibb received FDA approval to market Taxol®. Meanwhile companies have been attempting to synthesise this complex molecule. Rhone-Poulenc is currently marketing a synthetic analogue, Taxotere®, which is water-soluble and appears to be more potent than the natural compound. Meanwhile many research groups are investigating the uses of taxols in the treatment of advanced ovarian and breast cancer, especially those resistant to drugs such as doxorubicin and cisplatin (95), as well as small-cell and non-small-cell lung cancer, and neck and head cancers (96).

7. Acknowledgements

The authors wish to thank T. Reynolds, E. Porter, P. Green, N. Veitch and G. Kite for their help in the production of this chapter.

8. References

1. J.B. Harborne, *Introduction to Ecological Biochemistry*, 4th edition (Academic Press, London, 1993).
2. S.S. Rehr, D.H. Janzen and P. P. Feeny, *Science* **181** (1973) 81.
3. S. Teramoto and A. Komamine, in *Biotechnology in Agriculture and Forestry* Vol. 4 *Medicinal and Aromatic Plants* I, ed. Y.P.S Bajaj (Springer-Verlag, New York, 1988) p. 209.
4. P.A. Cox in *Phytochemistry of Plants used in Traditional Medicine*, eds K. Hostettmann, A. Marston, M. Maillard and M. Hamburger (Clarendon Press, Oxford, 1995) p.1.
5. J. H. Cardellina II and M.R. Boyd, in *Phytochemistry of Plants used in Traditional Medicine*, eds K. Hostettmann, A.Marston, M. Maillard and M. Hamburger (Clarendon Press, Oxford, 1995) p.81.
6. P.P. Principle, in *Economic and Medicinal Plant Research* Vol. 3, eds H. Wagner and N.R. Farnsworth (Academic Press, London, 1990) p.1.
7. N.R. Farnsworth, in *Ethnobotany and the Search for New Drugs*, eds. G.T. Prance, D.J. Chadwick and J. Marsh (Ciba Foundation 185, England, 1994) p. 42.
8 M.A. Rouhi, *Chemical & Engineering News* **75** (14) (1997) 14.
9 N.R. Farnsworth, O. Akerele, A.S. Bingel, D.D. Soejarto and Z.G.Gyo, *Bull. WHO* **63** (1985) 965.
10 W.H. Lewis and M.P. Elvin-Lewis, in *Ethnobotany and the Search for New Drugs*, eds G.T. Prance, D.J. Chadwick and J. Marsh (Ciba Foundation 185, England, 1994) p. 60.
11. G.J. Martin, in *Ethnobotany and the Search for New Drugs*, eds G.T. Prance, D.J. Chadwick and J. Marsh (Ciba Foundation 185, England, 1994) p. 229.
12. O. Phillips and A.H. Gentry, *Econ. Bot.* **47** (1993) 15.
13. O. Phillips and A.H. Gentry, *Econ. Bot.* **47** (1993) 33.
14. D.A.Van den Berghe, A.J. Vlietinck and L. Van Hoof, in *Advances in Medicinal Research*, eds A.J. Vlietinck and R.A Dommisse (Wissenschaftliche Verlagsgesellschaft, Stuttgart, 1985) p. 47.

15. M.J. Balick, in *Ethnobotany and the Search for New Drugs*, eds G.T Prance, D.J. Chadwick and J. Marsh (Ciba Foundation Symposium 185, Wiley, England, 1994) p. 4.
16. N.R. Farnsworth, in *Natural Products and Drug Development*, eds P. Krogsgaard-Larsen, S.B. Christensen and H. Kofod (Baillière, Tindall and Cox, London, 1984) p.8.
17. S.R. King and M.S. Tempesta, in *Ethnobotany and the Search for New Drugs*, eds. G. T. Prance, D.J. Chadwick and J. Marsh (Ciba Foundation 185, England, 1994) p. 197.
18. L.E. Fellows, G.C. Kite, R.J. Nash, M.S.J. Simmonds and A.M. Scofield, in *Plant Nitrogen Metabolism*, eds J.E. Poulton, J.T. Romeo and E.E. Conn (Plenum Publishing, New York, 1989) p.395.
19. S.V. Evans, A.M.R. Gatehouse and L.E. Fellows, *Entomol. Exp. Appl.* **37** (1985) 257.
20. M.S.J.Simmonds, W.M. Blaney and L.E. Fellows, *J. Chem. Ecology* **16** (1990) 3167.
21. A.N.E. Birch, W.M. Robertson, I.E. Geoghegan, W.J. McGavin, T.J.W. Alphey, M.S. Phillips, L.E. Fellows, A.A. Watson, M.S.J. Simmonds and E.A. Porter, *Nematologica* **39** (1993) 521.
22. G.S. Jacob, *Current Opinion in Structural Biology* **5** (1995) 605.
23. R.J. Nash, A.A. Watson and N. Asano, *in Alkaloids: Chemical and Biological Perspectives*, Vol. II, ed. S.W. Pelletier (Pergamon, Oxford, UK, 1996) p.345.
24. L.D. Hohenschulz, E.A. Bell, P.J. Jewess, D.P. Leworthy, R.J. Pryce, E. Arnold and J. Clardy, *Phytochemistry* **20** (1981) 811.
25. S.M. Colgate, P.R. Dorling and C.R. Huxtable, *Aust. J. Chem.* **32** (1979) 2257.
26. R. J. Nash, L.E. Fellows, J.V. Dring, G.W.J. Fleet, A.E. Derome, T.A. Hamor, A.M. Scofield and D.J. Watkins, *Tetrahedron Lett.* **29** (1988) 2487.
27. R.J. Nash, P.I. Thomas, R.D. Waigh, G.W.J. Fleet, M.R. Wormald, P.D. deQ. Lilley and D.J. Watkins, *Tetrahedron Lett.* **35** (1994) 7849.
28. A. Welter, J. Jadot, G. Dardenne, M. Marlier and J. Casimir, *Phytochemistry* **15** (1977) 757.
29. G. A. Cordell, *Phytochemistry* **40** (1995) 1585.
30. G.M. Cragg, D.J. Neuman and K.M. Snader, *J. Nat. Products* **60** (1997) 52.
31. G.M. Cragg, M.R. Boyd, J.H. Cardellina II, M.R. Grever, S.A Schepartz, K.M. Snader and M. Suffness, in *Human Medicinal Agents from Plants*, eds A.D. Kinghorn and M.F. Balandrin (ACS Symposium Series 534, 1993) p.80.
32. M. O'Neil, Glaxo Wellcome, personal communications.
33. J. Lambert, World Bank, personal communications.
34. Anon, *Datamonitor* (Trade Source, UK, 1996).
35. J.B. Harborne and H. Baxter, *Phytochemical Dictionary* (Taylor & Francis, Bristol, PA, 1993).

36. J. Buckingham, D.C. Ayres, B.W. Bycroft, P.M. Gunstone, J.B. Harborne, E. Haslam, R.A. Hill, D.R. Kelly, F.J. Leeper, R.D.H. Mirray and I.W. Southon, *Dictionary of Natural Products* (Chapman & Hall, New York, 1994).
37. D.G. Corley and R.C. Durley, *J. Nat. Products* **57** (1994) 1484.
38. A.Y. Leung and S. Foster, *Encyclopedia of Common Natural Ingredients used in Food, Drugs, and Cosmetics* (John Wiley & Sons Inc, New York, 1996).
39. NAPRALERT: M.D. Loub, N.R. Farnsworth, D.D. Soejarto and M.L. Quinn, *J. Chemical Information and Computer Sciences* **25** (1985) 99.
40. C. Kosuge and M. Yokota, *Chem. Pharm. Bull.* **24** (1976) 176.
41. C. Konno, M. Shirasaka and H. Hikino, *Planta Med.* **35** (1984) 150.
42. R.J. Marles and N.R. Farnsworth, in *Economic and Medicinal Plant Research* Vol. 6, ed. H. Wagner and N.R. Farnsworth (Academic Press, New York, 1994) p.149.
43. A. Sendl, G. Elbl, B. Steinke, K. Redl, W. Breu and H. Wagner, *Planta Med.* **58** (1992) 1.
44. B. Mathew, *A Review of Allium section Allium* (Whitstable Litho Ltd, UK, 1996) pp.176.
45. H.D. Reuter, *Phytomedicine* **2** (1995) 73.
46. H.D. Reuter and A. Sendl, in *Economic and Medicinal Plant Research* Vol. 6, eds H. Wagner and N.R Farnsworth (Academic Press, New York, 1994) p. 55.
47. J. Robak and R. J. Gryglewski, *Biochem. Pharmacol.* **37** (1988) 837.
48. M. Yalpani, *Chemistry & Industry* **3** (1996) 85.
49. M.F. Balandrin, A.D. Kinghorn and N.R. Farnsworth, in *Human Medicinal Agents from Plants*, eds A.D. Kinghorn and M.F. Balandrin (ACS Symposium series 534, Washington, 1993) p.1.
50. G. Plewig, E. Holzle and P. Lehmann, *Curr. Probl. Dermatol.* **15** (1986) 254.
51. F.A. Wolff and T.V Tessa, *Clin. Pharmacokinet* **11** (1986) p.62.
52. Z. Shi-Yo and C. Kuo-Chang, in *Biotechnology in Agriculture and Forestry* Vol.7, *Medicinal and Aromatic Plants* II, ed. Y.P.S Bajaj (Academic Press, New York, 1989) p.1.
53. D.L. Klayman, in *Human Medicinal Agents from Plants*, eds A.D Kinghorn and M.F. Balandrin (ACS Symposium series Vol. 534, Washington, 1994) p.242.
54. P.I. Trigg. in *Economic and Medicinal Plant Research* Vol. 3, eds H. Wagner, H. Hikino and N.R. Farnsworth (Academic Press, London 1989) p.1.
55. N.B. Paniego, A.E. Maligne and A.M. Giulietti, in *Biotechnology in Agriculture and Forestry* Vol. 24, *Medicinal and Aromatic Plants* V, ed. Y.P.S. Bajaj (Springer-Verlag, Berlin, Heidelberg, 1993) p.64.
56. J.D. Phillipson, C.W. Wright, G.C. Kirby and D.C. Warhurst, in *Phytochemistry of Plants used in Traditional Medicine*, eds K. Hostettman, A. Marston, M. Maillard and M. Hamburger (Clarendon Press, Oxford, 1995) p.95.
57. E. Van Geldre, A.Vergauwe and E. Van den Eeckhout, *Plant Molecular Biology* **33** (1997) 199.
58. N. Weiner, in *The Pharmacological Basis of Therapeutics*, eds A.G. Gilman, L.S. Goodman, T. Rail and F. Murad (Macmillan, New York, 1995) p.130.

59. Y.P.S. Bajaj and L.K. Simola, in *Biotechnology in Agriculture and Forestry* Vol.15, *Medicinal and Aromatic Plants* III, ed. Y.P.S. Bajaj (Springer-Verlag, Berlin, 1991) p.1.
60. M. Potmesil and H. Pinedo, *Camptothecins: New Anticancer Agents* (CRC Press, Boca Raton, 1995) 149pp.
61. R.W. Zhang, Q.Y. Cai, J.R. Lindsey, Y.R. Li, B. Chambless and F.N.M. Naguib, *Int. Journal of Oncology* **10** (1997) 1147.
62. Q.Y. Cai, R. Lindsey and R.W. Zhang, *Int. Journal of Oncology* **10** (1997) 953.
63. Z.J. Liu and J. Adams, *Canadian Journal of Botany* **74** (1996) 360.
64. M. Lopezmeyer, C.L. Nessler and T.D. McKnight, *Planta Medica* **60** (1994) *558.*
65. J. Mann, *Murder, Magic and Medicine*, (Oxford University Press, Oxford, UK, 1992) pp. 232.
66. E.S. Onaivi, A. Chakrabarti and G. Chaudhuri, *Progress in Neurobiology* **48** (1996) 275.
67. T. Furuya and T. Yoshikawa, in *Biotechnology in Agriculture and Forestry* Vol. 15, *Medicinal and Aromatic Plants* III, ed. Y.P.S. Bajaj (Springer-Verlag, Berlin, Heidelberg, 1991) p.142.
68. I.H. Chu, J.A. Bodnar, R.N. Bowman and E.L. White, *J. Liquid Chromatography & Related Technologies* **20** (1997) 1175.
69. *British Pharmacopaeia*, (HMSO, London, 1993).
70. M. Pudney, in *Fifty Years of Antimicrobials: Past Perspectives and Future Trends*, eds P.A. Hunter, G.K. Darby and N.J. Russell (Cambridge University Press, UK, 1995) p.229.
71. D.U. Lee and K.C. Chang, *Archives of Pharm. Research* **19** (1996) 456.
72. A. Ikuta and H. Itokawa, in *Biotechnology in Agriculture and Forestry* Vol. 7, *Medicinal and Aromatic plants* V, ed. Y.P.S. Bajaj (Springer-Verlag, Berlin, Heidelberg, 1989) p.110.
73. F. Sato and Y. Yamada, *Phytochemistry* **23** (1984) 281.
74. W. Rucker, in *Biotechnology in Agriculture and Forestry* Vol. 4, *Medicinal and Aromatic Plants* IV, ed. Y.P.S. Bajaj (Springer-Verlag, New York, 1988) p.389.
75. D. Deepak, S. Srivastava, N.K. Khare and A. Khare, *in Progress in the Chemistry of Organic Natural Products* Vol. 69, eds W. Herz, G.W. Kirby, R.E. Moore, W. Steglich and Ch. Tamm (Springer, New York, 1996) p.71.
76. R. Major, *Science* **157** (1967) 1270.
77. J.Vesper, and K.D. Hansgen, *Phytomedicine* **1** (1994) 9.
78. P. Braquet, *Drugs of the Future* **12** (1987) 643.
79. C.A. Newall, L.A. Anderson and J.D Phillipson, *Herbal Medicines, A guide for healthcare professionals* (The Pharmaceutical Press, London, 1996) 296pp.
80. J. Reichling and R. Beiderbeck, in *Biotechnology in Agriculture and Forestry* Vol.15, *Medicinal and Aromatic Plants* III, ed. Y.P.S. Bajaj (Springer-Verlag, Berlin, 1991) p.142.
81. R. Carle and O. Isaac, *Dtsch Apotheker Z.* **125** Supp. 1 (1985) 2.

82. A.Y. Leung and S. Foster, *Encyclopedia of Common Natural Ingredients used in Food, Drugs and Cosmetics* (John Wiley & Sons Inc, New York, 1996) 649pp.
83. A. Schreiber, R. Carle and E. Reinhard, *Planta Medica* **56** (1990) 179.
84. M.E. Aggag and R.T. Yousef, *Planta Medica* **22** (1972) 140.
85. Monograph, *Matricariae flos* (Bundesanzeiger no. 228, Dec. 1984).
86. Ch. Franz and C. Kirsch, *Dtsch Apotheker Z.* **125** Suppl. 1 (1985) 20.
87. N. Preaux, *Isolement et structure d'alcaloides indoliques extraits des Ochrosiinees; contribution a la chimiotaxonomie de cette sous-tribu* (Thèse D. Stat. Sci. Phys, Univ. Paris VI, 1976) 174pp.
88. Anon, *Documentation celiptium* (Pasteur Inst., Paris,1982) pp. 123
89. R. Porter and M. Teich, *Drugs and Narcotics in History* (Cambridge University Press, Cambridge, UK., 1995) p. 227.
90. S. Benyhe, *Life Sciences* **55** (1994) 969.
91. Y. Fujii, in *Biotechnology in Agriculture and Forestry* Vol. 15, *Medicinal and Aromatic Plants* III, ed. Y.P.S. Bajaj (Springer-Verlag, Berlin, Heidelberg, 1991) p.362.
92. M.C. Wani, H.L. Taylor, M.E. Wall, P. Coggin and A.T. McPhail *J. Am. Chem. Soc.* **93** (1971) 2325.
93. M.E. Wall and M.C. Wani, in *Alkaloids: Chemical and Biological Perspectives*, Vol. 9, ed. S.W. Pelletier (Pergamon, UK, 1995) p.2.
94. H.N. ElSohly, E.M. Croom, W.J. Kopycki, A.S. Joshi and J.D. McChesney, *Phytochemical Analysis* **8** (1997) 124.
95. A. Sulkes, U. Beller, T. Peretz, J. Shacter, G. Hornreich, C. McDaniel and B. Winograd, *Israel Journal of Medical Sciences* **30** (1994) 70.
96. R. Foa, L. Norton and A.D Seidman, *Int. Journal of Clinical & Laboratory Research* **24** (1994) 6.

Chapter 6

DISEASE PREVENTION AND PLANT DIETARY SUBSTANCES

GARY WILLIAMSON, MICHAEL J. C. RHODES AND ADRIAN J. PARR

1. Introduction

Unlike the situation with animals, the non-mobile lifestyle of higher plants means that they are intrinsically able to avoid neither the attentions of predators and pathogens, nor the effects of stressful environmental conditions. During the course of evolution they have thus developed a plethora of physical and chemical defence systems to help minimise the influence of these factors (see e.g. 1 and Harborne, Chapter 1 of this volume). Plants are thus typified by the extensive range of chemicals, so called "secondary metabolites", that they contain not as part of the basic biochemistry of life, but as part of the way in which they interact with their environment. Man, in his choice of food crops, is acting as a typical herbivorous predator, and thus the chemical defences of plants are directly relevant to human diet and health. The obvious factor of potential poisoning, where human illness is a direct result of consuming food having a deleterious level of toxic metabolites, has received much attention over the years. Some aspects of this issue are touched upon in Section 2 below. Rather more recently it has also become clear that there are potentially beneficial effects of many secondary metabolites. Some may exert a protective role which presumably reflects similar protective actions within the parent plant material (e.g. anti-oxidants), while other compounds possibly have favourable "side effects" in man, unrelated to the evolutionary reason why they are present in plants. The issue of protective agents will be discussed in Sections 3 and 4.

2. Health Benefits From Diminished Levels of Toxicants

Just because a food has traditionally been eaten for an extended period of time does not necessarily mean that it is completely safe. As mentioned above, all plant-based foods will contain some level of secondary metabolites involved in deterring predators and pathogens. While the action of some of these will be against very specific targets, so that there is little or no biological action in man, others will have a more general effect (1). In a mixed diet, the amounts of individual metabolites consumed are often so small that their potential toxicity can be neglected, especially if food is prepared in ways which remove or inactivate potential toxins (e.g. 2). However this may not always be the case; for example procedures for preparation are open to abuse, and particular environmental conditions may just occasionally lead to substantially elevated toxicant levels. Also, the health risks from long-term chronic exposure to low levels of toxicants, as distinct from the better-understood features of acute poisoning, are still not completely clear. As the life-span of individuals increases with modern social and technological advances, so the risks associated with chronic exposure become greater. Moreover, during the domestication of wild species to yield improved cultivars, man has often, consciously or subconsciously, selected for forms with lower toxicant content, but with growing demands for varieties with increased natural pest resistance or drought resistance, breeding programmes may have to return to wild species as a source of some of these factors. An elevation in toxicant levels might then occur - indeed it might well be the basis behind some of the characters that are being selected for! A small number of crops in the Western diet have been identified in which the levels of potential toxicants in current and any newly-developed varieties are sufficient to merit monitoring, and research is under way to examine whether a deliberate reduction in toxicant levels may be of benefit. A complete eradication of toxicants may however not be feasible, or even desirable; toxicant-free crops could perhaps be highly pest-sensitive (and thus not growable in the absence of intensive artificial pesticide treatments), or else the potential toxicants may at low doses actually contribute favourably to attributes such as flavour. It is now also becoming clear that certain so-called toxicants may also possess potential beneficial effects which outweigh the toxicant activity under many circumstances.

Two of the more toxic classes of metabolites which might potentially occur in the average Western diet are the glycoalkaloids and the furanocoumarins. Certain groups within the population may also be exposed to other, more unusual, toxins - for instance the pyrrolizidine alkaloids found in comfrey (3), which has from time to time been promoted for use in herbal teas. Because of the more constrained nature of the food supply, the diet of many developing countries may also contain many additional potential toxicants, such as the cyanogenic glycoside linamarin present in cassava, the dietary staple of many tropical countries.

2.1. Glycoalkaloids

These are complex conjugates of sugars and nitrogenous analogues of steroids, found primarily in members of the plant family Solanaceae. A number of recent reviews (e.g. 4, 5) deal with these types of compounds in some detail. Although a wide range of chemical structures have been described, the most relevant in a food context are α-solanine [1] and α-chaconine [2] from potatoes. In the potato tuber, α-solanine and α-chaconine generally occur at highest concentration in the surface peel, with the content of the inner flesh often being rather low (6). Other parts of the plant, and in particular the shoots of sprouting tubers, typically show a much higher glycoalkaloid content than tubers (6), so explaining why these are not used for food. Glycoalkaloids are heat-stable, and so are generally not removed by cooking (7).

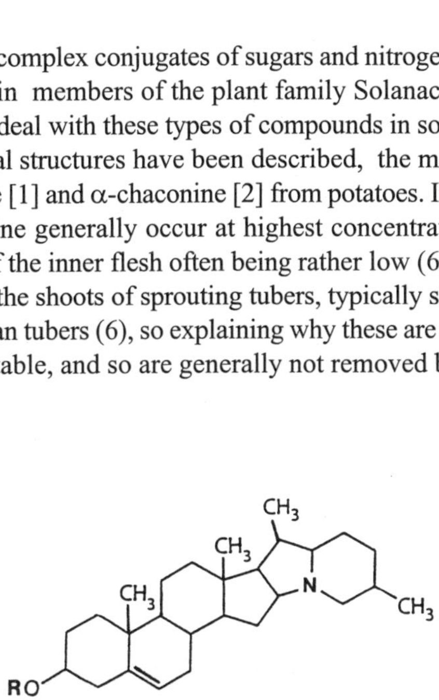

α-solanine [1]
R = rhamnose-galactose--
glucose′

α-chaconine [2]
R = rhamnose-glucose--
rhamnose′

Glycoalkaloids (GAs) exhibit a number of toxic effects, of which an anti-cholinergic activity (8, 9) and the ability of many to disrupt cellular membranes (10) may be most important. Although some compounds are less active than others, there is often a synergistic effect between different structures. For instance, the effect on membrane permeability of a mixture of α-solanine and α-chaconine is greatest at a relative molar proportion of approximately 1 solanine to 1.5 chaconine (10), which is close to that actually found in potatoes (6). The levels of GAs present in tubers are sensitive to a range of environmental factors, and several situations - e.g. high levels of nitrogen fertilisation (11), insect attack on the above-ground parts (12), exposure of tubers to light, physical damage or other stresses (7) - may lead to an elevation of total GA. There have thus been many reports of acute poisoning resulting from consumption of potatoes with elevated GA levels, but fortunately fatalities are rare

(13). The closeness of even normal levels to the dosage at which acute toxic effects become apparent, and continuing debate over possible long-term teratogenic effects (e.g. 14), has however meant that all new potato varieties developed from breeding programmes are now screened for GA content, with an acceptable upper limit of 20 mg GA / 100 g being recognised by many countries (7, 12), though a slightly lower upper limit is in place in some others. Some previously well-established varieties have also subsequently been removed from commercial cultivation following the discovery of conditions under which unacceptably high levels of GAs can be accumulated (see e.g. 15).

Apart from the potato, the other main GA-containing foodcrop is the tomato, in which α-tomatine is found throughout the plant. The GA content is high in small green fruits, but this declines as the fruit ripens, with the fully-ripe tomato typically having only a low or negligible level of α-tomatine (16,17). In the case of this foodstuff, GAs are thus not usually an issue, despite α-tomatine itself being able to disrupt membranes (18).

2.2. Furanocoumarins

These compounds, sometimes also referred to as furocoumarins or as psoralens, are phenolics of often relatively simple structure, examplified by psoralen itself [3]. They occur most typically in members of the Apiaceae (= Umbelliferae) and also in the Rutaceae (which includes the citrus fruits), though they are not completely restricted to these families. They are generally stable to boiling or microwaving (19).

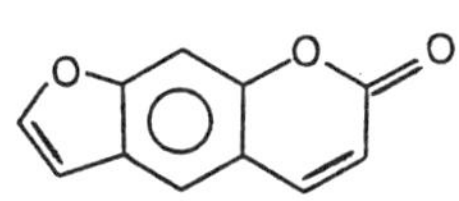

Psoralen [3]

Many furanocoumarins (FCs) are able to intercalate into DNA and, in the presence of long-wave UV light (so-called "UV-A"), then become covalently crosslinked, either at one end or at both ends, onto pyrimidine bases (20). This would have direct effects on gene expression and also may have mutagenic potential. Effects due to photo-adduct formation with proteins may also be important (21), though these are currently only very poorly researched. Actions of FCs which are independent of light, or which occur in tissues not exposed to light, have also been described (e.g. 22, 23). Photo-sensitisation of skin following contact with FC-containing plants is a well-known phenomenon outside the foodcrops area; for example the Giant Hogweed, *Heracleum mantegazzianum,* is a potent photosensitising agent (24). Rather less well-known is the fact that varieties of several edible umbelliferous crops, e.g. parsnip, celery and parsley, also contain readily detectable levels of FCs (19, 25, 26). While other members of the family, such as carrot, show only very low levels under most conditions (27), many umbellifers respond to pathogen attack, and to environmental stress, by the rapid synthesis of FCs (27, 28). Accidental inclusion of infected material in the food chain might thus increase the potential threat to health, even where crops with a low basal FC level are involved. In practice there is still relatively little information available on the true prevalence of dietary FC-mediated illness. Athough consumption of single high doses of celery roots (measured as containing over 8 mg of FC in one case) did not seem to lead to any obvious deleterious effects (29, 30), there was a suggestion that safety margins could be low (29). Moreover, problems of photo-sensitisation amongst celery field workers and packers, and greengrocers who have a long-term exposure to FC-containing material, is a well-recognised phenomenon (e.g. 31, 32).

2.3. Miscellaneous Toxicants

In addition to the toxicants described above, there are many other potentially deleterious plant secondary metabolites which might sometimes be encountered in the Western diet, for instance certain cyanogenic glycosides, glucosinolates and aminoacid analogues (1). The toxicities of these compounds are often either lower than that of the GAs and FCs, their toxic activities in man being offset by independent beneficial properties, the levels encountered are low, or else the sources are only very minor and unpredictable components of the diet. For the purposes of this article, they are thus considered as being of secondary importance as toxicants. In a wider global context, however, it should be remembered that there are a number of compounds found in staple foodstuffs which can have potentially major deleterious effects on health if the crops are eaten in exceptionally large amounts, in the wrong condition, or without adequate processing to remove the toxins. Examples include (i) linamarin and other cyanogenic glycosides from cassava, which in addition to having potentially toxic actions in their own right, also release free cyanide upon hydrolysis during digestion (reviewed in 33), (ii) β-*N*-oxalyl-L-2,3-diaminopropionic acid (β-ODAP), present in the drought-tolerant legume *Lathyrus sativus* and involved in the disease neurolathyrism (34), and (iii) ipomeamarone and 4-ipomeanol, highly-toxic furanoterpenoid phytoalexins which are produced in diseased sweet potato, *Ipomoea batatis* (35).

2.4. Anti-nutritional Factors

In addition to the overtly-toxic chemicals found in plants, there are also a range of metabolites best classified as antinutritional in that, while being of low intrinsic toxicity, they either reduce the acceptability of otherwise highly-nutritious foodstuffs or else reduce the availability of certain of their nutrients. For the purpose of this article two classes of compound, the saponins and the raffinose oligosaccharides, will be discussed in detail, though others such as the proteinase inhibitors (36) should not be overlooked.

2.4.1. Raffinose oligosaccharides

The presence of raffinose oligosaccharides (RFOs) in legume seeds is associated with tolerance of desiccation and also with seed storability (37). These compounds are α-galactosides of sucrose and are laid down in the seed at the time of onset of desiccation. They appear to protect seed proteins from denaturation during this process. They arise by successive transfer of galactose units from galactinol to a sucrose acceptor. In peas, raffinose, stachyose and verbascose with respectively one, two and three galactose units, accumulate. Peas contain up to 4.6% of their dry weight as RFOs; the comparable figures for soybeans, lupin and cowpeas are 4.0, 5.4 and 5.3% respectively. In peas, about 50 % of RFOs are accounted for by verbascose, 37% by stachyose and only 13% by raffinose (38). Nutritionally these compounds are important because of their flatulence-producing property which is generally acknowledged as a major obstacle to increased consumption of legumes, which in other respects are of high nutritional value (39). These oligosaccharides are not digested in man and in monogastric animals such as pigs and poultry due to a lack of an α-1,6 galactosidase in the intestinal mucosa to effect their breakdown. Thus the unhydrolysed, undegraded RFOs pass from the small bowel into the large intestine where they are metabolised by gut bacteria to produce carbon dioxide, methane and hydrogen. Although other food components may contribute to flatulence, RFOs are the major factors. Thus there are opportunities to modify the RFO level and composition of legume seeds to reduce the flatulence-inducing effect and hence improve the nutritional contribution of legume seeds while retaining sufficient RFOs to carry out their important roles in seed maturation.

2.4.2. Saponins

Saponins are complex glycosides of triterpenes or steroidal aglycones. The sugars may be attached by either ether or ester bonds and a wide variety of sugars may be involved with mono- and bis-desmosides formed often with oligosaccharide substituents. This diverse class of compounds is found in many human foods including legumes (soya, peas, beans), and in potatoes, asparagus and in *Allium* vegetables (onions and garlic). Soybean has a complex composition of saponins; the principal aglycones are soyasapogenols A, B and E and at least 10 different glycosylated forms have been described (40). Kudou *et al.* (40) were able to show that in their native state the saponins based on the soyasapogenol B aglycone were conjugated to a 2,3-dihydro-2,5-dihydroxy-6-methyl-4H-pyran-4-one (DDMP) moiety through the C-22 hydroxyl of the aglycone. Ruiz *et al.* (41) showed that the pattern of saponins in lentils and chick peas was much simpler than that of soybean. Here one compound dominated, which is identical to soyasaponin VI of soya and which also contained a DDMP substituent at C-22. This DDMP unit is lost from soyasaponin VI in the chick pea during cooking, when it is degraded to maltol and is leached from the tissue into the cooking water. A review of the chemical and biological significance of DDMP-saponins has recently been published (42).

The ingestion of saponins in both man and animals has been associated with both deleterious and beneficial effects. DDMP saponins have antioxidant properties and many saponins have antitumour and antiviral activities. There are also reports that saponins may have hypocholesterolemic properties. However, the most important property of saponins is their ability to interact with cell membranes and potentially to influence the uptake and hence bioavailability of other food components. Gee *et al.* (43) showed that saponins were able to depolarise the potential difference (PD) generated across the gut wall of rats during Na^+-dependent glucose transport. They showed that there were significant differences between saponins in their ability to depolarise the gut wall and the activity was dependent on the pattern of glycosylation of the saponin, the aglycones being inactive. Subsequent studies have shown that this loss of PD is associated with a permeabilisation of the gut membrane in which pores are formed (44). Although this has direct toxicological implications, there is also the potential for saponins to promote the bioavailability of other food components which might be excluded by an intact membrane. The overall significance of dietary saponins remains uncertain but their interactions with other food components may ultimately prove their most interesting property.

3. Health Benefits From Increased Comsumption of Phytochemicals

3.1. Antioxidants

The term "antioxidant" is almost impossible to define, except for the chemical definition of an agent that prevents oxidation. Many attempts have been made, but the most useful definition is probably by Halliwell and Gutteridge (45): "any substance that, when present at low concentration compared to that of an oxidisable substrate, significantly delays or inhibits oxidation of that substrate". The term antioxidant is a much used (and abused) term, and it is important to understand that an effective antioxidant in one assay system is not necessarily an effective antioxidant in another. For this reason, there is no single assay which can possibly define the total efficacy of an antioxidant. In assessing an antioxidant, it is important to concentrate on the mechanism of action, since there are many ways to generate free radicals, many ways in which a free radical can be quenched by an antioxidant, many factors which can influence this reaction, and an almost infinite number of possible targets of the free radicals. Antioxidant action arises from a combination of several distinct chemical events, which include: metal chelation; hydrogen donation from phenolic groups; partition coefficients; oxidation to a non-propagating radical; redox potential; enzyme inhibition. Furthermore, the health implications of a putative antioxidant depend upon how well it is absorbed by the body and how it is metabolised.

Free radicals have been implicated in over 100 diseases, including cancer, atherosclerosis, rheumatoid arthritis, inflammatory bowel disease and cataract. This large range of targets has sometimes had the effect of reducing the credibility of the antioxidant hypothesis since many of these diseases involve the production of free radicals during the course of illness, and there is only proof that free radicals are involved in the pathogenesis of a few diseases; these, however, include two of the major causes of death in Western society, heart disease and cancer, and also a major factor which influences the quality of life, cataracts. Antioxidants are thought to decrease the risk of these diseases by lowering the concentration of free radicals and so reducing the likelihood that one of these radicals will cause an initiation event. This is a deceptively simple hypothesis. The efficacy of an antioxidant *in vivo* depends not only on the free radical scavenging activity of the compound, but also on the changes in the putative antioxidant on processing/cooking, the bioavailability and absorption from the gut, the distribution in the blood and uptake by target organs, and the metabolism and excretion of the compound.

Epidemiological evidence shows that increased levels of fruit and vegetables in the diet reduce the risk of cancer and heart disease. This has been the subject of several comprehensive reviews (e.g. 46, 47) and will not be covered here. The evidence shows that proportions of the population with a high (>3 portions/day) consumption of fruit and vegetables have a 2-4 fold lower risk of contracting cancer of the intestinal tract. However, epidemiology is too blunt a tool to indicate which components of fruit and vegetables have the most important health implications.

3.1.1. Classes of antioxidants in plants and plant foods

Vitamin C or ascorbic acid is highly water-soluble and is found in a wide range of fruits and vegetables. The best dietary sources include green pepper, strawberries, oranges and blackcurrants, which contain approximately 60-100 mg per "portion" (100 g). Vitamin E (tocopherol) is found in high levels in wheat germ, peanuts, and avocado. There are several tocopherols in the diet, but *in vivo* the most important is the α form. This vitamin partitions into membranes and is almost completely insoluble in water. The carotenoids are highly-coloured compounds which are abundant in sweet potato, carrots, spinach, red pepper, kale, tomatoes and broccoli. Several forms exist, including β-carotene, lycopene and lutein. The flavonoids are a very broad class of phenolic compounds; in the diet the most commonly encountered flavonoids are the flavonol, quercetin (which is present in foods such as onions, apples, and broccoli, as glycosidic derivatives), flavanols (found in tea) and anthocyanins (found in berry fruits). Hydroxycinnamic acids are also dietary antioxidants and are present at very high levels in some foods. Chlorogenic acid is caffeic acid linked to quinic acid and is found in coffee and many fruits. Ferulic acid is a major component of wheat bran, but is covalently linked into the cell wall. Dimers of hydroxycinnamic acids, such as curcumin in turmeric and phenylindan dimers in roasted coffee, are amongst the most potent antioxidants.

3.1.2. Mechanism of action of antioxidants

3.1.2.1. Metal chelation

Transition metals readily produce free radicals in the presence of hydrogen peroxide. The latter is present at a steady state level in many organs and is continuously generated by normal metabolic processes such as flavoprotein oxidases. The reaction between Fe^{2+} and hydrogen peroxide produces hydroxyl radicals by the well-studied Fenton reaction; hydroxyl radicals react readily with all biological molecules. The strategy of biological organisms to prevent the Fenton reaction is to sequester all iron,

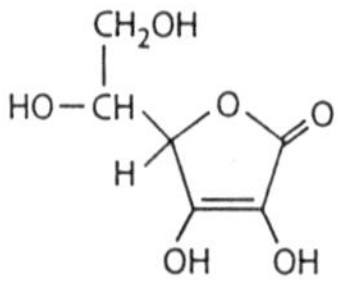

Ascorbic acid
(vitamin C)

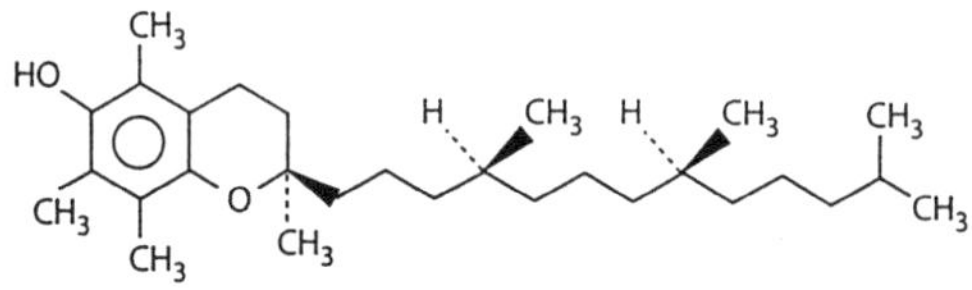

α-tocopherol
(vitamin E)

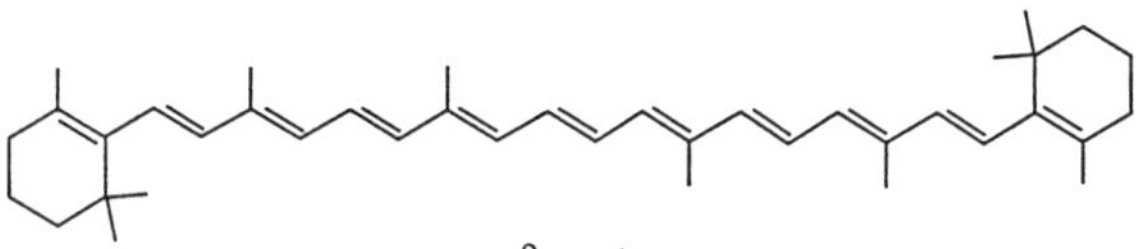

β-carotene

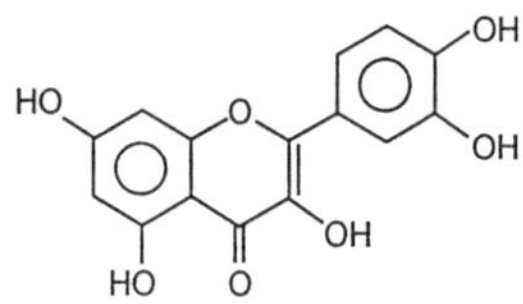

quercetin
(a flavonol)

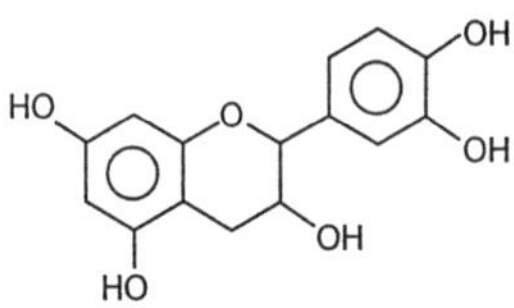

epi-catechin
(a flavanol)

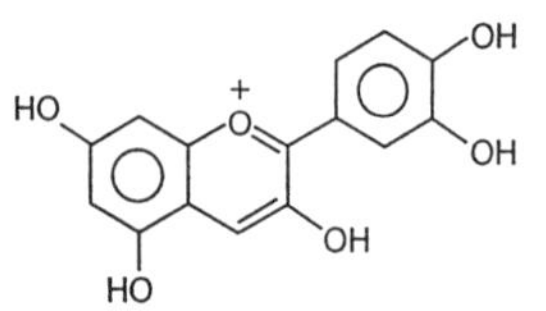

cyanidin
(an anthocyanin)

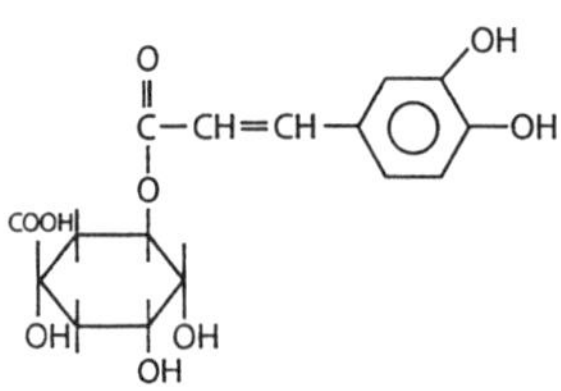

chlorogenic acid
(a hydroxycinnamate ester)

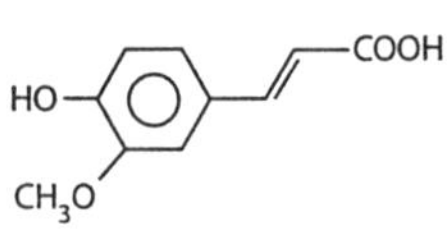

ferulic acid
(a hydroxycinnamate)

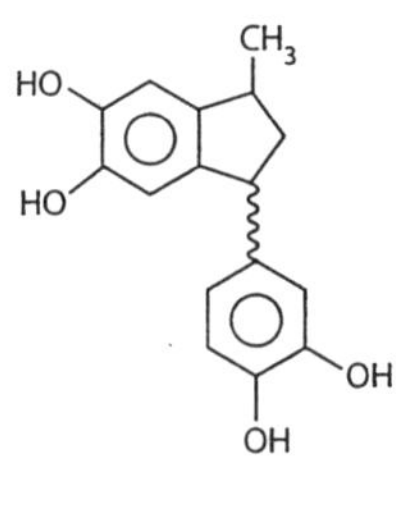

Tetraoxygenated phenylindan
(caffeic acid dimer)

Figure 1. Chemical structures of representative antioxidants.

which is achieved intracellularly by ferritin, and by iron-binding proteins such as transferrin in extracellular fluids. The concentration of free iron in healthy blood isvery low. However, during stress such as septic shock, inflammation and tissue damage, iron can be released from its stores and will then be available to generate free radicals. Iron is chelated by many phenolics, but tight binding requires vicinal hydroxyl groups (48), such as those found on the B ring of quercetin and on caffeic acid (Figure 1). In general, the iron-chelating properties of phenolics and other compounds are relevant in the gut lumen, but are probably not relevant in blood or other tissues because of the highly efficient scavenging of iron by specific proteins.

3.1.2.2. Hydrogen donation/radical scavenging

Peroxidation of lipids occurs both *in vivo* and *in vitro* and gives rise to cytotoxic and reactive products. These products disrupt the normal functioning of a cell and can give rise to damaged or modified DNA (49). Lipid peroxidation has been extensively studied by many researchers. The general mechanism requires oxygen, a free radical initiating species (R, derived from the Fenton reaction, UV light, etc) and the presence of an unsaturated double bond on the lipid (L).

$LH + R^{\bullet} \Rightarrow L^{\bullet} + RH$ Initiation

$L^{\bullet} + O_2 \Rightarrow LO_2^{\bullet}$ Propagation

$LO_2^{\bullet} + LH \Rightarrow LOOH + L^{\bullet}$ Propagation

These steps give rise to a self-propagating reaction with the generation of free radicals and lipid hydroperoxides (LOOH). Hydrogen-donating antioxidants can react with lipid peroxyl radicals in a termination reaction, which breaks the cycle of generation of new radicals:

$$LO_2^{\bullet} + AH \Rightarrow LOOH + A^{\bullet}$$

where A represents an antioxidant. In the case of vitamin E, the resulting radical has a much longer half life than the lipid peroxyl radical and so is much more stable. This therefore increases the chance of occurrence of a true termination reaction:

$A^{\bullet} + A^{\bullet} \Rightarrow A\text{-}A$ Termination

In the case of the flavonoids, hydroxyl groups are the most important for hydrogen donation (50). In the case of the flavonols, hydroxyl groups at the 2', 3' and 4' positions increase the stability of radicals (51).

3.1.2.3. Partition coefficients

Partition coefficients measure the relative solubility of compounds in two solvents. The most usual method is to determine the concentration of compounds in octanol and water and to express the distribution as log $([A]_{octanol}/[A]_{water})$ (e.g. 53). A higher value indicates greater hydrophobicity. Vitamin E and carotenoids are highly hydrophobic. Vitamin C is very hydrophilic. Flavonoids and hydroxycinnamates are intermediate, but the partition coefficient depends strongly on the chemical structure of the flavonoid and on the presence of covalently-linked sugars. Partition coefficients for several antioxidants are shown in Table I. Generally, hydrophobic compounds enter cells more readily than hydrophilic ones. However, many specific transport mechanisms exist, such as the ascorbate transporter (54), and these use active transport to take up molecules into cells.

Table I. Partition coefficients expressed as log(P) for representative phenolic antioxidants (from reference 52)

Antioxidant	log(P)
Ferulic acid	0.55
Quercetin	1.26
Caffeic acid	0.56
Catechin	1.31

3.1.2.4. Redox potential

The oxidation-reduction potential defines the ability of a compound to donate or accept reducing equivalents, and consequently is a fundamental parameter in defining antioxidant action. Values are expressed in volts (V) and represent a form of the equilibrium constant, as defined by the Nernst equation, between oxidized and reduced couples:

$$E_h = E'_o + (2.303\ RT)/nF \log [\text{electron acceptor}]/[\text{electron donor}]$$

where E_h is the observed electrode potential, R is the gas constant (8.31 $J.deg^{-1}.mol^{-1}$),

T is the temperature (K), n is the number of electrons being transferred, F is the Faraday constant (96.4 $kJ.V^{-1}$) and E'_o is the standard redox potential (pH = 7.0, T = 25°C and all concentrations at 1 M). The standard redox potentials allow prediction of the way in which electrons will flow from one redox pair to another under standard conditions. A redox couple with a lower E'_o will reduce (i.e. transfer electrons to) a redox couple with a higher E'_o value. Redox reactions can be slow in the absence of a catalyst or enzyme. In practice, and especially in biological systems, the transport of reducing equivalents only occurs if compounds can make contact, which depends on the presence of appropriate enzymes, the partition coefficients, transport mechanisms, etc.

The redox potentials for some antioxidants have been measured by cyclic voltametry (55). The first redox potentials for quercetin and trolox (a water soluble analogue of vitamin E) are similar (-0.11 V), and for chlorogenic acid the first redox potential is -0.23V. In general, radical scavengers are oxidised at relatively low oxidation potentials, and there is often good correlation between redox potential and antioxidant activity.

3.1.2.5. Enzyme inhibition

Some enzymes, such as certain cytochrome P450 isoforms, lipoxygenases, cyclo-oxygenases and xanthine oxidase are potentially pro-oxidant and can generate radicals. Many flavonoids and hydroxycinnamic acids are effective inhibitors of enzymes. The literature on this area will not be extensively discussed, but as an example, quercetin inhibits 5-lipoxygenase with I_{50} = 3.5 mM (56), and the C-7 hydroxyl of flavonoids is essential for inhibition of xanthine oxidase (51).

3.1.3. Protection of targets from free radical attack

3.1.3.1. Protection of circulating lipoproteins

Protection of LDL (low density lipoprotein in the plasma) from oxidative damage is an important mechanism of antioxidant action. Vitamin E, and possibly other antioxidants, are incorporated into the LDL particle and act to prevent oxidative modification of the lipid and protein components (57). If LDL is oxidatively modified, then it binds to a different receptor on the surface of epithelial cells and may initiate foam cell formation leading to atherogenesis and coronary heart disease (CHD) (58). Antioxidants may prevent this from occurring, and so reduce the risk of CHD.

3.1.3.2. Inhibition of carcinogenesis

Antioxidants also act to block carcinogenesis. Free radicals and reactive oxygen species give rise to changes within a cell which can lead to modified DNA. The free radicals can be generated either from external sources such as cigarette smoke, burnt food and UV light or from normal cellular metabolism. Phorbol esters, such as 12-*O*-tetradecanoyl-phorbol-13-acetate (TPA), hydrogen peroxide and cytokines (such as tumour necrosis factor-a) generate pro-oxidant states in cells and may activate several signal transduction mechanisms (Figure 2). These lead to induction of proto-oncogenes such as *c-fos* and *c-jun* which have a variety of cellular effects; *c-fos* is transiently increased during normal growth, but a prolonged increase can stimulate a cell to initiate carcinogenesis. It is not possible to cover this complex area in any detail, and the reader is referred to some comprehensive reviews (59, 60).

Plant phytochemicals and antioxidants can act to modulate some of these signal transduction pathways. For example, using TPA-induced production of intracellular free radicals in HL-60 cells as a model, the effect of various compounds on this reaction was examined (61). Caffeic acid, α-tocopherol succinate and indole-3-carbinol (derived from breakdown of indolyl glucosinolate) were effective inhibitors of intracellular free radical production, whereas quercetin, catechin and vitamin C were not. Inhibition of tyrosine kinase activity is also an indicator for compounds which modulate signal transduction pathways. Quercetin, catechin, indole-3-carbinol and vitamin C were effective, whereas caffeic acid and α-tocopherol succinate were not (61).

3.2. Other Phytochemicals

Carcinogenesis is a multistage process during which a normal cell becomes a tumorous cell as a result of specific mutations in certain genes. Two such genes may be important: the *apc* and the *p53* genes. The *apc* gene is mutated in almost all cases of colon cancer and plays a central role in modulating cellular activity and signal transduction. Mutations in the *apc* gene lead to a high chance of initiation of carcinogenesis (62). Another important gene is the *p53* gene. The gene is mutated in a wide variety of human tumours. In normal cells, agents that damage DNA induce *p53*, and this over-expression leads to growth arrest of cells, or to apoptosis. This ensures that the cell is prevented from developing with damaged DNA. A mutation in *p53* allows the cell to continue functioning even with damaged DNA, and so the chances of carcinogenesis are increased (63).

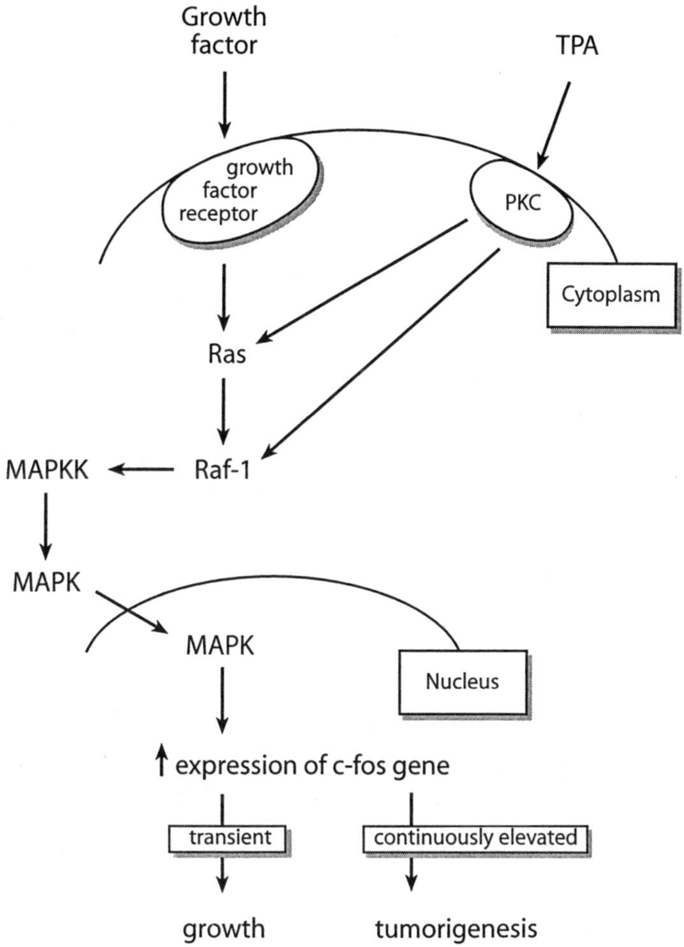

Figure 2. Mechanism of action of phorbol esters.

Protection of cells from DNA damage therefore decreases the chances of mutations in genes such as the tumour suppressor genes. Metabolism of carcinogens and excretion decrease the chance that these agents can modify DNA. The role of diet in this complicated process is potentially throughout the whole of the carcinogenic process (Figure 3). To simplify the possible mechanism of action, anticarcinogens can be classified into two groups, blocking and suppressing, depending on the point of action. Some compounds can both block and suppress. In rodents, it has been shown that increasing the levels of detoxifying enzymes (a blocking action) decreases the incidence of chemically-induced carcinogenesis (64). Also, certain dietary components, such as phenethyl isothiocyanate found in crushed watercress, blocked tumour formation at some sites by almost 100% (65).

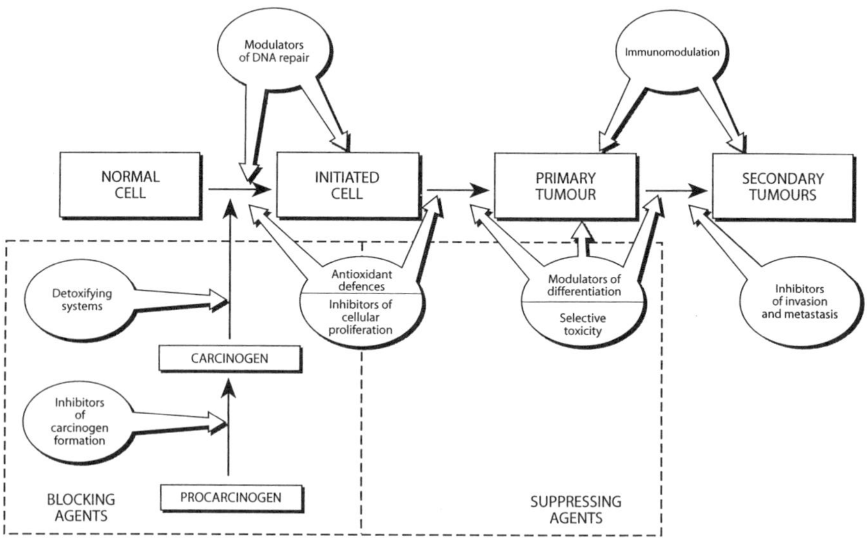

Figure 3. Mechanisms and sites of interaction whereby plant dietary substances may inhibit the carcinogenic process (from 66).

3.2.1. Blocking agents

Investigation of the anticarcinogenic properties of naturally-occurring compounds was pioneered by Wattenberg, who proposed that blocking agents were compounds that were most effective when given before or during treatment with carcinogens in animal models (64). The main action of blocking agents is to stimulate the carcinogen detoxifying enzymes and to inhibit enzymes which have the potential to activate pre-carcinogens into carcinogens. Activating (phase I) enzymes include some forms of cytochrome P450, and detoxifying (phase II) enzymes include glutathione S-transferase, quinone reductase and UDP-glucuronosyl transferase. Induction of glutathione transferase in rodent models correlates with a reduction in risk from carcinogens (66). Agents which induce phase II enzymes have been reviewed by Talalay and co-workers (67), but in the diet the most significant are the breakdown products of glucosinolates and cysteine sulphoxides.

Glucosinolates (Table II) are found in *Brassica* vegetables, such as broccoli, Brussels sprouts, cabbage and cauliflower. Glucosinolates are not themselves blocking agents (68), but on disruption of the plant tissue the glucosinolates are hydrolysed by the endogenous plant enzyme, myrosinase, to a wide range of breakdown products; these include isothiocyanates, which are potent inducers of phase II enzymes (69). The most effective class of isothiocyanates is the aliphatic sulphinyl derivatives, of which sulphoraphane is a well known example (70). Certain isothiocyanates also inhibit phase I enzymes (71), which reduces the rate of carcinogen activation. Some glucosinolate breakdown products also reduce the rate of transcription of cytochrome P450 1A (72). The combination of increasing phase II detoxification and reducing phase I activation increases the likelihood that a carcinogen will be metabolised and excreted without the formation of DNA adducts or formation of reactive oxygen species. This decreases the probability of an initiation event in carcinogenesis. Cysteine sulphoxides are found in *Allium* vegetables and, as for glucosinolates, are broken down by endogenous enzymes. The resulting compounds induce phase II enzymes, with the most effective being diallyl disulphide (73).

Table II. Commonly-occurring glucosinolates in *Brassica* vegetables

Glucosinolate side chain	Common name	Sources containing high amounts
Prop-2-enyl	Sinigrin	Brussels sprouts, cabbage
But-3-enyl	Gluconapin	Brussels sprouts, chinese cabbage
Pent-4-enyl	Glucobrassicanapin	Chinese cabbage
4-Methylsulphinylpropyl	Glucoiberin	B. sprouts, broccoli, cabbage
Indolyl-3-methyl	Glucobrassicin	B. sprouts, broccoli, cabbage
2-Hydroxy-but-3-enyl	Progoitrin	B. sprouts, turnip
2-Phenethyl	Gluconasturtiin	Watercress

In humans, the effect of phenethylisothiocyanate (PEITC, a breakdown product of phenethyl glucosinolate) consumption on metabolism of carcinogens was examined

(74). Smokers were used as a group who expose themselves to carcinogens on a regular basis. Consumption of PEITC-rich watercress altered the metabolism of 4-(methylnitrosamino)-1-(3-pyridyl)-1-butanone (NNK), a tobacco carcinogen. The levels of detoxified metabolites in the urine of volunteers was increased by PEITC consumption, which indicates an increased rate of detoxification and a reduced rate of activation of the carcinogen, NNK.

3.2.2. Suppressing agents

This concept was also introduced by Wattenberg, and refers to compounds that inhibit the appearance of tumours, even after administration of a carcinogen. Certain isothiocyanates, especially benzyl and allyl-isothiocyanates, are suppressing agents (66). Suppressing agents may act by modifying intracellular signalling, inhibition of oncogene expression, or modification of polyamine or oestrogen metabolism. As an example, sinigrin in Brussels sprouts breaks down to allylisothiocyanate (AITC), and this compound selectively kills certain tumour cells, e.g. the HT29 cell line, rather than the differentiated derivative of HT29 (75). For a full review on suppressing agents, the reader is referred to reference 66.

4. Phytoestrogens and Disease Prevention

Plants contain a number of classes of compound which are structural analogues of the natural oestrogens and can, to some extent, mimic their activity. These compounds may interact with the oestrogen receptors in human cells and act as either agonists or antagonists of hormone-dependent responses in these cells. At least two classes of oestrogen receptors have been characterised in animal cells (76). Interesting speculations have been made on the evolutionary basis of this endocrine activity of plant-derived compounds (77). Insights into this relationship have come from sequence analysis of related genes in plants and animals which suggests that key enzymes which regulate steroid biosynthesis in animals and those leading to phytoestrogen formation in plants may have a common ancestral form. Oestrogenic principles have been isolated from a wide range of plants including both monocots and dicots and these represent a wide range of classes of compounds including isocoumarins, flavonoids, hop acids, and terpenes (78). However, by far the most abundant phytoestrogens are compounds derived from phenylpropanoids, the isoflavonoids and lignans.

4.1. *Isoflavonoid Phytoestrogens*

There are two main classes of isoflavonoid phytoestrogens, the isoflavones and the coumestans which are almost exclusively found in plants of the family Leguminosae. The principal dietary source of isoflavones is the soybean seed (*Glycine max*). Soybean contains high levels of isoflavones, up to 4.2 mg / g; these compounds occur in the seed as conjugates of three related aglycones, daidzein, genistein [4] and glycitein. The conjugates are the 7-glucosides of the aglycones and the glucose moiety may be acylated with malonic acid.

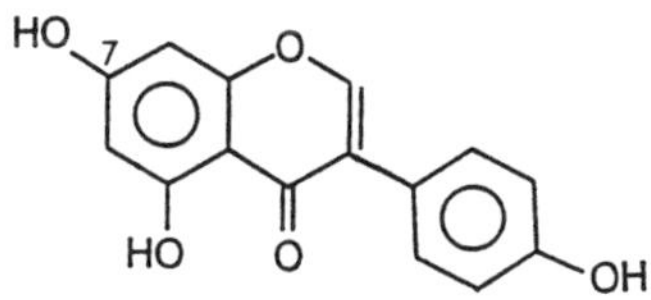

[4] genistein

Typically 30-50 % of the glucose moieties are malonated. Although isoflavone aglycones are found in soybean seeds their level is very low (2-3% of the total) (79). In soy products, processing can lead to an overall loss of isoflavone content due to leaching or to microbial fermentation. Heat treatments can lead to decarboxylation of the malonated isoflavones and the appearance of the equivalent acetyl derivatives (79). Some fermented products, even though they have lower levels of isoflavones compared to the fresh seed, have a relatively high level of aglycones. Other legumes contain isoflavones but generally at much lower levels than in soybean; other major important sources are the chick pea, *Cicer arietinum*, and black bean, *Phaseolus vulgaris* (80).

Coumestans are structurally related to the isoflavones. They are only present in low quantities in most plant foods and are probably insignificant in most diets. However, as will be described later, they have a relatively high oestrogenic activity compared with isoflavones and in some less conventional foods may be present at high levels. The best example of this is the alfalfa sprout, which can contain up to 46 mg / kg fresh weight.

4.2. Lignans

Lignans are phenylpropanoid dimers derived from coniferyl alcohol. The dietary significance of lignans was recognised in studies of compounds commonly detected in urine and identified as enterolactone and enterolactol (81). It was deduced that these compounds arose by microbial degradation of plant lignans. The occurrence of lignans in plant foods is not fully documented, but they are known to be abundant in linseed and relatively so in cereal seeds. In linseed, the precursor of the enterolactone is the diglucoside of secoisolariciresinol, a plant lignan. This compound, if exposed to human faecal bacteria *in vitro*, is degraded via deglucosylation, dehydroxylation and demethylation to yield enterolactone and enterodiol. Another widely-distributed plant lignan is matairesinol, which undergoes similar microbial degradation to form enterolactone. Much more knowledge of the occurrence of lignans in food is required before their significance in the overall diet can be assessed.

4.3. Biological Activity of Isoflavones and Lignans

Evidence from epidemiological studies suggests that diets rich in phytoestrogens, principally based on soy-rich diets, offer protection against hormone-dependent cancers in man (82). In populations, typically in the Far East where soy-products form a significant part of the diet, the incidence of such hormone-dependent diseases including breast and prostate cancer is significantly lower than in the West. A comparative study of the serum concentrations of isoflavones in Japanese men compared to a Finnish group eating typical diets for these two countries showed that the level of isoflavones in the blood was 7-110 times higher in the Japanese men (83). The compounds detected included conjugates of daidzein and genistein but also the microbial degradation products of these compounds including equol. It is thus likely that isoflavones are readily taken up into the blood stream and this is confirmed in recent direct intervention studies in which soya products were fed to volunteers and the kinetics of appearance of isoflavones and their degradation products in blood followed over a 16 hour period (84). Similar feeding studies in which linseed was fed to humans also showed a rapid appearance of lignan degradation products, enterolactol and enterodiol, in urine (85).

Isoflavones have anticancer properties in animal models of carcinogenesis and they inhibit growth of many cancer cell lines in vitro (86). Feeding diets rich in soy for 1 month to premenopausal women increased the length of the follicular phase and delayed menstruation. In this study, the plasma level of estrodiol during the follicular phase was reduced by 9.6% (87). Isoflavones bind to the oestrogen receptors but at much lower affinity compared with steroidal hormones. Typically these affinities are 3 orders of magnitude lower than that of steroidal hormones; the relative oestrogenic activities are of the order 17-β estradiol 100; coumestan 0.202 ; genistein 0.084; equol 0.061; daidzein 0.013 (88). Adlercreutz *et al.* (83) in their study of the levels of isoflavones in the blood of Japanese men concluded that these compounds were present in sufficient concentrations to affect oestrogen-sensitive cells such as prostatic cancer cells.

Although isoflavones and lignans have this ability to bind to the oestrogen receptors, which may in part explain the biological activity of these compounds, they have other activities which may be relevant. They are both antioxidants and are enzyme inhibitors. Of particular interest is the activity of genistein as a specific inhibitor of tyrosine kinase (89). It may thus influence protein phosphorylation reactions and in turn signal transduction pathways which are known to mediate growth responses in cells. Although it is now widely argued that ingestion of phytoestrogens is beneficial to health by offering protection against hormone-dependent diseases such as breast and prostrate cancer, there are some possible negative aspects, particularly in the exposure of infants to high levels of phytoestrogens in baby feeds. This concern was highlighted by studies in New Zealand (90) which calculated that exposure in infants fed soy-based formulae could reach 4 mg / kg body weight / day, greater than the amount reported to cause hormonal effects in pre-menopausal women (about 0.73 mg / kg body weight / day). Little is known of the effects of exposure to such levels in humans but exposure of infants to soy-based feeds is probably best avoided in Western countries where parents are not adapted to a soy-rich diet.

5. Summary

The basic concept of the diet as providing simply a supply of minerals, vitamins and nutrients clearly requires modification in view of the large numbers of non-nutritive, but highly biologically-active, secondary metabolites present in plant-based foods. Individual metabolites may have toxic actions, positive protective effects or even subtle combinations of the two. Foodstuffs, by their nature or quality, thus have the potential to influence both short-term and long-term human health and well-being via mechanisms in addition to the simple effects on nutrient status. On the consumer side, the importance of plant-derived protective factors is clearly emphasised. Whilst there has been some move in certain market sectors towards the sale of specific foods or partially-purifed components from them as "nutraceutical" dietary supplements, the wide range of such factors found in nature means that it is likely that a simple varied diet, rich in vegetables, fibre and fruit, will yield the most benefit. On the agricultural side there are also far-reaching aspects to arise from the growing understanding of the biological action of plant secondary products in man. Breeding programmes aimed at reducing levels of potential toxicants present in many important foodcrops of the Third World should have a major impact on global health issues. In the Western World these issues are no longer quite so directly important, but when developing new commercial crop varieties selection for agronomically-desirable traits (such as yield, pest resistance etc.) clearly must not be undertaken without wider reference to possible changes in the secondary metabolite pattern of these varieties. Although assays for levels of selected toxicants have already been included in the evaluation of new varieties of certain foodcrops, as research continues slowly to clarify the relative importance of specific factors, perhaps this idea should be broadened to include also assays for protective factors. As mentioned above (Section 2), the domestication of many foodcrops has also involved an only partly intentional selection for varieties with modified secondary metabolite profiles. With a growing understanding of the relevance of the different types of compounds, both in terms of the interaction of the plant with its environment, and the consequences to human health of the presence of these metabolites, maybe now is the time to consider also the development of new varieties with specifically optimised secondary metabolite levels. Even in Western agriculture, this might involve deliberate reductions in toxicant levels, though as these compounds must fulfil a role within the plant, a total abolition may not be feasible. Similarly, an increased level of certain protective factors might also be desirable, though as many of these can show undesirable side effects or even toxicity at high levels, again there may be limits as to the changes that are practicable. Not only does classical plant breeding have a role to play here, but as our knowledge of the biochemistry and genetics of secondary metabolite formation increases, so increasingly do the more targetted approaches of modern molecular biology (see Chapters 2 and 9 of this volume),

6. References

1. A. J. Parr and M. J. C. Rhodes, in *Crop Protection Agents from Nature: Natural Products and Analogues*, ed. L. G. Copping (Royal Society of Chemistry, Cambridge, 1996) p. 301.
2. S. Srivastava and S. Khokhar, *J. Sci. Food Agric.* **71** (1996) 50.
3. C. E. Couet, C. Crews and A. B. Hanley, *Nat. Toxins* **4** (1996) 163.
4. J. A. Maga, *Food Rev. Int.* **10** (1994) 385.
5. M. Friedman and G.M. McDonald, *Crit. Rev. Plant Sci.* **16** (1997) 55.
6. M. Friedman and L. Dao, *J. Agric. Food Chem.* **40** (1992) 419.

7. J. A. Woolfe, *The Potato in the Human Diet* (Cambridge University Press, Cambridge U.K., 1987).
8. R. J. Bushway, S. A. Savage and B. S. Ferguson, *Am. Potato J.* **64** (1987) 409.
9. J. G. Roddick, *Phytochemistry* **28** (1989) 2631.
10. J. G. Roddick and A. L. Rijnenberg, *Phytochemistry* **26** (1987) 1325.
11. N. I. Mondy and C. B. Munshi, *J. Agric. Food Chem.* **38** (1990) 565.
12. J. J. Hlywka, G. R. Stephenson, M. K. Sears and R. Y. Yada, *J. Agric. Food Chem.* **42** (1994) 2545.
13. S. C. Morris and T. H. Lee, *Food Technol. Aust.* **36** (1984) 118.
14. J. R. Rayburn, J. A. Bantle, C. W. Qualls Jr. and M. Friedman, *Food Chem. Toxic.* **33** (1995) 1021.
15. K-E. Hellenas, C. Branzell, H. Johnsson and P. Slanina, *J. Sci. Food Agric.* **68** (1995) 249.
16. J. G. Roddick, *Phytochemistry* **13** (1974) 9.
17. M. Friedman, C. E. Levin and G. M. McDonald, *J. Agric. Food Chem.* **42** (1994) 1959.
18. J. G. Roddick and R. B. Drysdale, *Phytochemistry* **23** (1984) 543.
19. G. W. Ivie, D. L. Holt and M. C. Ivey, *Science* **213** (1981) 909.
20. G. D. Cimino, H. B. Gamper, S. T. Isaacs and J. E. Hearst, *Ann. Rev. Biochem.* **53** (1985) 1151.
21. I. M. Schmitt, S. Chimenti and F. P. Gasparro, *J. Photochem. Photobiol.* **B27** (1995) 101.
22. A. E. Desjardins, G. F. Spencer and R. D. Plattner, *Phytochemistry* **28** (1989) 2963.
23. A. Gawron and B. Pawlikowska, *J. Pharm. Pharmacol.* **45** (1993) 1087.
24. K. Lagey, L. Duinslaeger and A. Vanderkelen, *Burns* **21** (1995) 542.
25. J. T. Trumble, J. G. Millar, D. E. Ott and W. C. Carson, *J. Agric. Food Chem.* **40** (1992) 1501.
26. R. C. Beier, G. W. Ivie and E. H. Oertli, *Phytochemistry* **36** (1994) 869.
27. O. Ceska, S. K. Chaudhary, P. J. Warrington and M. J. Ashwood-Smith, *Phytochemistry* **25** (1986) 81.
28. R. C. Beier and E. H. Oertli, *Phytochemistry* **22** (1983) 2595.
29. J. Schlatter, B. Zimmerli, R. Dick, R. Panizzon and C. Schlatter, *Food Chem.Toxicol.* **29** (1991) 523.
30. N. Gral, J. C. Beani, D. Bonnot, A. M. Mariotte, J. L. Reymond and P. Amblard, *Ann. Dermatol. Venereol.* **120** (1993) 599.
31. R. C. Beier, *Rev. Environ. Contam. Toxicol.* **113** (1990) 47.
32. E. Finkelstein, U. Afek, E. Gross, N. Aharoni, L. Rosenberg and S. Halevy, *Int. J. Dermatol.* **33** (1994) 116.
33. B. P. Kamalu, *Int. J. Food Sci. Nutr.* **46** (1995) 65.
34. P. S. Spencer, D. N. Roy, A. Ludolph, J. Hugon, M. P. Dwivedi and H. H. Schaumburg, *Lancet* No.8515 (1986) 1066.
35. M. R. Boyd and B. J. Wilson, *J. Agric. Food Chem.* **20** (1972) 428.

36. K. A. Wilson, in *Antinutrients and Natural Toxicants in Foods*, ed. R.L. Ory (Food & Nutrition Press, Westport, Conn., 1981.)
37. M. Horbowicz and R. L. Obendorf, *Seed Science Research* **4** (1994) 385.
38. P. M. Dey, *Methods in Plant Biochemistry 2* (Academic Press, London, 1990) p. 189.
39. B.O. De Lumen, *Food Structure* **11** (1992) 33.
40. S. Kudou, M. Tonomura, C. Tsukamoto, T. Uchida, T. Sakabe, N. Tamura and K. Okubo, *Biosci. Biotech. Biochem.* **57** (1993) 546.
41. R. G. Ruiz, K. R. Price, E. Arthur, M. E. Rose, M. J. C. Rhodes and G. R. Fenwick, *J. Agric. Food Chem.* **44** (1996) 1526.
42. R. G. Ruiz, K.R. Price, Y. Yoshiki, K. Okubo, M. J. C. Rhodes and G. R. Fenwick, *J. Agric. Food Chem.* (1997) in press.
43. J. M. Gee, K.R. Price, C. L. Ridout, I. T. Johnson and G. R. Fenwick, *Toxic. in vitro* **3** (1989) 85.
44. J. M. Gee, G. M. Wortley, I. T. Johnson, K. R. Price, A. A. J. J. L. Rutten, G. F. Houben and A. H. Penninks, *Toxicol. in vitro* **10** (1996) 117.
45. B. Halliwell and J. M. C. Gutteridge, *Free Rad. Biol. Med.* **18** (1995) 125.
46. G. Block, B. Patterson and A. Subar, *Nutr. Cancer* **18** (1992) 1.
47. K. A. Steinmetz and J. D. Potter, *Cancer Causes and Control* **2** (1991) 325.
48. I. Morel, G. Lescoat, P. Cogrel, O. Sergent, N. Pasdeloup, P. Brissot, P. Cillard and J. Cillard, *Biochem. Pharmacol.* **45** (1993) 13.
49. E. H. Cao, X. Q. Liu, L. G. Wang, J. J. Wang, and N. F. Xu, *Biochim. Biophys. Acta.* **1259** (1995) 187.
50. N. Salah, N. J. Miller, G. Paganga, L. Tijburg, G. P. Bolwell and C. Rice-Evans, *Arch. Biochem. Biophys.* **322** (1995) 339.
51. N. Cotelle, J. L. Bernier, J. P. Catteau, J. Pommery, J. C. Wallet and E. M. Gaydou, *Free Radical. Biol. Med.* **20** (1996) 35.
52. M. Foti, M. Piattelli, M. T. Baratta, and G. J. Ruberto, *Agric. Food Chem.* **44** (1996) 497.
53. W. A. Bruggeman, J. van der Steen and O. Hutzinger, *J. Chromatogr.* **238** (1982) 335.
54. R. W. Welch, Y. H. Wang, A. Crossman, J. B. Park, K. L. Kirk and M. Levine, *J. Biol. Chem.* **270** (1995) 12584.
55. M. Born, P. A. Carrupt, R. Zini, F. Bree, J. P. Tillement, K. Hostettmann and B. Testa, *Helv. Chim. Acta* **79** (1996) 1147.
56. M. J. Laughton, P. J. Evans, M. A. Moroney, J. R. S. Hoult and B. Halliwell, *Biochem. Pharmacol.* **42** (1991) 1673.
57. A. Schmitt, R. Salvayre, J. Delchambre and A. Negresalvayre, *Br. J. Pharmacol.* **116** (1995) 1985.
58. L. Hegyi, J. N. Skepper, N. R. B. Cary and M. J. Mitchinson, *J. Pathol.* **180** (1996) 423.
59. H. J. Rahmsdorf, *J. Molecular Med. Imm.* **74** (1996) 725.
60. M. Karin, *J. Biol. Chem.* **270** (1995) 16483.

61. S. Sharma, J. D. Stutzman, G. J. Kelloff and V. E. Steele, *Cancer Res.* **54** (1994) 5848.
62. R. L. White, *Pathologie Biologie* **45** (1997) 240.
63. A. J. Levine, J. Momaid and C. A. Finlay, *Nature* **351** (1991) 453.
64. V. L. Sparins, P. L. Venegas and L. W. Wattenberg, *J. Nat. Cancer Inst.* **3** (1982) 493.
65. L. O. Dragsted, M. Strube and J. C. Larsen, *Pharmacol. Toxicol.* **72** (1993) S116.
66. I. T. Johnson, G. Williamson and S. Musk, *Nutrition Res. Rev.* **7** (1994) 175.
67. P. Talalay, *Adv. Enzyme Reg.* **28** (1989) 237.
68. N. Tawfiq, R. K. Heaney, J. A. Plumb, G. R. Fenwick, S. R. R. Musk and G. Williamson, *Carcinogen.* **16** (1995) 1191.
69. G. H. Posner, C. Cho, J. V. Green,Y. Zhang and P. Talalay, *J. Med. Chem.* **37** (1994) 170.
70. Y. Zhang, P. Talalay, C. Cho and G. H. Posner, *Proc. Natl. Acad. Sci. U.S.A.* **89** (1992) 2399.
71. S. M. Hamilton and R. W. Teel, *Anticancer Res.* **16** (1996) 3597.
72. H. Wang, S. Griffiths, and G. Williamson, *Cancer Lett.* **114** (1997) 121.
73. V. L. Sparnins, G. Barany and L. W. Wattenberg, *Carcinogen.* **9** (1988) 131.
74. S. S. Hecht, F. L. Chung, J. P. Richie Jr., S. A. Akerkar, A. Borukhova, L. Skowronski and S. Carmella, *Cancer Epidemiol. Biomarkers Prev.* **4** (1995) 877.
75. S. R. R. Musk, P. Stephenson, T. K. Smith, P. Stening, D. Fyfe and I. T. Johnson, *Nutr. Cancer* **24** (1995) 289.
76. W. G. N. Gray, E.E. Biswas, N. R. Bashirelahi and S. B. Biswas, *Proc. Natl. Acad. Sci. U.S.A.* **91** (1994) 11502.
77. M. E. Baker, *Proc. Soc. Expt. Biol. Med.* **208** (1995) 131.
78. K. R. Price and G.R. Fenwick, *Food Additives and Contaminants* **2** (1985) 73.
79. H-J. Wang and P.A. Murphy, *J. Agric. Food Chem.* **42** (1994) 1666.
80. A. A. Franke, L. J. Custer, C. M. Cerna and K. K. Narala, *J. Agric. Food Chem.* **42** (1994) 1905.
81. K. D. R. Setchell and H. Adlercreutz, *Role of the Gut Flora in Toxicity and Cancer* (Academic Press, London, 1988) p 315.
82. C. Herman, T. Aldercreutz, B. R. Goldin, S. L. Gorbach, K. A. V. Hockerstedt, S. Watanabe, E. K. Hamalainen, M. H. Markkanen, T. H. Makela, K. T. Wahala, T. A. Hase and T. Fotsis, *J. Nutr.* **125** (1995) 757S.
83. H. Adlercreutz, H. Markkanen and S. Watanabe, *The Lancet* **342** (1993) 1209.
84. P. Creeke *et al.*, in preparation.
85. J. W. Lampe, M. C. Martin, M. S. Kurzer, H. Adlerceutz and J. L. Slavin, *Am. J. Clin. Nutr.* **60** (1994) 122.
86. S. Barnes, C. Grubbs, J. Carlson and K.D. R. Setchell, *Mutagens and Carcinogens in the Diet* (Wiley-Liss, New York, 1990) p 239-253.
87. A. Cassidy, S. Bingham and K. D. R. Setchell, *Am. J. Clin. Nutr.* **60** (1994) 333.
88. L. Markiewicz, J. Garey, H. Adlercreutz and E. Gurpide, *J. Steroid Biochem. Mol. Biol.* **45** (1993) 399.

89. T. Akiyama, J. Ishida, S. Nakagawa, H. Ogawara, S-I. Watanabe, N. Itoh, M. Shibuya and Y. Fukami, *J. Biol. Chem.* **262** (1987) 5592.
90. C. Irvine, M. Fitzpatrick, I. Robertson and D. Woodhams, *New Zealand Med. J.* **108** (1995) 208.

Chapter 7

BIOTRANSFORMATIONS

MAURICE C.R. FRANSSEN AND NICHOLAS J. WALTON

1. Introduction

1.1. Definition and scope

Biotransformations are uses of enzymes, cells or organs (ideally, biocatalysts) to perform chemical reactions in which the starting substances and products are of comparable chemical complexity. They are distinct from *biosyntheses*, in which relatively complex products are assembled essentially *de novo* by whole cells, tissues, organs or organisms from simple starting substances, such as carbon dioxide, ammonia or glucose, and from *biodegradations*, conceptually the converse process, in which complex substances are broken down again to simple ones. Aspects of plant secondary product biosynthesis are covered in Chapter 2 of this book. The detoxification of xenobiotics by plants has recently been reviewed (1,2).

De novo biosynthesis of plant secondary products by cell and organ cultures is not the principal focus of this chapter and the reader is referred elsewhere for specific reviews of this field (3-7). Instead, a more general emphasis will be placed on the discovery and use of isolated enzymes or whole cells to perform particular types of chemical reactions, for example where these are not easily achievable by non-biocatalytic means. These reactions do not necessarily have current technological application and some are included because they are catalysed by recently-characterised or mechanistically-interesting enzymes. Some examples are given of multi-step processes catalysed by cell extracts *in vitro* or by whole cells or organs in culture. They can illustrate the potential of biotransformations to generate existing or novel substances, help to establish a biosynthetic pathway (see Chapter 2) or point the way towards genetic manipulation work. This chapter does not aim to be comprehensive; rather it is meant to show some of the diversity of biocatalytic capabilities available in plants.

During the early years of synthetic organic chemistry, great efforts were made to replicate natural biosyntheses in the laboratory. For example, the blue dye indigo, traditionally obtained from woad, *Isatis tinctoria*, following a microbial fermentation/biotransformation, and from *Indigofera* spp., was first synthesised by a commercially-viable chemical route in the late 19th century and the synthetic product then rapidly displaced the traditional one (8). Around the same time, natural plant products began to be used as lead structures for the development of synthetic or semi-synthetic drugs with modified properties. This activity continues and expands (see Chapter 6). The wheel has now turned full-circle and, from seeking to replicate or replace the natural chemistry, chemists now seek to utilise biology to achieve reactions which conventional chemistry may less readily bring about or from which there may be some other commercial or environmental advantage. For instance, indigo can now be produced in the laboratory by genetically-engineered *Escherichia coli* (9, 10).

Properties common to both plant cells and enzymes as biocatalysts are first considered in general terms. The rest of the chapter is then devoted to selected illustrative examples in two sections: in the first, cell and organ cultures (and some extracts derived from them) are considered; in the second, a number of specific bioconversions with isolated enzymes from a variety of plant sources are discussed.

1.2. Biocatalysis

A catalyst is defined as a compound or substance which enhances the rate of a particular chemical reaction without being consumed or incorporated into the final product. It does so by providing an alternate pathway for the reaction in which $\Delta G^{\ddagger}$ is less than it would be without the catalyst (11). Catalysts do not change ΔG. Typically, a good catalyst is present in the reaction mixture in only small amounts, when compared to the concentrations of the reactants. *E.g.*, the rate of esterification of acetic acid with ethanol is increased considerably when just a few drops of concentrated sulphuric acid are added to the reaction mixture [eq. {1}]:

$$CH_3C(O)OH + CH_3CH_2OH \xrightleftharpoons{H+} CH_3C(O)OCH_2CH_3 + H_2O \quad \{1\}$$

As depicted in the equation above, both the forward and the backward reaction are acid catalysed. This is another typical feature of a catalyst: a catalyst does not change the position of the reaction equilibrium, it only helps to attain the equilibrium faster.

Nature produces an enormous variety of chemical compounds and it is therefore logical that Nature uses its own catalysts for their production. These biological catalysts, or *biocatalysts*, are generally enzymes, although catalytic activity of RNA (12) and antibodies (13) are known. In biotechnology, the word biocatalyst is used both for isolated enzymes and whole cells. Of course, in the latter case, the enzymes inside the cells are the real biocatalysts that are responsible for the chemical transformations ("biotransformations").

The most important features of biocatalysts are (14):
1. they are very efficient;
2. they are environmentally acceptable;
3. they operate under mild conditions;
4. they are selective;
5. they are cheap.
These advantages will be discussed below.

1. Biocatalysts are very efficient catalysts. Typically the rates of enzyme-mediated processes are accelerated, compared to those of the corresponding nonenzymatic reactions, by a factor of 10^8. The acceleration may even exceed a value of 10^{12}, which is far above what chemical catalysts can achieve. Because of this enormous rate enhancement, biocatalysts are generally used at 10^{-3} - 10^{-4} mole %, as compared to 0.1 - 1 % for usual chemical catalysts.

2. Biocatalysts are environmentally acceptable because they are natural compounds, and are, therefore, fully biodegradable. Biocatalysts are attractive for the food industry because products made by biocatalysis can be labeled "natural", leading to improved consumer acceptance.

3. Biocatalysts operate under mild conditions. Optimum conditions are usually room temperature or slightly higher, an aqueous medium at near-neutral pH and ambient pressure. This is quite different from usual organic chemistry were either high temperatures and pressures (especially in industry) or very low temperatures are used (in the laboratory when working with aggressive reagents). The use of harmful or environmentally-unfriendly organic solvents is not necessary in biocatalytic reactions, although it is known nowadays that many organic solvents and biocatalysts are compatible with each other (15).

4. Biocatalysts are selective. The goal of a chemical synthesis is to obtain the highest possible yield of the desired compound. In many cases, reactants have several reaction possibilities and a mixture is obtained instead of a pure compound. A catalyst can specifically enhance the rate of *one* of the possible reactions, leading to a much cleaner reaction mixture. Biocatalysts have been designed by Nature to do just this: to convert a substrate into *one*, well-defined product. This striking ability of biocatalysts has drawn the attention of chemists during the last twenty years leading to thousands of publications and some excellent books (14,16,17) and reviews (18, 19) on the application of biocatalysts to selective organic reactions. Three different kinds of selectivity can be distinguished:

- *Chemoselectivity*: the ability to react with only one type of functional group when several groups with comparable reactivities are present in the substrate. For instance, esters and acetals can both be cleaved by acid treatment, but an esterase hydrolyses esters only;
- *Regioselectivity*: the ability to distinguish between identical functional groups or positions at different positions in the substrate;
- *Stereoselectivity*: the ability to transform an achiral compound into one and only one stereo-isomer of a chiral compound, or the ability to specifically react with with one isomer of a mixture of stereo-isomers.

Most enzymes are very selective, although it should be kept in mind that, in order to be useful, enzymes should both be very selective *and* accept a wide range of substrates, which is a *contradictio in terminus*.

5. Biocatalysts are cheap catalysts because they can be produced by fermentation. This is generally true for microbial cells or enzymes but not for plant cells (*vide infra*).

Of course, biocatalysts have also some disadvantages. They are sensitive to denaturating conditions like extreme pH values, heat and specific organic solvents. At the end of the reaction, they are difficult to recover from the reaction medium without inactivation. They frequently suffer from inhibition phenomena which hampers conversions in concentrated solutions. However, in many cases it is possible to circumvent these problems by, e.g., buffering, immobilisation or working in two-phase systems using hydrophobic organic cosolvents.

1.3. Plant Biocatalysis Versus Microbial Biocatalysis

Up to now, the large majority of enzymes have been obtained from microbial sources. The reason for this is that microbial cells have doubling times in the range of hours, so they can be produced in huge amounts by fermentation. Furthermore, it is relatively easy to treat them in such a way (*e.g.*, by genetic engineering) that they produce the desired enzymes in copious amounts. A minor fraction of enzymes is obtained from plant sources (14). Undifferentiated plant cells have doubling times of days and produce the desired enzymes mostly in minute amounts. Quite often, the required enzymes are not formed at all because the corresponding genes are only turned on in differentiated tissues (see Chapter 2).

Nevertheless, the plant kingdom is an important source for the chemist and the biotechnologist because plants produce a unique variety of chemicals. The rationale for this is simple: there is no way plants can escape from their predators, so they have to defend themselves by chemical means. Plants make compounds with the most fantastic chemical structures, as is shown below in Figure 1: the antitumour drug Taxol® [1] (from *Taxus brevifolia*), the insect feeding deterrent azadirachtin [2] (from the Indian neem tree *Azadirachta indica*) and the analgesic morphine [3] (from *Papaver somniferum*).

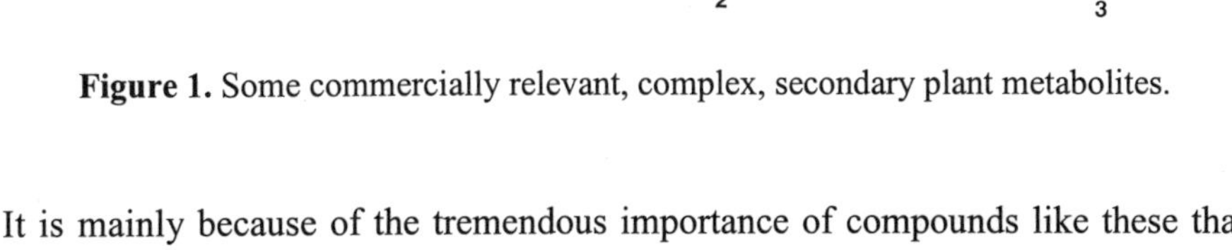

Figure 1. Some commercially relevant, complex, secondary plant metabolites.

It is mainly because of the tremendous importance of compounds like these that science has been, and still is, interested in the application of plant cells and enzymes, irrespective of the disadvantages they sometimes have.

Many different enzymes are involved in the biosynthetic pathways of complex plant metabolites like the ones depicted above. The enzymes catalyse, among others, stereo- and regioselective hydroxylations, (ep)oxidations, esterifications and cyclisations. However, one should keep in mind that not all reactions in plants are enzyme-catalysed! *E.g.*, the alkaloid nitraramine [2] (see Scheme 1), which contains

7 stereogenic centres, is present in the plant *Nitraria schoberi* as a racemate. Recent investigations showed that this complex alkaloid can be formed spontaneously at pH 7 from the achiral precursor [1], which in its turn can be biosynthesised from lysine in several steps (20). Its postulated formation is depicted in Scheme 1.

Reactions like the one depicted below prove that caution has to exercised in postulating the role of enzymes in biosynthetic routes. Also the isolation of a *chiral* metabolite might, in principle, be due to spontaneous, nonenzymatic reactions starting from an achiral precursor followed by specific enzyme-catalysed metabolism of *one* of the enantiomers into a different compound (21).

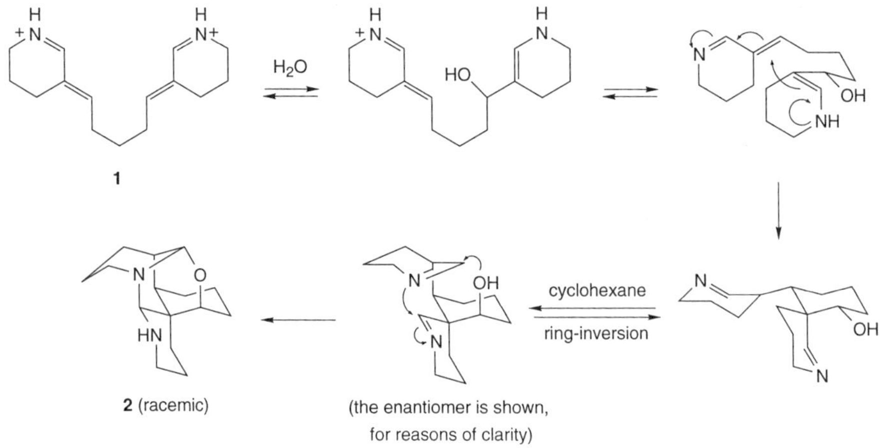

Scheme 1. Spontaneous formation of racemic nitraramine [2] from a simple achiral precursor.

1.4. Immobilisation of Biocatalysts

In order to be useful in synthetic biotransformations, it is a prerequisite that the applied biocatalyst is reusable and stable. This can be achieved by binding the biocatalyst to a solid support or entrapment in a matrix. These so-called immobilised biocatalysts are easy to recover after the reaction by filtration. Immobilisation often leads to increased stability, though this is not a general rule. However, immobilisation frequently results in lower reaction rates, mostly due to diffusional limitation.

Apart from the improved stability and facilitated downstream processing, immobilisation of plant cells has two additional advantages (22). Firstly, immobilised plant cells are in physical contact with each other, which induces some differentiation of the cells, in comparison to calli. It has been observed many times that differentiated plant cells have a (much) higher production of secondary metabolites. Secondly, not all cells in these immobilised cell preparations are equally accessible to nutrients etc., which leads to concentration gradients of chemical compounds. Gradients influence the expression of biosynthetic pathways essential to secondary metabolite production and accumulation. This, again, often leads to improved productivity of immobilised cells with respect to cells which are freely suspended (23).

A few systems which are often applied for the immobilisation of enzymes and (plant) cells are given below.

- Adsorption

Enzymes can be adsorbed to insoluble supports. The forces that are responsible for the binding between the enzyme and the support are hydrogen bonds, dipole-dipole interactions and hydrophobic interactions. The immobilisation procedure is very mild and consists of simple mixing of the enzyme and the support and rotating them together slowly overnight. A disadvantage is the fact that the enzyme is not tightly bound to the support and may detach at high ionic strengths. Commonly used supports are polypropene (e.g. Accurel™) and diatomaceous earth (Celite).

Solid surfaces can also be used for the adsorption of cells, including plant cells (24).

- Ionic binding

When the optimum pH of an enzyme is not too close to its isoelectric point, enzymes are charged and may be bound to oppositely charged solid supports. The procedure is similar to the one described above and is equally mild. The enzymes detach at high ionic strengths and at pH values where the support and the enzyme have the same charge. Popular is DEAE-cellulose, well-known from ion-exchange chromatography of proteins.

- Covalent binding

The best way to overcome leaching of enzyme is to link it covalently to a solid support, controlled-pore glass (CPG) being one of the best. However, the enzyme has to undergo a chemical reaction in that case. This often leads to some inactivation of the enzyme. Many different methods exists; for an overview the reader is referred to the old but still very useful books of Gutcho (25) and Zaborsky (26).

- Entrapment

A very attractive way for the immobilisation of whole cells is entrapment. In this technique, the cells are brought into a "cage" which is accessible to substrates but from which the cells themselves cannot escape. This cage can be an organic polymer like polyurethane or a biopolymeric gel like alginate or κ-carrageenan. Generally, immobilisation yields are high when using entrapment, but the growth rate and the production rate of the cells diminishes because of diffusional limitation of the nutrients and substrates.

This is not completely valid for plant cells because rapidly growing cells frequently do not produce much, if any, secondary metabolite (27). Minimal growth rate favours secondary metabolite production. Therefore, culturing plant cells in a matrix can be beneficial. The cells are encouraged to grow together in a partially organised multicellular system, where the physical and chemical gradients that are brought about approach those existing in the differentiated plant. The achievement of the stationary phase of the cell growth cycle, where cell division is minimal, for the production of secondary metabolites can be induced by placing the cells in stress conditions. The cells can then be maintained in the non-replicating state which will prevent them outgrowing the supporting immobilisation matrix. After some time it is necessary to revitalise the cell culture by reintroducing the standard growth media. Some entrapment media are treated below.

Plant cells have large dimensions, compared to microorganisms, and possess a rigid, thick cell wall. This makes plant cells sensitive to shear forces, which implies that care has to be taken during their immobilisation (28). Entrapment in gels is the most popular way to immobilise whole cells, including plant cells, because it is a cheap, simple and reproducible technique with mild conditions during the immobilisation (28). In particular, alginate is a commonly used matrix. Alginate is a polysaccharide which can be isolated from several kinds of macroalgae. It consists of regularly alternating units of β-(1→4)-D-mannuronic acid and α-(1→4)-L-guluronic acid. Its sodium salt is readily soluble in water, but its calcium salt is insoluble. For immobilisation, a suspension of plant cells is mixed with a sodium alginate buffer and dropwise extruded through an orifice or a hollow needle. The formed drops are caught in a calcium chloride solution. As soon as the drops fall into the calcium chloride solution, sodium and calcium ions begin to exchange. An insoluble calcium alginate gel is formed at the outer surface of the drop, resulting in a spherical particle. The bead is further hardened by remaining in the salt solution, so that the calcium ions can be transported to the centre of the bead until an equilibrium state has been achieved. A procedure for the upscaling of this immobilisation technique is available (29).

κ-Carrageenan is another useful algal polysaccharide, although of a less well defined composition. Immobilisation in this matrix is carried out in the same way as described for alginate, but now ammonium or potassium ions are needed for gelification. A drawback of both polysaccharides is the fact that complexing agents (like phosphate in the case of calcium alginate) lead to dissolution of the gel. When this is a problem, other entrapment gels like agarose and gelatin can be applied.

A very effective method is the entrapment of cells in a three-dimensional reticulate, e.g. plastic foam, pan scrubbers and steel knit mesh. Once cells are lodged into the voids of the reticulates they may multiply until the aggregate is entrapped. When this method of immobilisation is used, cell growth and cell production can be combined in one reactor, with time intervals (27). Polyurethane reticulates are most frequently used for this purpose. The immobilisation procedure is very simple; the cells migrate into the foam without any force, so the viability of the cells is not influenced by changes in temperature or ionic strength.

A totally different way of immobilisation is the entrapment of cells in membrane reactors. Shuler *et al.* (30) defined five possible configurations, namely hollow fibre units, flat plate systems, spiral wound units, tubular membrane reactors and multimembrane reactors. Membrane reactors are based on the fact that plant cells are physically separated from the liquid production medium by a membrane, allowing substrates, nutrients and products freely to pass. In comparison to plant cells immobilised in gels, membrane systems allow better control of fluid dynamics and flow distribution, direct access to the cells during actual operation and an easier scale-up (30). Another important advantage is the fact that the membrane protects the plant cells against intrusion of foreign cells, which makes sterile operation very easy (31). Kim *et al.* (32) used a dual hollow fibre reactor to grow their *Lithospermum erythrorhizon* cells (see Figure 2). With this system, they reached very high biomass and secondary metabolite concentrations: up to 325 g/l dry biomass and 221 mg/l.day of products. This cell density is one of the highest ever reported. High cell density is an important parameter to increase the overall volumetric productivity of plant cell bioreactors.

Nevertheless, membrane bioreactors do have some disadvantages, like any other immobilisation technique: they are expensive, liable to fouling and may suffer from gas transfer problems.

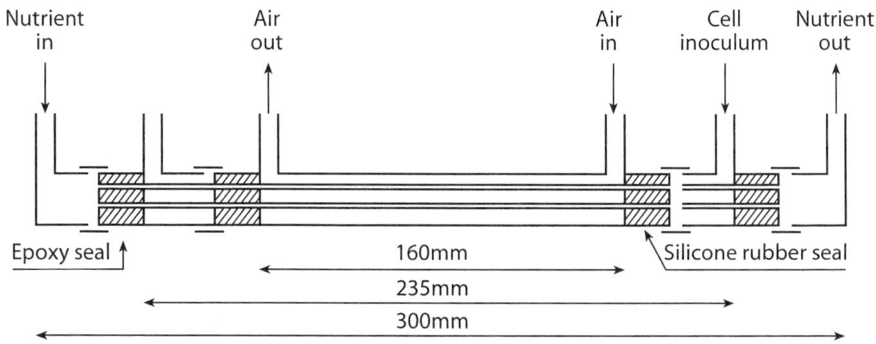

Figure 2. Dual hollow fibre bioreactor, as used by Kim *et al.* (32) for the immobilisation of *Lithospermum erythrorhizon* cells. Reproduced with permission.

- General remarks

As will be clear from this overview, many different systems exist for the immobilisation of plant enzymes and plant cells. It is not possible to give general guidelines on which support/matrix to choose; this remains a matter of trial and error. Although each method has its own advantages and disadvantages, it is nearly always possible to find an immobilisation method which gives a stable preparation with acceptable loss of activity.

2. Bioconversions Using Plant Cell or Organ Cultures

2.1. Some General Considerations

Whole plant cells or organs - whether immobilised or not - are in principle capable of performing a wider range of bioconversions than isolated plant enzymes. Such bioconversions may be multi-step, amounting to partial biosynthetic pathways. They may be catalysed by enzymes not yet successfully isolated or characterised. Cells also possess accessory enzymes for the provision and recycling of cofactors such as NAD so that the need for any additional cofactor is in principle obviated. These are advantageous characteristics of whole cells or organs.

Other properties, however, may be disadvantageous in relation to enzymes. First, cells and organ cultures are subject to risks of microbial contamination which can only be contained by rigorous adherence to sterility. Secondly, although cells and cultures are not subject to denaturation (though they may well be subject to mechanical shear damage), the sustained use of cells or cultures as biocatalysts implies, *sensu stricto*, that their growth should be restricted, particularly if immobilisation is used. This usually requires starving the cells of some essential component of the medium, such as phosphate; and this need not necessarily maintain the cells in the best physiological condition for the bioconversion to occur. Sometimes, the cells will need to be grown up initially in a medium different from that employed to perform the bioconversion. Next, substrates may be toxic, or the cells may be impermeable. This may limit the potential for bioconversion reactions. Terpenes and other components of essential oils are often toxic at mM concentrations, and it may be necessary to introduce them continually into the flask or bioreactor at low concentrations to avoid phytotoxic effects. Release of the product into the medium is essential if reuse of the cell or organ culture is contemplated, a very important consideration for a commercial process. In principle, this might be manipulated by the use of an external "product sink", such as an Amberlite™ XAD resin, which can continuously deplete the concentration of product in the medium (33), or in the case of volatile products by pervaporation or pertraction (34), but obviously it is essential for the substrate not to behave similarly. Direct approaches to the product-release problem, seeking to permeabilise cells with low concentrations of detergents, for example dimethylsulphoxide, have on the whole been unpromising (35). Finally, though a gross generalisation, the behaviour of cell cultures tends to be less reproducible than that of either enzymes or microorganisms.

Variation and instability are notorious problems in plant cell culture and may impose a requirement for continual screening to maintain highly-productive cell lines (3-7). This is likely to be of lesser importance where cultures are used to perform biotransformations than where they are used to carry out biosyntheses *de novo*. Some of the variation is genetic and involves chromosomal changes and instability; some in contrast may be epigenetic. In favourable cases, cell lines have maintained their desirable biosynthetic characteristics over a period of years, comparing well in stability and reproducibility with microbial cultures. Differentiation enhances genetic stability, however, and this is one advantage of organ cultures, particularly those established by genetic transformation with strains of the crown-gall bacterium, *Agrobacterium tumefaciens* or the related "hairy root disease" bacterium, *A.*

rhizogenes. The genetic transformation results in shoot or root cultures which, unlike untransformed plant cell or organ cultures, are independent of added plant-growth substances including auxins and cytokinins. In principle and with regular subculturing they can be maintained indefinitely on a simple sucrose- and mineral-salts-based medium. "Hairy root" cultures in particular have been studied in many laboratories (3, 36, 37) and have been established from a range of plant genera. In general, their biosynthetic characteristics seem to resemble those of normal plant roots. A further attractive characteristic associated with these cultures is that the *Agrobacterium* strains used to establish them are genetic vectors which can be used to introduce foreign genes.

Biotransformations with cell or organ cultures can be separated into two broad - and by no means separate - categories. The first includes applications of a characteristic biosynthetic capacity of the plant - a "partial biosynthesis", or a reaction sequence using an analogue of a normal biosynthetic intermediate. The second consists of apparently non-specific reactions. Many plant cells have the capacity to biotransform or derivatise uncharacteristic substances. In some cases, too, "characteristic" substances may be biotransformed in an "uncharacteristic" way - for example if a substance formed endogenously is provided in a high or toxic concentration exogenously; a good example is the formation of norharman by *Cinchona* cultures supplied with high concentrations of tryptophan (38). Both categories of reaction are potentially useful and are discussed here. For the sake of completeness and comparison, some reactions corresponding to *in vivo* bioconversions but occurring *in vitro* in extracts of cell cultures are also included.

2.2. Pathway Biotransformations

Biotransformations of pathway intermediates have been studied for many years. For instance, digitoxin (Figure 3) and its 12-hydroxy derivative, digoxin, are both produced by plants of *Digitalis* spp. and remain important cardioactive drugs. Since digoxin is the preferred drug for pharmaceutical application, yet is present in the plant at lower concentrations than digitoxin, there is interest in the biotransformation of digitoxin to digoxin. The study of *Digitalis* cultures for the production and biotransformation of digitoxin and its aglycone, digitoxigenin, was amongst the very early applications of plant cell culture (39). The 12-hydroxylation of digitoxin, including its scale-up in bioreactors, has continued to be studied in several laboratories. Levels of 0.6 mmol per litre of digoxin have been reported in 300-litre airlift bioreactors supplied with digitoxin at 0.65-0.8 mmol per litre (40); approximately 80% of the digoxin produced was in the medium. Digitoxin 12β-hydroxylase is a cytochrome P450 monoxygenase (see Section 3.8).

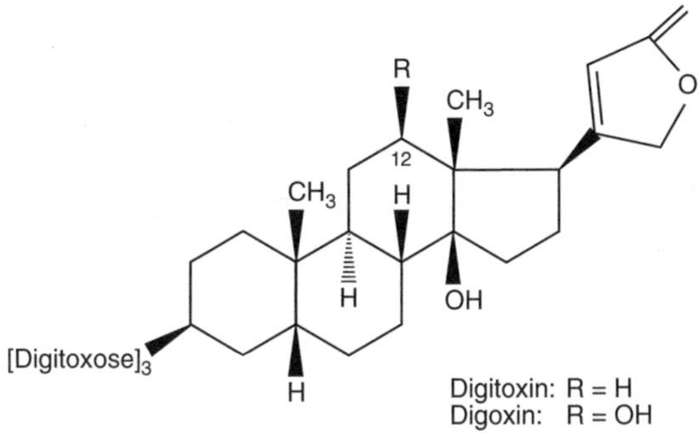

Figure 3. Digitoxin and digoxin.

An entirely different example of a pathway biotransformation is provided by the conversion to raumacline (and other minor products) of the monoterpenoid indole alkaloid, ajmaline, in cell suspension cultures of the Indian medicinal plant, *Rauwolfia serpentina* (41). The biotransformation could be mimicked on a gram scale by borohydride reduction of ajmaline to dihydrochanoajmaline, followed by a flavin-mediated photooxidation to raumacline.

In agreement with this chemical transformation, the oxidative conversion of the intermediate compound, dihydrochanoajmaline, to raumacline could be demonstrated both *in vivo*, with cell suspension cultures, and *in vitro*, using cell-wall preparations in the presence of hydrogen peroxide. Elucidation of the details of the overall biotransformation was achieved by studies of cell-free preparations *in vitro*. In contrast to the chemical route, no reductive conversion of ajmaline to dihydrochanoajmaline could be demonstrated and, instead, the identity of the intermediate was demonstrated to be 21-hydroxyraumacline, produced by an initial oxidation of ajmaline - i.e. a reversal of the order of reductive and oxidative steps demonstrated in the chemical route. Subsequently, a (slow) biotransformation of 21-hydroxyraumacline to raumacline by cell suspension cultures was demonstrated. The second reaction involved in the overall biotransformation from ajmaline has been demonstrated and the enzyme responsible, 21-hydroxyraumacline reductase, purified and characterised. This example is illustrative on two counts: first, it indicates that a *chemical* conversion is not necessarily a reliable guide to an analogous *bio*-conversion; secondly, it illustrates that the ability to demonstrate the bioconversion of a putative intermediate - in this case, of dihydrochanoajmaline to raumacline - is no guarantee that that conversion occurs *in vivo* as part of an overall bioconversion or biosynthesis.

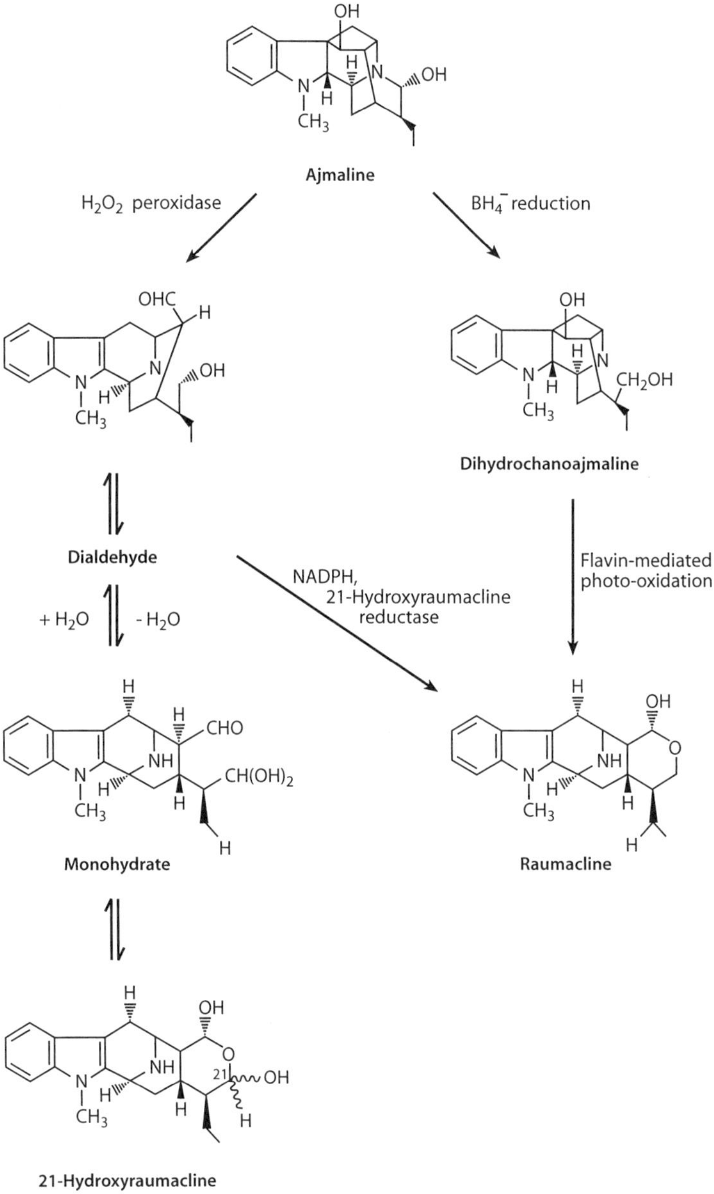

Figure 4. The conversion of the indole alkaloid, ajmaline, to raumacline.

The biotransformation of benzylisoquinoline alkaloids first studied some years ago in cell cultures of *Papaver somniferum* (42,43) illustrates that useful biotransformation capacities may be present even if the complete biosynthetic pathway is absent. These cultures were unable to produce morphinans. Nevertheless, codeinone could be bioconverted to codeine. The biotransformation of codeinone could be carried out using cells immobilised for relatively long periods (a month or more) in calcium alginate, either in flasks or in a bioreactor. The lack of morphinan biosynthesis *de novo* was reflected in the inability to metabolise the key intermediate (-)-reticuline; however, its enantiomer, (+)-reticuline, was found to be biotransformed to (-)-scoulerine by the "berberine-bridge" enzyme, which catalyses a biochemically-unique reaction of the benzophenanthridine alkaloid pathway (Figure 5).

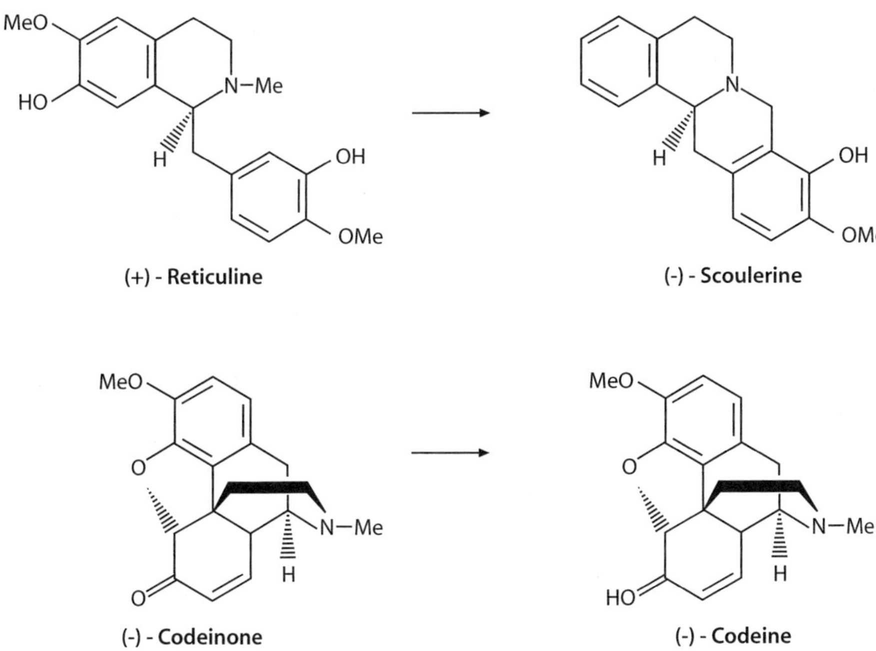

Figure 5. Biotransformations of the benzylisoquinoline alkaloid pathway.

Peroxidase-catalysed reactions, and in particular, radical-coupling reactions, are prominent in secondary product biosynthesis, most notably in the formation of lignin, lignans and suberin (see Chapter 2). Although the details of their enzymology and stereospecific control often remain incompletely characterised, reactions of this type can be utilised to perform *in vivo* biotransformations of biotechnological interest. The properties and uses of isolated (halo)peroxidases are discussed in Section 3.6. A good example of their application in cell culture is given by the case of the lignan, podophyllotoxin, the precursor of the semi-synthetic anti-cancer drug, etoposide. This compound is derived chemically from podophyllotoxin by a two-step conversion to 4'-demethylepipodophyllotoxin, followed by a five-step reaction series to attach the carbohydrate unit (44). The extraction of podophyllotoxin from its source plant, *Podophyllum peltatum*, could in principle be obviated if a biosynthetic precursor, synthesised chemically by a cost-effective procedure, could be supplied to *P. peltatum* cell cultures able to catalyse as a biotransformation the final steps of the overall biosynthetic pathway. Dibenzylbutanolides are putative biosynthetic precursors of podophyllotoxin (45) and its derivatives and can be produced chemically from aromatic aldehydes (46). Kutney and his co-workers demonstrated that a cell line of *P. peltatum* active in the biosynthesis of podophyllotoxins was able to maintain repeated biotransformation, by oxidative coupling, of the butanolide [1] to the podophyllotoxin analogue [2] (Figure 6) in a bioreactor, for a total of more than 15 cycles, each of 24 h or more (46, 47). The yield was around 50%. This result indicates the feasibility of a biotransformation route to podophyllotoxin using a chemically-synthesised precursor, although the optical activity of the product was not identical to that of natural podophyllotoxin.

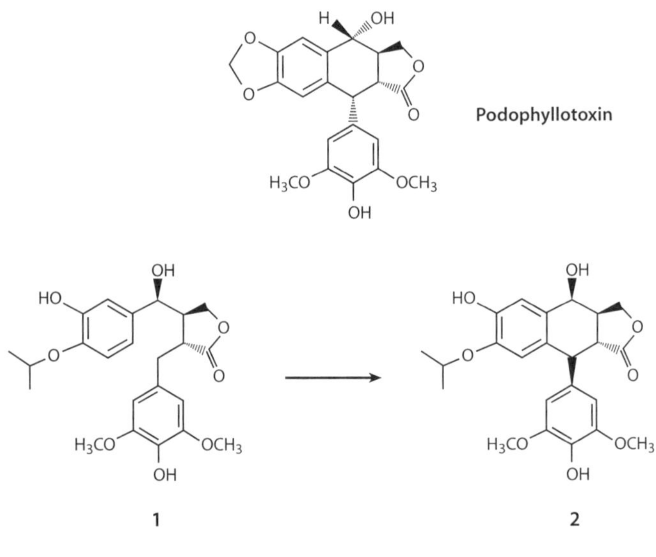

Figure 6. The butanolide analogue [1] and the podophyllotoxin analogue [2] produced from it by peroxidative ring-closure.

Though a "partial biosynthesis", or a close analogue of one, this reaction also provided an example of a non-specific biotransformation, since the reaction could be performed by a cell suspension culture of *Nicotiana sylvestris* (47). This was attributed to the extracellular secretion of peroxidase(s) into the culture medium and the reaction was easily demonstrated using culture broth from which the cells had been removed. Under appropriate conditions, an 87% conversion could be demonstrated within 40 minutes. It seems probable that the reaction could be performed using whole cells, cell-free extracts or culture broth from cell suspension cultures of a range of species, owing to the widespread distribution - even ubiquity - of peroxidases amongst plants (see Section 3.6). The use of cell-free extracts from cultures of *Cassia didymobotrya*, for example, has also been demonstrated (48).

A peroxidase is responsible for the coupling of the monoterpenoid indole alkaloids, catharanthine and vindoline, to form the precursor of the pharmaceutically-important - but non-abundant - anti-leukaemic bis-indole alkaloid, vinblastine, in the Madagascar periwinkle, *Catharanthus roseus* (Figure 7). With an immobilised enzyme preparation from a cell culture of *C. roseus*, a 90% coupling of catharanthine and vindoline could be achieved within 15-30 minutes (49). More protracted incubations in the presence of NADH led to a sequence of two intermediate compounds, followed finally by vinblastine itself. In this case, however, study of the *in vitro* biotransformation was used as the basis for devising an entirely chemical

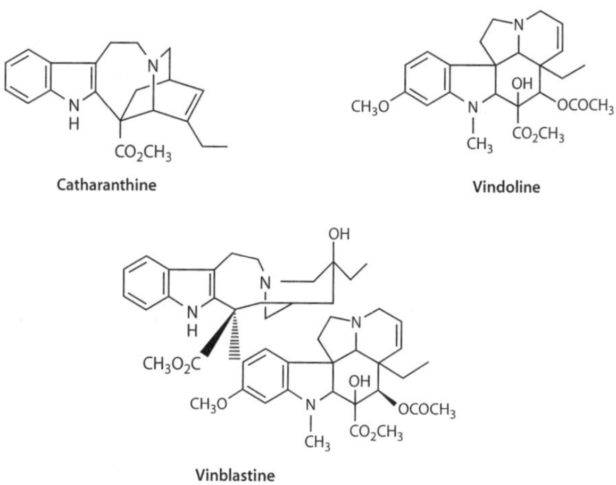

Figure 7. The indole alkaloids, catharanthine and vindoline, and the bis-indole alkaloid, vinblastine, derived from them.

route (47,50). Catharanthine was first oxidised to catharanthine *N*-oxide and the coupling with vindoline was achieved with trifluoroacetic anhydride as a catalyst. The subsequent NADH-dependent reduction was achieved with a dihydronicotinamide analogue of the physiological reductant, NADH. The final two reactions to vinblastine, an oxidation and a reduction, were brought about by the use of ferric chloride and sodium borohydride, respectively. The overall yield of the five sequential reactions, occurring as a "one-pot" synthesis, was 40%.

Triptolide and tripdiolide are diterpene triepoxides of the Chinese medicinal plant, *Tripterygium wilfordii*. They show, in particular, strong immunosuppressive activity. As in the case of the biotransformation to produce podophyllotoxin from a synthetic butanolide, so it was of interest to discover whether a cell line of *T. wilfordii* was capable of the biotransformation - in this case, epoxidation - of a putative metabolic precursor accessible by chemosynthetic means (51). Isotriptophenolide, produced chemically by a multi-step sequence from dehydroabietic acid, was fed for 4 days to 15-21-day-old cell cultures of *T. wilfordii*. Several biotransformation products were produced, the two principal compounds being the mono-epoxides shown in Figure 8; unfortunately, however, no detectable amounts of the di- or tri-epoxides appeared to be formed (52).

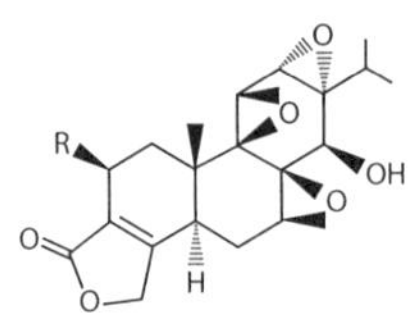

Triptolide (R = H)
and
Tripdiolide (R = OH)

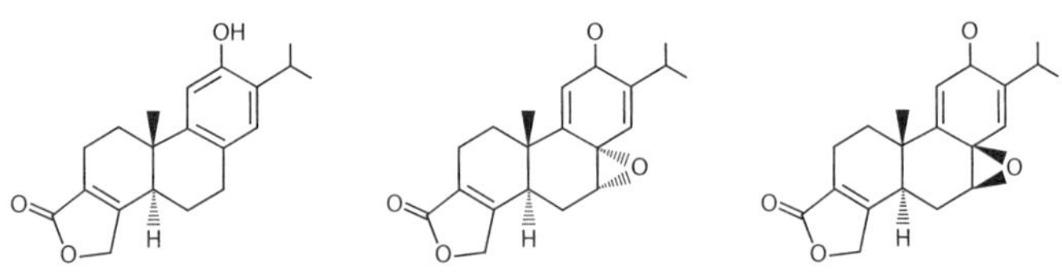

Isotriptophenolide 1 2

Figure 8. The diterpenoid triepoxides, triptolide and tripdiolide, and the monoepoxide bioconversion products [1] and [2] of isotriptophenolide.

Biotransformations of metabolite analogues have been studied in transformed ("hairy") root cultures of solanaceous species. When cadaverine (1,5-diaminopentane) was fed to "hairy root" cultures of *Nicotiana* species, the formation of the usual alkaloid, nicotine, derived from putrescine (1,4-diaminobutane), was inhibited and that of the minor - often undetectable - alkaloid, anabasine, was markedly stimulated (Figure 9; 53). The basis of this effect appeared to lie in the competition between cadaverine and *N*-methylputrescine and their respective metabolites as substrates for the enzymes of the nicotine biosynthetic pathway.

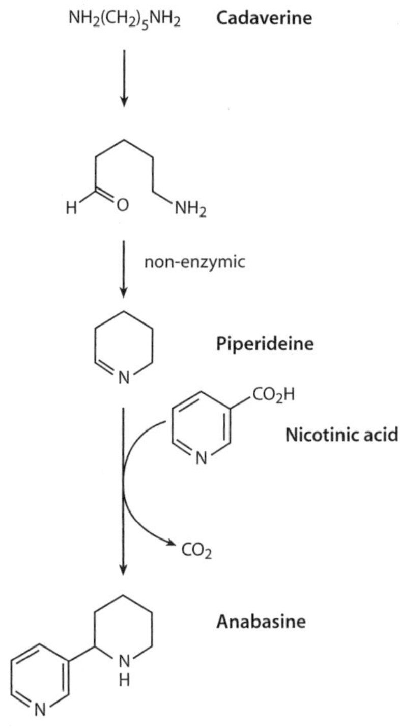

Figure 9. The bioconversion of cadaverine to anabasine.

Because cadaverine is a product of the decarboxylation of the ubiquitous amino acid, lysine (*cf.* putrescine from ornithine), its production could be engineered by the introduction and expression of a heterologous gene for lysine decarboxylase (*ldc*). This was demonstrated by Fecker *et al.* (54), who expressed an *ldc* gene from the bacterium, *Hafnia alvei*, in *Nicotiana* "hairy root" cultures and showed a 10-fold increase in cadaverine production and a 3-fold increase in anabasine. The effects of the introduced gene were further enhanced by directing the expressed protein to the leucoplast where lysine is synthesised (55). This example illustrates how a biotransformation can suggest a genetic-manipulation approach, increasing the supply of an unavailable or severely rate-limiting precursor in order to express a "latent" biosynthetic pathway.

The isomer of cadaverine, *N*-ethylputrescine, is also biotransformed by "hairy root" cultures of *Nicotiana* and in this case gives the anticipated higher homologue of (*S*)-nicotine, *N*-ethyl (*S*)-nornicotine (56). When fed at 1 mM concentration, *N*-ethylputrescine was converted to *N*-ethyl (*S*)-nornicotine in 6% yield. The initial stages of tropane alkaloid biosynthesis are closely similar, if not identical, to those of nicotine biosynthesis and in an analogous study, *N*-ethylputrescine was found to be metabolised by "hairy root" cultures of a *Brugmansia candida x aurea* hybrid active in the biosynthesis of tropane alkaloids (H.D. Boswell and coworkers, unpublished work). Analogues of the normal metabolic intermediates throughout the pathway were detected by GC/MS, including an analogue of the most distant metabolite, the tropane alkaloid, scopolamine. The metabolism of higher *N*-alkylputrescines was less extensive and led to the formation only of analogues of relatively "early" intermediates of the pathway. The main value of such multi-step analogue biotransformations at present is as a means of exploring the plasticity of secondary metabolism and the substrate specificity of as-yet-uncharacterised enzymes; their biotechnological potential remains unproven.

2.3. "Non-specific" Biotransformations

The number of individual biotransformations reported to be carried out by plant cell cultures is large - considerably larger than the number of species reported to perform them. Most, however, fall into a relatively small number of defined classes. A comprehensive review of these reactions was prepared some years ago (57). In the account that follows, some of the major classes of "non-specific" biotransformations will be summarised according to reaction type, with illustrative examples.

Hydroxylations - Many plant cell cultures can hydroxylate exogenous substances. These reactions commonly exhibit regio- and stereo-selectivity, enhancing their potential value to the chemist, and several compounds can often be formed from a single substrate. Partly on account of their importance in the fragrance and flavour industries, the hydroxylation and further biotransformation of terpenoids has been particularly well studied, using cell cultures of essential-oil-producing species and of other plants, such as *Nicotiana tabacum* (57). Stereospecificity was shown, for example, by the formation only of *trans* isomers from hydroxylation occurring at C-4 of β-terpineol and its acetate and by the predominant formation of a *trans*-diol as a result of hydroxylation of the endocyclic double bond of α-terpinyl acetate (Figure 10). Enantioselectivity was apparent in the hydroxylation of enantiomers of α-terpineol and its acetate: hydroxylation at the 6-position took place principally with the (4*R*)-isomers. In contrast, diol formation at the double bond of α-terpinyl acetate occurred mainly with the (4*S*)-isomer. It was suggested that the difference in enantioselectivity between the hydroxylation occurring allylic to the double bond (α-terpineol) and at the double bond itself (α-terpinyl acetate) implied that different enzyme systems were responsible (57, and *loc. cit.*).

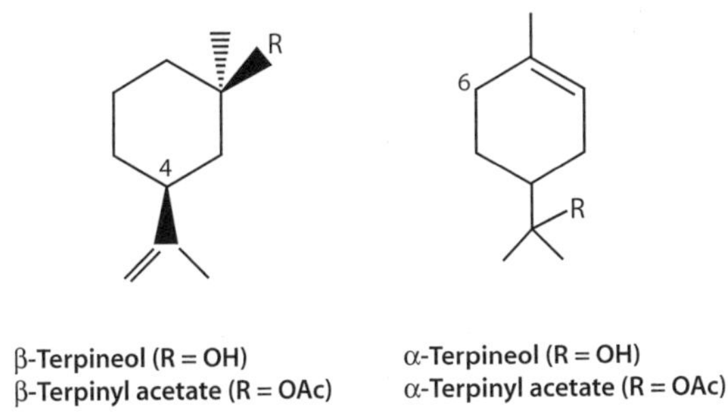

Figure 10. α- and β-Terpineol and their acetate esters.

Eucalyptus plants contain an essential oil, a principal component of which is 1,8-cineole. Cell cultures of *Eucalyptus perriniana* can biotransform a range of monoterpenoids, including 1,8-cineole itself, (+)- and (-)-menthol, (-)-borneol, (+)- and (-)-fenchone and (+)-camphor. The initial step in the biotransformation of 1,8-cineole, (-)-borneol, the fenchones and (+)-camphor is hydroxylation of a methylene carbon. Analysis of the pattern of hydroxylation of each of these substrates revealed that hydroxylation occurred principally on the opposite side of the molecule to the pre-existing oxygen function and it was suggested that this might reflect the structure of the binding site of the hydroxylase(s) responsible (58). In view of the natural occurrence of 1,8-cineole, these reactions in *E. perriniana* cultures might also be considered "pathway biotransformations". These hydroxylations appear to be carried out by cytochrome P450 monooxygenases. Further reference to cytochrome P450 enzymes is made in Section 3.8.

Glycoside formation - Glycoside formation increases the hydrophilicity of many substances, reducing their ability to dissolve in membrane lipids, and is of substantial physiological importance. It is now clear that glycosides, especially of glucose - glucosides - play a mainstream role in secondary metabolism - for example in the biosynthesis of lignin (59) and anthocyanins (60). The ability of plant cell cultures to form β-D-glucosides of exogenous substances is widespread and, as with hydroxylation, is not limited to substances known to be characteristic of the plant species in question. Cell cultures of *Rauwolfia serpentina* glucosidate a variety of simple phenolic substances (61). The extent of the reaction can be substantial: for example, cell cultures of *Gardenia jasminoides* were able to glucosidate about 70%

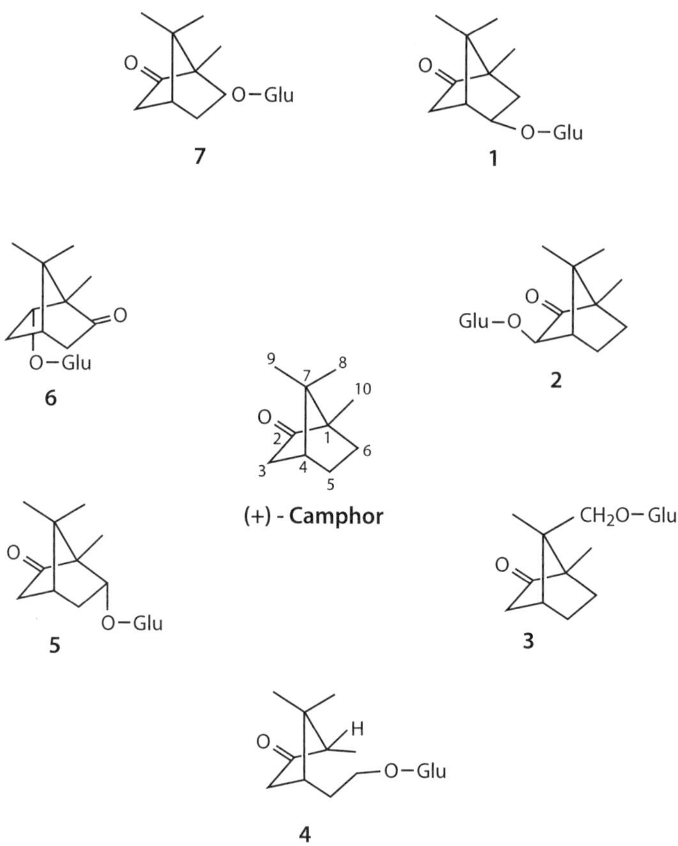

Figure 11. Bioconversion products of (+)-camphor in cell cultures of *Eucalyptus perriniana*.

of added salicyl alcohol to salicin within four days (62). When hydroxylation of an exogenous substance occurs, glucoside formation very frequently follows. In many cases, the initial hydroxylated intermediate may not be detected at all. Figure 11 shows the range of products which were obtained following the feeding of (+)-camphor to cell cultures of *E. perriniana* (63).

The putative hydroxylated precursors of the seven glucosides formed were not found in the cell culture. The biotransformation of 1,8-cineole by the same cell line similarly led to a range of β-D-glucosides (Figure 12), without any accumulation of the initial hydroxylated products (64). As indicated above, hydroxylation and glucoside formation occurred at positions opposite to the oxygen function. The total extent of biotransformation was 15-20%.

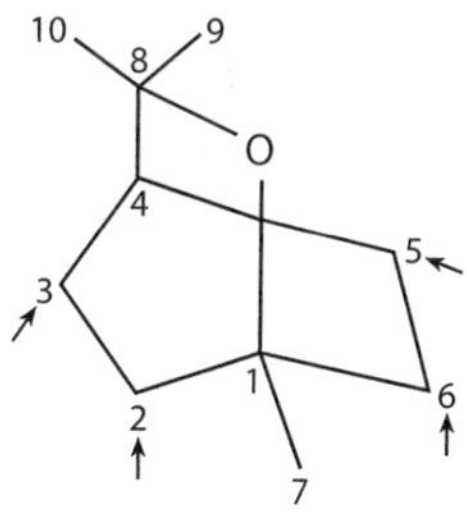

Figure 12. β-Glucoside formation following hydroxylation of 1,8-cineole by cell cultures of *E. perriniana*. Arrows indicate positions derivatised.

A potential biotechnological application of glucoside formation is that it affords a means of masking the properties of an aglycone, which can subsequently be regenerated by glucosidase action as required. For example, cell suspension cultures of *Coffea arabica* were found to be able to β-D-glucosidate capsaicin, a medium-chain fatty-acyl amide of vanillylamine (Figure 13) which is the pungent principle of chilli peppers (65). The extent of conversion was only about 7%, and the β-D-glucoside was not stable in the culture. The pungency of this compound was estimated by a taste panel to be about 100-fold less than that of capsaicin itself and its solubility in water was considerably increased. Treatment *in vitro* with a β-glucosidase released capsaicin, restoring the property of pungency. Several other plant cell cultures, including those of the chilli pepper, *Capsicum annuum*, were found in this work to be incapable of glucosidating capsaicin, but it has recently been shown that capsaicin can be β-glucosidated by cell cultures of *C. annuum* and that small quantities of the compound might also be present in *Capsicum* fruits (G. Calva-Calva, unpublished work). As indicated above (61,62), many simple phenolic substances can be rapidly β-glucosidated by plant cell cultures. Vanillin β-D-glucoside is a naturally-occurring compound in freshly-harvested *Vanilla* pods which, prior to "fermentation", contain virtually no free vanillin (4-hydroxy-3-methoxybenzaldehyde); vanillin accumulation is the result of deglucosidation, apparently occurring as the result of the action of a specific β-D-glucosidase (A. Narbad, unpublished work). Vanillin fed to *Nicotiana plumbaginifolia* cells was found to be biotransformed principally to vanillin-β-D-glucoside; however, in addition, the two possible β-D-glucosides of vanillyl alcohol were found, indicating reduction of the aldehyde group of vanillin (66). Almost all of the vanillin supplied was converted to intracellular glucosides. This example is of particular analytical interest, since the biotransformations were followed *in vivo* using inverse-correlated 2D ^{1}H-^{13}C NMR spectroscopy at the natural abundance of ^{13}C. The 2D NMR method was able to detect bioconversions in the millimolar range and demonstrated a significant advance in sensitivity and acquisition time relative to natural abundance ^{13}C NMR.

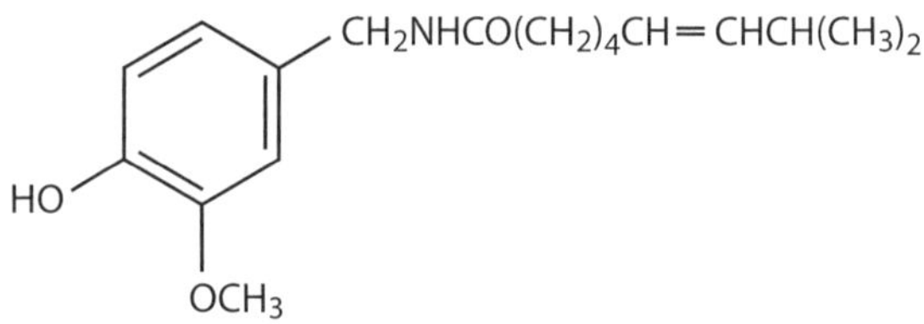

Figure 13. Capsaicin, the pungent principle of chilli peppers.

The applications of plant glucosidases *in vitro*, including their application in low-water media to carry out the reverse of their normal reaction, i.e. glucoside formation, is discussed below in Section 3.5.

Redox reactions - Hydroxylation is frequently also succeeded by oxidation, as shown for example in the biotransformation of Δ^2-carene by dissociated callus cultures of *Myrtillocactus geometrizans* and *N. tabacum* (67).

Cultures of both plant species biotransformed Δ^2-carene into a pair of diastereomeric alcohols (Figure 14) and in the case of *Myrtillocactus* these alcohols were substantially oxidised to the corresponding ketones. Such oxidations are very

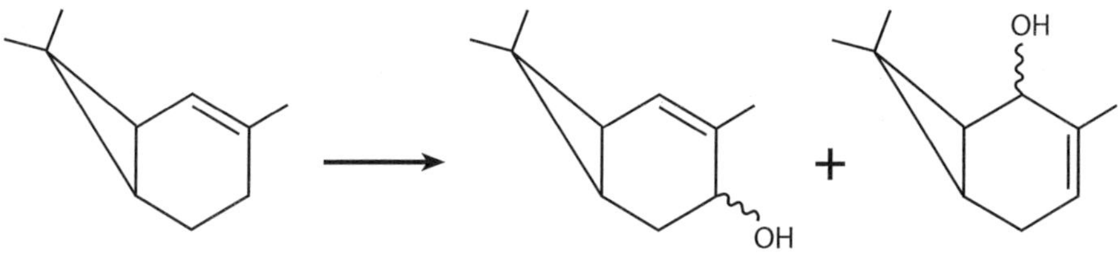

Δ^2-Carene Diastereomeric Carenols

Figure 14. Diastereomeric alcohols formed from $\boldsymbol{\Delta}^2$-carene.

frequently enantioselective. For instance, cell cultures of *N. tabacum* converted (1*R*,2*S*,4*R*)-borneol present in (*RS*)-borneol or (1*R*,2*R*,4*R*)-isoborneol, present in (*RS*)-isoborneol into (+)-camphor (Figure 15), leaving behind the unchanged substrates, (1*S*,2*R*,4*S*)-borneol or (1*S*,2*S*,4*S*)-isoborneol; optical purities of the product and the remaining starting material were in excess of 90% (57 and *loc. cit.*).

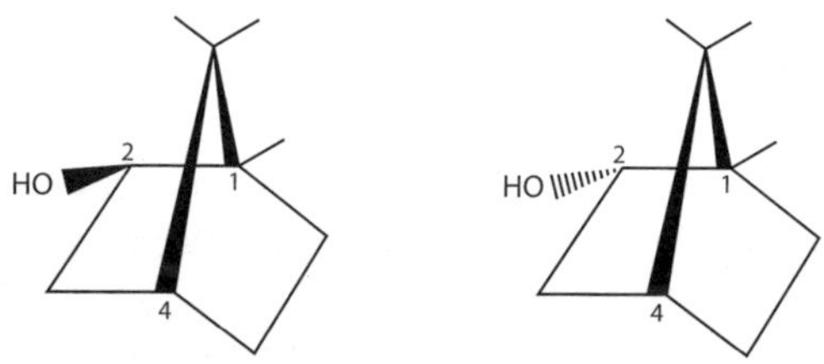

(1*R*,2*R*,4*R*)-Isoborneol (1*R*,2*S*,4*R*)-Borneol

Figure 15. Enantiomers of borneol and isoborneol converted into (+)-camphor.

Conversely, stereospecificity is also shown in the reduction of aldehydes and ketones to alcohols. For example, the reduction of (1*R*,4*R*)- and (1*S*,4*S*)-carvomenthones by cultured cells of *N. tabacum* gave rise to (1*R*,2*S*,4*R*)- and (1*S*,2*R*,4*S*)-neocarvomenthol, respectively (Figure 16), indicating hydrogen attack from the *re*-face of the carbonyl group to produce the alcohols with the *S*-chirality at the carbon atom bearing the hydroxyl group (68,69).

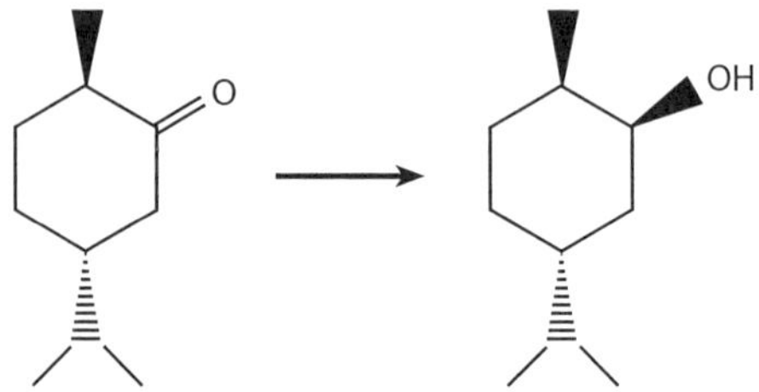

(1*R*,4*R*)-Carvomenthone (1*R*,2*S*,4*R*)-Neocarvomenthol

Figure 16. Carvomenthones and neocarvomenthols.

An interesting example of diastereo- and enantioselective ketone reduction is given by the reduction of the β-keto ester, ethyl 2-methyl-3-oxobutanoate (70). Yields of 85-100% of the β-hydroxy ester were obtained with cultures of *Marchantia polymorpha*, *Catharanthus roseus* and *Glycine max*. The diastereoselectivity varied with the culture; the *Marchantia* culture generated predominantly the *anti*-isomer (96%), whilst the *Glycine* culture produced chiefly the *syn*-isomer (92%). In each case, there was also a high degree of enantioselectivity, since the *anti*-isomer produced by the *Marchantia* culture was almost exclusively (2*S*,3*S*) and the *syn*-isomer produced by the *Glycine* culture was *ca*. 97% (2*R*,3*S*). In this instance, the biotransformations were suggested to offer superior diastereoselectivity compared to that available by chemical reduction using BINAP-Ru complexes.

A number of examples of the reduction of carbon-carbon double bonds have also been reported and regiospecificity and stereospecificity demonstrated. Cultured cells of *N. tabacum* reduced (4*R*)- and (4*S*)- carvone (71), but of the two carbon-carbon double bonds available in the molecule, only that adjacent to the carbonyl group was reduced, to give (1*R*,4*R*)-dihydrocarvone and (1*R*,4*S*)-isodihydrocarvone, respectively (Figure 17). Other carbon-carbon double bond reductions, especially of terpenoids, have been studied in cell cultures of *Mentha*, *Medicago* and *Catharanthus* (58, and *loc. cit.*). Where a carbonyl group is additionally present in the molecule (as in carvone), this is frequently reduced also (*vide supra*).

Hydrolysis - Hydrolyses are of interest because, if enantioselective, they offer the possibility for the resolution of racemates. The hydrolysis of a range of terpenoid esters, including α-terpinyl, bornyl, isobornyl and isopinocampheyl acetates, has been studied in cell cultures of *N. tabacum*. Enantiomers with an *R*-configuration at the

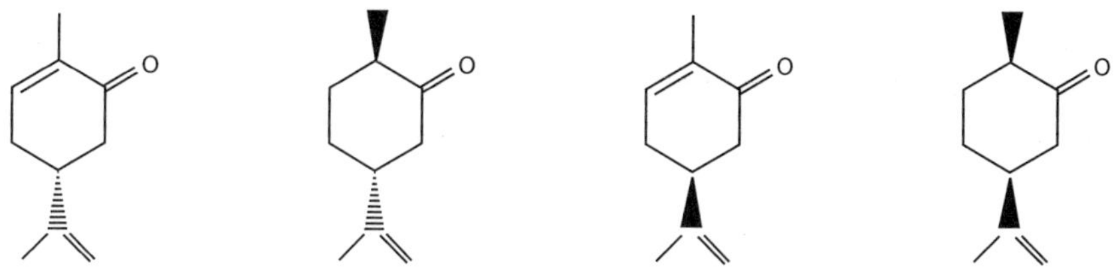

(4*R*)-Carvone (1*R*,4*R*)-Dihydrocarvone (4*S*)-Carvone (1*R*,4*S*)-Isodihydrocarvone

Figure 17. Reduction of (4*R*)- and (4*S*)-carvone.

carbon atom to which the acetoxyl group was attached were found to be preferentially hydrolysed (72). Comparable results were found with the hydrolysis of (*RS*)-1-phenylethyl acetate and related esters by cell cultures of *Spirodela oligorrhiza* (73). In principle, cell cultures might be found to hydrolyse many other conjugates; however, hydrolyses of many substances are achievable *in vitro* using preparations of esterases, lipases, peptidases and glycosidases, from a variety of source organisms.

3. Enzymic Bioconversions

3.1. Introduction

In this section, some (potentially) useful reactions with isolated enzymes, free or immobilised, derived from plant sources are given. It is not at all a comprehensive treatise, but rather a collection of examples showing the potential of enzymes from the plant kingdom. The examples are listed by the enzyme.

3.2. Papain

The latex of the green fruit and and leaves of papaya (*Carica papaya*) contains a high concentration of the thiol protease papain. It is generally used for tenderising meat and bating skins (74). However, papain is also capable of performing selective biotransformations.

Many proteases selectively react with the α-carboxylic acid amide group of acidic amino acids like L-glutamic acid and L-aspartic acid, leaving the side-chain acid group untouched. This is an example of a regioselective reaction. The "natural" reaction of papain is the splitting of peptide bonds, but ester linkages are also cleaved in some cases. A remarkable example in which the terminal ester group of an unnatural amino acid is hydrolysed is shown below in Scheme 2 (75). The product, monoester [2], is used for the production of unnatural peptides which are more stable to physiological breakdown.

Because of the principle of microscopic reversibility, enzymes catalyse both forward and backward reactions - i.e., a protease is not only able to *hydrolyse* peptide bonds but also to *form* them, provided the water concentration is low enough to shift the equilibrium in Scheme 3 to the left:

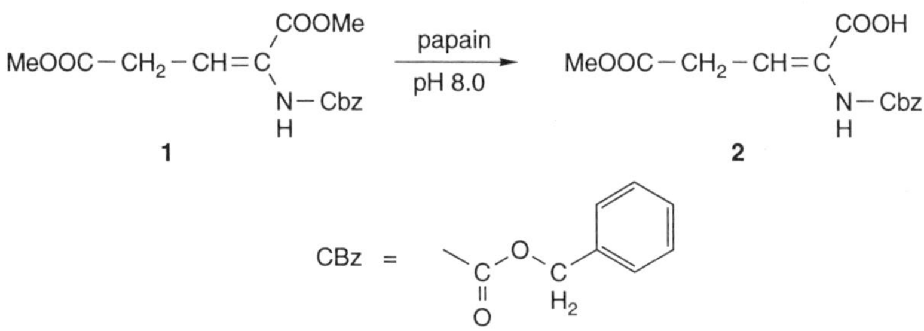

Scheme 2. Papain-mediated regioselective hydrolysis reaction.

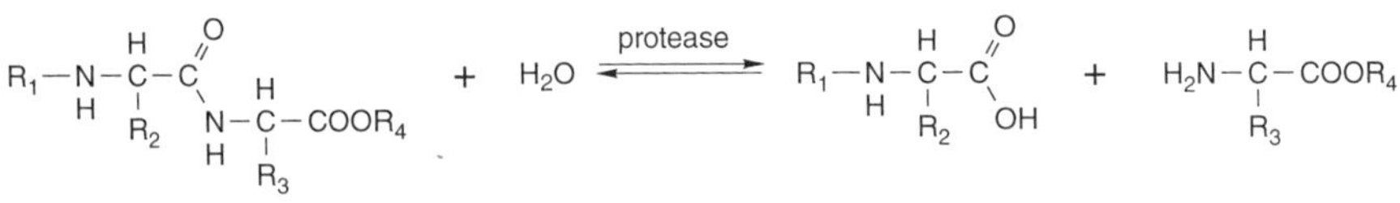

Scheme 3. Protease-catalysed formation and hydrolysis of peptides

Using this strategy, Boc-protected glycine and L-phenylalanine methyl ester were coupled to the protected dipeptide Boc-Gly-PheOMe by papain immobilised on Celite by adsorption or by covalent attachment (76). The reaction equilibrium was shifted in the desired direction by performing the reaction either in a water-trichloroethylene biphasic mixture or in a homogeneous mixture of water and acetonitrile.

3.3. Oxynitrilases

Oxynitrilases catalyse the addition of hydrogen cyanide to aldehydes. The enzyme also accepts ketones as substrates, but these reactions are seldom of practical value. The obtained cyanohydrins are chiral; oxynitrilases are stereoselective enzymes and usually produce only one enantiomer (see Scheme 4). The configuration of the product obtained depends on the enzyme source.

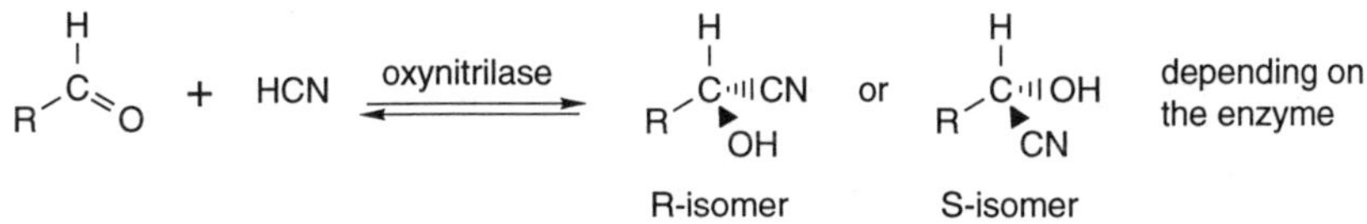

Scheme 4. Preparation of chiral cyanohydrins by oxynitrilases.

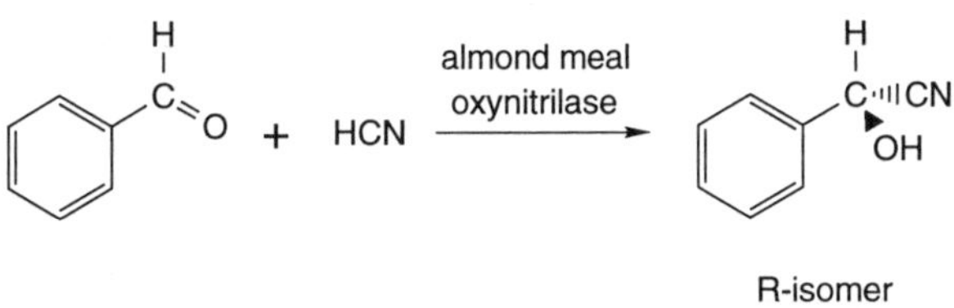

Scheme 5. Stereoselective formation of benzaldehyde cyanohydrin by almond meal oxynitrilase.

The *R*-specific enzyme is found in several Rosaceae, almonds being a particularly good source. The enzyme is involved in the biosynthesis of amygdalin (Figure 18 [1]) in bitter almonds. Amygdalin is the disaccharide of the cyanohydrin of benzaldehyde; enzymatic or chemical hydrolysis of the glycosidic linkage yields mandelonitrile [2] which is in equilibrium with benzaldehyde [3] and toxic hydrogen cyanide. Almonds are such a rich source for the oxynitrilase that adding ground, defatted almonds to the reaction mixture of aldehyde and HCN suffices to get a clean reaction (77, 78).

S-Specific oxynitrilases have only been described recently. Useful sources are millet (*Sorghum bicolor*) (79) and the gum-tree (*Hevea brasiliensis*) (80). The latter enzyme is more useful as it also accepts aliphatic aldehydes.

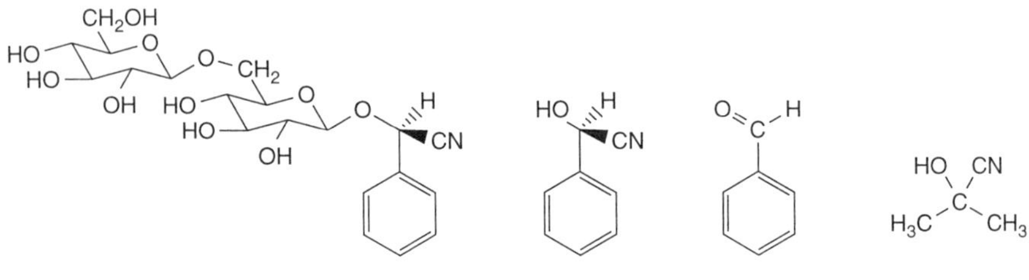

1 2 3 4

Figure 18. Structures of amygdalin [1], mandelonitrile [2], benzaldehyde [3] and acetone cyanohydrin [4].

The cyanohydrins obtained by these enzymes are very useful building blocks for chiral chemical products (14, 19) like α-hydroxy acids, α-hydroxy ketones (77, 81), vicinal diols (81) and ethanolamines (81, 82). Acetone cyanohydrin (Figure 18 [4]) has been used as a safe equivalent to hydrogen or potassium cyanide (83); in solution, it is in equilibrium with acetone and hydrogen cyanide which is used by the enzyme.

3.4. Cyclases

Germacrane sesquiterpenes are believed to be important intermediates in the biosynthetic pathway towards guaiane-, eudesmane- and other types of sesquiterpenes (see Figure 19). Compounds with these skeletons have been detected in the sprouts and roots of chicory *(Cichorium intybus)* and are responsible for the bitter taste of this vegetable (84, 85). One of the most abundant bitter compounds in chicory is lactucin (Figure 19), which is an oxidised guaiane lactone.

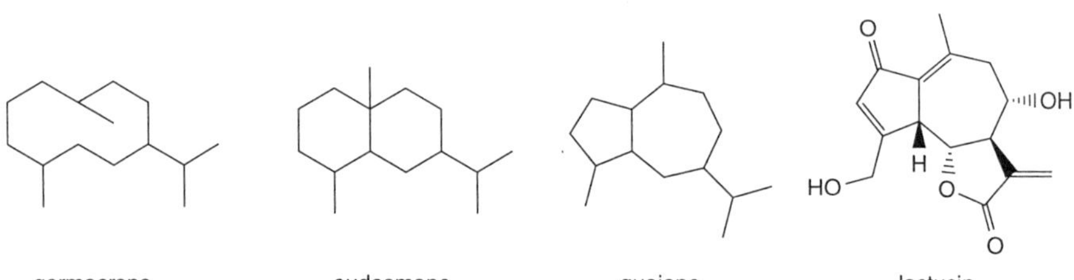

germacrane eudesmane guaiane lactucin

Figure 19. Chemical structure of the germacrane, guaiane and eudesmane skeleton, and the chicory bitter principle lactucin.

Since guaianes, eudesmanes and germacranes co-occur in chicory, it is quite likely that they share common biosynthetic precursors. Very recently, a cyclase was detected in chicory which selectively cyclises the germacrane derivative germacrone 4,5-epoxide ([1], Scheme 6) into neoprocurcumenol [3] (86). The reaction starts, presumably, by the enzyme-mediated protonation of the epoxide group in [1], followed by ring closure leading to carbocation [2]. This can give, in principle, three different alkenes when [2] is deprotonated, or an alcohol when [2] takes up a water molecule. However, the enzyme selectively removes the proton at C1 leading to [3] which has its carbon-carbon double bond at the same position as lactucin, proving the biomimetic character of this conversion.

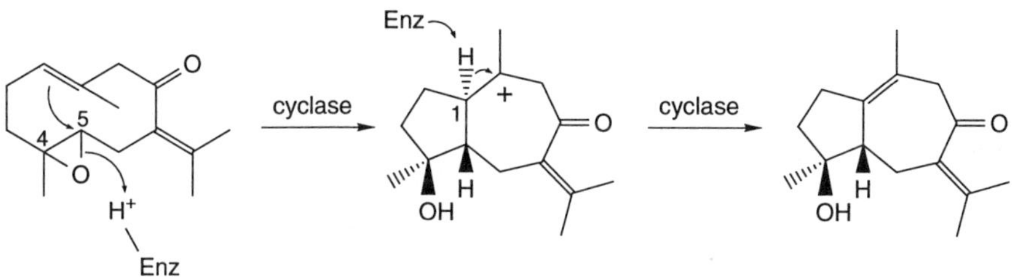

1 2 3

Scheme 6. Biotransformation of germacrone 4,5-epoxide [1] into neoprocurcumenol [3] by chicory root cyclase. All compounds are racemates.

There is evidence now that the chicory root cyclase transforms 1,5-germacranes (87, 88) ([1], see Figure 20) and 1,5-germacrane-1,10-epoxides (87) ([2]) into eudesmanes, whereas guaianes are formed exclusively from 1,5-germacrane-4,5-epoxides (86) ([3]). The cyclase seems to have a broad substrate specificity, since 1,6-cyclodecadienes ([4]) are also accepted as substrates (89). Therefore, these enzymes may be very useful for the smooth cyclisation of cyclic dienes and their epoxides.

3.5. Glycosidases

Glycosides are interesting compounds both from a commercial and a scientific point of view, as already pointed out in Section 2.3. Glycosides of higher alcohols, particularly 1-heptyl- and 1-octyl-β-glucoside, are marketed as biologically degradable non-ionic surfactants, whereas glycolipids are used as emulgators in the

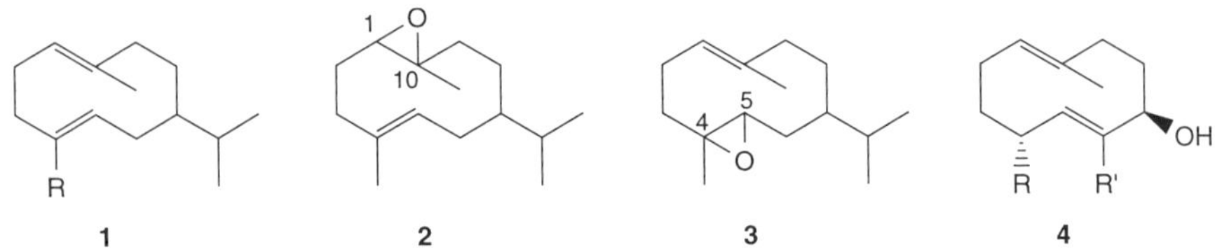

Figure 20. Germacrane and germacrane-like substrates for the cyclase from *Cichorium intybus.*

food industry. Glycosides of some ω-hydroxy carboxylic acids are intermediates in the flavour and fragrance industry (90) and 1-octylglucoside is being used in research towards lipid vesicles (91).

Chemical glycosidation methods have several disadvantages (92):
1. They are multistep processes because of the necessity to attach and remove protective groups;
2. The synthetic routes are not applicable to compounds having either acid- or base-labile groups;
3. In many cases, it is difficult to obtain selectively the α- or the β-isomer.

Enzymatic methods do not have these disadvantages. Two classes of enzymes are able to form glycosidic linkages: glycosyl transferases (EC 2.4) and glycosidases (EC 3.2). The glycosyl transferases are *in vivo* responsible for the formation of glycosides, whereas the glycosidases are responsible for their selective hydrolysis. However, since the latter reaction is reversible, glycosidases can also be used for the synthesis of glycosides. Glycosidases have several advantages with respect to glycosyl transferases, see Table I.

Glycosyl transferases transfer a glycoside moiety from a specific donor (an oligosaccharide or an nucleoside diphosphate glycoside) to an aglycone acceptor (Equation {2}):

$$\text{sugar-O-A} + \text{R-OH} \xrightarrow{\text{glycosyl transferase}} \text{sugar-O-R} + \text{A-OH} \qquad \{2\}$$

A = nucleoside diphosphate or oligosaccharide, R-OH = alcohol

Glycosidases are not only able to hydrolyse specific glycosidic linkages but can also transfer the glycosyl group to an alcohol acceptor (Equations {3}, {4})

$$\text{sugar-O-B} + H_2O \underset{}{\overset{\text{glycosidase}}{\rightleftharpoons}} \text{sugar-OH} + \text{B-OH} \quad \{3\}$$
$$\text{sugar-O-B} + \text{R-OH} \rightleftharpoons \text{sugar-O-R} + \text{B-OH} \quad \{4\}$$

B = mono- or oligosaccharide, alkyl- or aryl group; R-OH = alcohol

Table I. Properties of glycosyl transferases and glycosidases

	glycosyl transferases	glycosidases
Availability	not commercially available or very expensive	cheap and available in large quantities
Specificity	very specific towards both glycosyl donor and -acceptor	low specificity towards the glycosyl acceptor
Substrates	oligosaccharides, nucleotide diphosphate, glycosides , etc.	mono- and disaccharides, alkyl- and arylglycosides
Stability	lack of knowledge prevents general conclusions	very thermostable enzymes are known; generally resistant toward organic solvents

Reports on enzyme-catalysed production of glycosides deal mainly with glycosidases, because of the disadvantages of glycosyl transferases as mentioned in Table I and the very high price of nucleoside diphosphate sugars. Glycosides can be synthesised by Equation {3} (backward reaction) or Equation {4} (forward reaction). Equilibrium {3} can be shifted to the left side in media with low water content (93).

Several glycosidase-mediated one- and two-phase systems for glycosidation have been described. The most often used enzyme is the β-glucosidase from almond meal. In the literature, the attention is always focused on the yields based on sugar, which are acceptable. In contrast to this, the yields with respect to the aglycone are always very low and have to be calculated by the readers themselves. The groups of Vulfson (94, 95), Chalid (96) and Adlercreutz (93) have used the β-glucosidase from almonds for the glucosidation of simple aliphatic alcohols like 1-hexanol or 1-octanol. Yields are 20-30% based on the sugar and 0.5% based on the alcohol. Since only the β-isomer is formed, this enzymatic procedure is very attractive for the stereoselective glycosidation of cheap alcohols.

3.6. (Halo)peroxidases

Haloperoxidases catalyse the halogenation of a variety of organic compounds using hydrogen peroxide and halide ions as substrates (97, 98). The general overall reaction is given in Equation {5}:

$$AH + H_2O_2 + H^+ + X^- \longrightarrow AX + 2\,H_2O \qquad \{5\}$$

where AH is the organic substrate, X^- is chloride, bromide or iodide ion and AH is the halogenated product. The enzymes are called chloro-, bromo- or iodoperoxidases, depending on the smallest halide they can oxidise. This means that chloroperoxidases are able to oxidise chloride, bromide and iodide ions, whereas bromoperoxidases only accept bromide and iodide. Fluoride is in some cases an inhibitor for these enzymes, but is never a substrate.

Some 30 years ago, haloperoxidases were thought to be unique enzymes: their halogenating capacities were unusual and they were considered to be rare. Nowadays, haloperoxidases are known from more than 100 sources including mammals, birds, terrestial plants, algae, fungi and bacteria, clearly showing their wide occurrence.

Algae are a particularly rich source of haloperoxidases (99). The enzymes are involved in the biosynthesis of a variety of halogenated natural compounds. The halometabolites that are produced by these species range from simple chloro/ bromomethanes (100) to complex, chiral halogenated hydrocarbons like [1] (Figure 21) (from *Plocamium sp.*, a red alga).

The biological function of these halometabolites is still obscure. It is believed that they play a role in the defence mechanism of their producers. In this respect, "defence" should be divided in two separate strategies: 1) the excretion of a halogenated metabolite by an organism in order to kill or deter potential enemies; or 2) the accumulation of the toxin to make its producer inedible for possible predators.

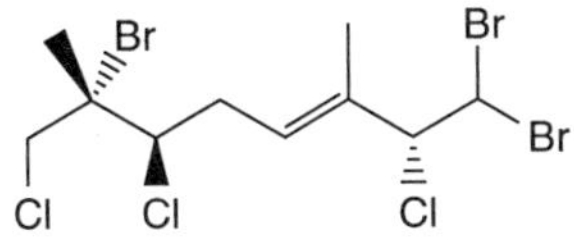

1

Figure 21. Halogenated metabolite from *Plocamium sp.* (a red alga).

Compounds involved in the first strategy include simple brominated phenols, which possess antibacterial and antialgal activities: these compounds have been found in a number of algal species. The second strategy is nicely illustrated by the life of the sea hares (*Aplysia*). These molluscs feed on the red seaweeds *Laurencia* and *Plocamium*. In doing so, they accumulate the halogenated monoterpenes (like [1]) present in these algae and become unpalatable themselves. At any rate, algae produce a vast amount of chlorinated and brominated compounds. The biological production of chloromethane is estimated to be 2-5 megaton per annum, which is at least one order of magnitude higher than its industrial production (101). A relation between algal halomethane production and the reduction of the stratospheric ozone density has been suggested by Wever *et al.* (102, 103).

The prosthetic groups of the halogenating catalysts, the haloperoxidases, are mostly haem or vanadium. The way these enzymes work is still not completely understood, although the reaction mechanism of the haem enzymes has been quite well documented (104).

The haem-containing haloperoxidases possess an essential iron-protoporphyrin IX complex in their active pocket. Both the iron (III) ion and the porphyrin group play an important role, as is shown in Scheme 7. When the native enzyme reacts with hydrogen peroxide (or other peroxy compounds like ethyl hydroperoxide or peracetic acid) an enzyme intermediate (commonly called "Compound I") is formed which is two oxidising equivalents above the native state. The oxidation equivalents reside in the iron ion which has a formal charge of +4, and in the porphyrin group which is transformed into a radical cation. Compound I is unstable (half-life of minutes or less) and has a characteristic green colour. Compound I reacts with oxidisable organic compounds (*vide infra*) or with halide ions, forming the very unstable (and still disputed) Compound EOX. In this reaction, two electrons are transferred from the halide ion to the iron (IV) and the porphyrin radical cation, thereby oxidising the halide into an X^+ species. Compound EOX rapidly decomposes, presumably liberating an oxidised halide species as the active halogenating agent (105). The exact nature of the halide is difficult to determine because X^+ does not exist freely in water, but is bound as HOX, X_2 or X_3^-.

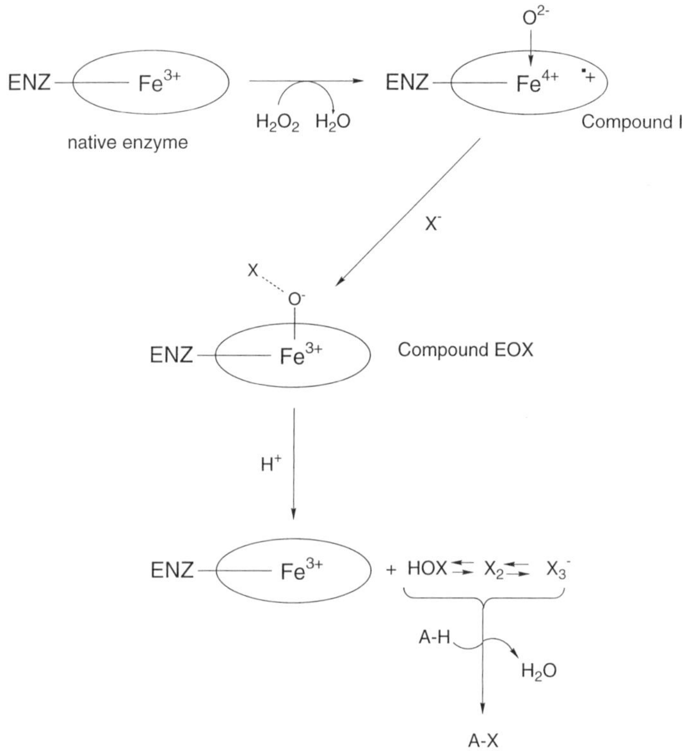

Scheme 7. Halogenation mechanism of haem-containing haloperoxidases. The protein part of the enzyme is represented by ENZ, the haem group is depicted as Fe^{x+} (x=3 or 4) in the centre of an ellipse, for reasons of clarity.

The vanadium-containing enzymes are also believed to produce hypohalous acids (or equivalents thereof) as the actual halogenating agents (106, 107). This implies that haloperoxidase-mediated regio- or stereoselective halogenation reactions are not to be expected. Nevertheless, the enzymes can be useful to the organic chemist because they produce HOX in a continuous way, avoiding high concentrations of this reactive agent. This may lead to improved yields when compared to reaction in which HOX is added directly (108). A few examples are given below.

The vanadium-containing bromoperoxidase from the red alga *Corallina pilulifera* was used for the conversion of styrene [1] (Scheme 8) into its bromohydrin [2]. The isolated yield of the racemic product was 69% (109). Various coralline algae (e.g., *Corallina pilulifera*, *C. officinalis* and *Amphiroa zonata*) gave upon treatment of

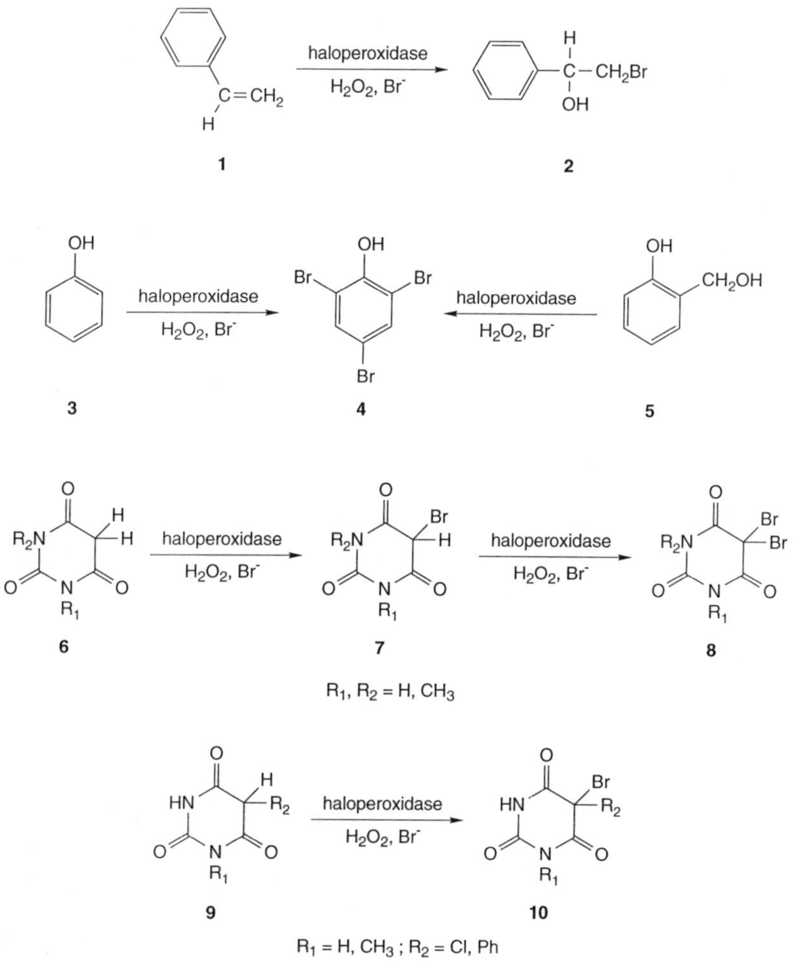

Scheme 8. Some halogenation reactions with algal haloperoxidases. See text for details.

phenol [3] and 2-hydroxymethyl phenol [5] with bromide and hydrogen peroxide, 2,4,6-tribromophenol [4] (110). The vanadium bromoperoxidase from the brown alga *Ascophyllum nodosum* is able to convert barbituric acid and some of its derivatives ([6], [9]) into their respective 5-bromo- ([7], [10]) and/or 5,5-dibromo derivatives [8] in good yields (106).

(Halo)peroxidases are also abundant in higher plants. The best characterised peroxidases are the ones obtained from horseradish roots, soybean, turnip and Japanese radish roots (109). These enzymes generally perform one-electron oxidations leading to dehydrogenated products or oligomers. These reactions are considered as not very relevant for organic chemists. However, this situation may change upon the discovery of the stereoselective sulphoxidation and epoxidation reactions which are catalysed by some mutants of horseradish peroxidase.

Initial studies on the sulphoxidation of thioanisole [1] (Scheme 9) showed that the haem-containing horseradish peroxidase (HRP) gave the corresponding (*S*)-sulphoxide [2] in moderate enantiomeric excess (e.e.) (112), whereas [1] is converted into the (*R*)-sulphoxide [3] in very high e.e. upon treatment with the chloroperoxidase from the mould *Caldariomyces fumago* and t-butyl hydroperoxide (113).

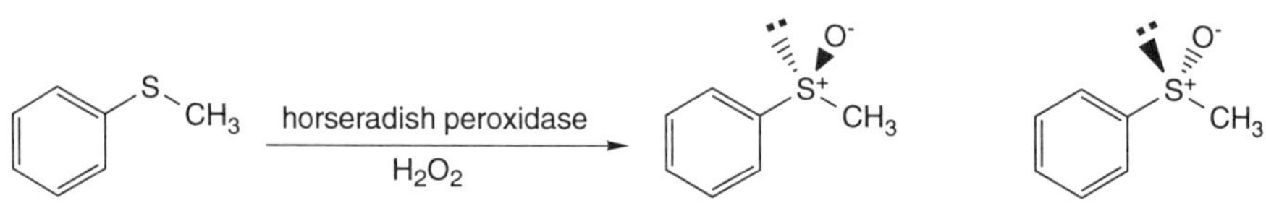

1 **2** **3**

Scheme 9. Peroxidase-catalysed sulphoxidation of thioanisole [1].

The enantioselectivity of the HRP-mediated reaction could be strongly increased by replacing amino acid residue 41, a phenylalanine, by a leucine (called the F41L mutant) (114). This is explained as follows. At the distal side of the haem prosthetic group of peroxidases, three conserved amino acid residues are found: an arginine, a histidine (His 42 in HRP) and a tryptophan or a phenylalanine (Phe 41 in HRP). There is evidence that Phe 41 blocks the access of the substrate to the $Fe^{IV} \leftarrow O^{2-}$ species which is present in Compound I. When a sulphide like [1] enters the enzyme active site, the sulphur atom will donate its electrons to the iron ion via the porphyrin ring, thereby becoming positively charged. The sulphur cation will then, more or less randomly, take up water from the environment to give the corresponding sulphoxide with low e.e. However, when the phenylalanine is replaced by the much less bulky leucine group, the substrate may interact with the $Fe^{IV} \leftarrow O^{2-}$ group leading to direct, stereoselective transfer of oxygen from the iron-oxo complex to the sulphur. In this way, unselective HRP has been transformed into a very useful stereoselective biocatalyst.

When the genetically altered HRP is able to oxidise sulphides into their corresponding sulphoxides, it is tempting to investigate whether this enzyme can also epoxidise unfunctionalised alkenes in a stereoselective way, since this reaction has been a "hot topic" among synthetic chemists for several years already (115). The haem-containing chloroperoxidase from the mould *Caldariomyces fumago* was shown to able to epoxidise a variety of alkenes in a stereoselective manner (116), whereas native HRP is unreactive towards these substrates (117). However, both the F41L and the related threonine mutant (F41T) of HRP effectively catalyse the stereoselective epoxidation of styrene and some of its methylated derivatives (see Table II) (118). Although the reaction is rather slow, several byproducts like phenylacetaldehyde and benzaldehyde are produced and the stereochemistry of the carbon-carbon double bond is not always conserved, this reaction is promising and once more shows the usefulness of plant enzymes.

3.7. Lipoxygenase

Lipoxygenase is a non-haem iron-containing dioxygenase which catalyses the incorporation of dioxygen into suitable unsaturated substrates. During catalysis, the oxygen atoms of O_2 remain bound to each other, which implies that the enzymatic products are hydroperoxides (ROOH) or endoperoxides (ROOR'). Since the enzyme is very stereoselective and the peroxides can be easily reduced, lipoxygenase can be used for the production of chiral alcohols. Its favoured substrates are fatty acids having 1,4-diene units; the products are in that case 1,3-diene-5-peroxy acids. In Scheme 10 an example is given of the soybean lipoxygenase-mediated oxidation of linoleic acid. The reaction is not only very stereoselective but also highly regioselective, since only the 13-peroxy-9,11-diene is formed (119). The product is used as an intermediate in the flavour and fragrance industries.

Lipoxygenase from soybean has received the most attention in terms of a detailed characterisation, because of its early discovery, its ease of isolation and its stability

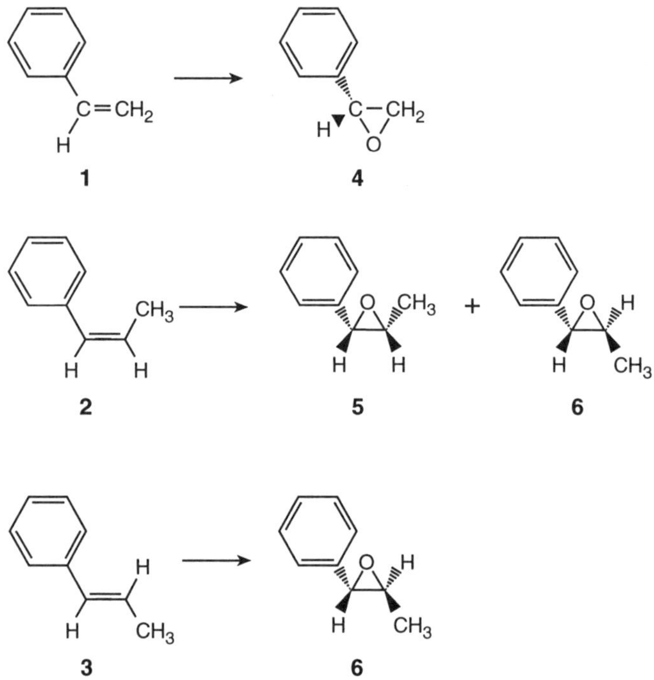

		Table II. Stereochemistries of the epoxides formed in the reactions of styrene [1], cis-β-methyl styrene [2] and trans-β-methylstyrene [3] with F41L and F41T HRP (118)	
		% e.e.	
substrate	epoxide	F41L	F41T
[1]	[4]	46	2
[2]	[5]	*	>99
	[6]	*	10
[3]	[6]	36	64

* no reaction

(4). The configuration of the newly formed oxygenated chiral centre is *S*, in most cases (120). The enzyme accepts a variety of unnatural substrates, but E,E-dienes and conjugated dienes are usually not oxidised.

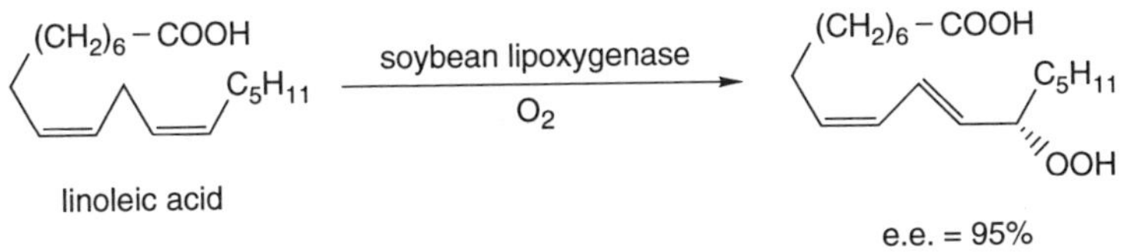

Scheme 10. Oxidation of linoleic acid by soybean lipoxygenase. Taken from (119).

Another soybean lipoxygenase-catalysed reaction of industrial interest is the oxidation of esters of coniferyl alcohol (like [1], Scheme 11) into vanillin [2] (121). However, this reaction is not a commercial process yet because of the relatively high price and the moderate stability of [1].

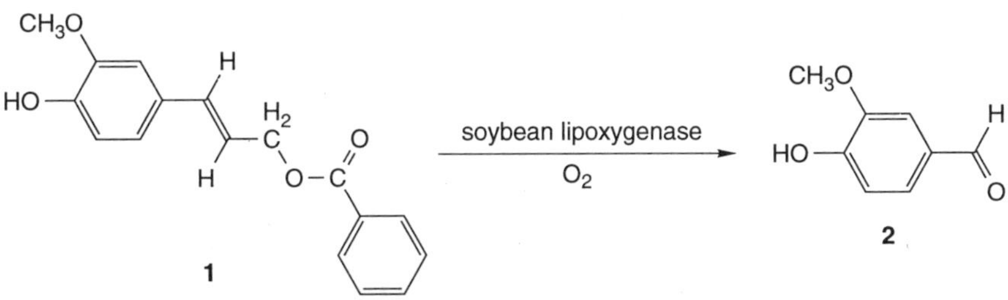

Scheme 11. Potential application of soybean lipoxygenase in the production of vanillin [2].

3.8. Cytochrome P450 Monooxygenases

Cytochrome P450 enzymes activate molecular oxygen and, typically but not invariably, use one oxygen atom in hydroxylating their substrates. The designation "P450" arises from the absorption peak at 450 nm seen in the carbon monoxide difference spectrum of reduced cytochrome P450.Two components are required for the complete reaction - a separate reductase component, which transfers reducing equivalents from NADPH and is a flavoprotein (or, in bacterial and mitochondrial P450 reductases, comprises a ferredoxin reductase and a non-haem iron protein); and the terminal cytochrome P450 itself, which accepts the reducing equivalents from the reductase, binds the substrate and activates the molecular oxygen. The initial stages of the catalytic cycle involve displacement of a coordinated water molecule from the ferric haem prosthetic group upon binding of the substrate molecule, a change in spin state of the iron and its reduction from the ferric state to the ferrous, followed by the binding of molecular oxygen. Limited direct information is available on the subsequent phases of the cycle, which include the cleavage of molecular oxygen, the release of water, the formation of an activated oxygen species, formally $[FeO]^{3+}$, and the hydroxylation of the substrate.

As a class, cytochrome P450 enzymes are remarkable both for the diversity of substances which are substrates and for the variation that exists - sometimes in the case of a single enzyme - in the reactions of an individual substrate. The substrates are generally lipophilic compounds and include both intermediates of normal cellular metabolism and a wide range of xenobiotic substances. It is impossible here to provide more than the briefest introduction; excellent comprehensive reviews of cytochrome P450 enzymes and their mode of action (122) are available, including two accounts devoted specifically to the plant enzymes (123,124), and the reader is referred to these for full details. (It is important to remember that cytochrome P450 enzymes are not alone in catalysing hydroxylation reactions; for example, a hydroxylation may be catalysed by an oxoglutarate-linked dioxygenase (125), or of course a hydroxyl group may be introduced by the action of a hydratase on a double bond.) Cytochrome P450 enzymes catalyse a wealth of important reactions - predominantly hydroxylations - in many or most plant secondary product pathways, including the central phenylpropanoid pathway (for example, cinnamic acid 4-hydroxylase; ferulic acid 5-hydroxylase), flavonoid and coumarin biosynthesis (for example, flavonoid-3′-hydroxylase; flavonoid-3′5′-hydroxylase; isoflavone synthase; psoralen synthase), terpenoid biosynthesis (notably geraniol / nerol 10-hydroxylase), alkaloid biosynthesis (including indole and benzylisoquinoline alkaloid biosyntheses) and the pathway of formation of cyanogenic glucosides. The cytochrome P450 terpene hydroxylases have been reviewed (126). One unusual example is the rubber-particle protein of guayule *(Parthenium argentatum)*, which has been identified as a cytochrome P450 enzyme with strong similarity to allene oxide synthetase, a P450 enzyme which converts lipid hydroperoxides to epoxides; the rubber-particle protein accounts for *ca.* 50% of the protein of the (highly-abundant) rubber particles, though its function remains unknown (127). Because of the range of substrates, the cytochrome P450 enzymes offer in principle enormous potential for the catalysis of biotransformation reactions *in vitro*. Unfortunately, however, isolation of these enzymes is often not facile. Their association with microsomal membranes generally demands solubilisation with detergents, often resulting in loss of activity. Successful reconstitution following purification requires a source of NADPH-cytochrome P450 reductase and, frequently, phospholipid, such as dilauroyl-phosphatidylcholine. The digitoxin 12β-hydroxylase of *Digitalis lanata* (see Section 2.2) has been purified and reconstituted, and successfully immobilised (128, 129).

Where a cDNA has been isolated and cloned for an identified cytochrome P450 enzyme, the protein can in principle be produced by heterologous over-expression and this may obviate some or most of the problems associated with isolation from the native source. This has been demonstrated elegantly in the case of berbamunine synthase, a cytochrome P450 enzyme of *Berberis stolonifera* which catalyses the stereospecific phenolic coupling of (*R*)- and (*S*)-*N*-methylcoclaurine to form the bisbenzylisoquinoline alkaloid, (*R,S*)-berbamunine (130); unusually, in this reaction there is no incorporation of oxygen into the product alkaloid (Figure 22). The cDNA was expressed in insect cell culture, using a baculovirus-based expression system in a cell line of *Spodoptera frugiperda*. Haemin was added to the culture medium to ensure an adequate supply of prosthetic group to support the over-expression. Following microsome isolation and solubilisation with sodium cholate, the heterologously-expressed berbamunine synthase could be purified to near-homogeneity by a single fractionation step on ω-aminooctylagarose, and to apparently-complete homogeneity by a subsequent hydroxylapatite step. The yield after the ω-aminooctylagarose step was approximately 5 mg per litre of cells and the authors calculated that to isolate an equivalent quantity using *B. stolonifera* cell suspension cultures would require 18,000 litres of culture. An interesting feature of the purified enzyme was that the stereochemistry of the reaction product depended not only upon the ratio of the (*R*)- and (*S*)- substrates but also upon the source of the NADPH-cytochrome P450 reductase provided. Other authors have reported a strategy for the functional expression in *Escherichia coli* of plant cytochrome P450 proteins as translational fusions with cytochrome P450 reductase; this seems likely to be useful in the characterisation of the increasing number of cytochrome P450 cDNA clones, isolated on the basis of sequence homology, to which a biochemical function remains to be assigned (131).

Although at the time of writing a three-dimensional structure has not yet been reported for a plant cytochrome P450, the structures of several soluble bacterial P450 enzymes (notably "$P450_{cam}$", a camphor 5-exo-hydroxylase from *Pseudomonas putida*) have been solved and sequence homologies indicate that there are clear structural similarities between them. A key objective is to discover the identity of the residues which most critically determine substrate specificity. Progress in this direction has been made with the CYP2 family of mammalian, drug-metabolising enzymes, by comparing their sequences with those of the bacterial P450 enzymes. Six

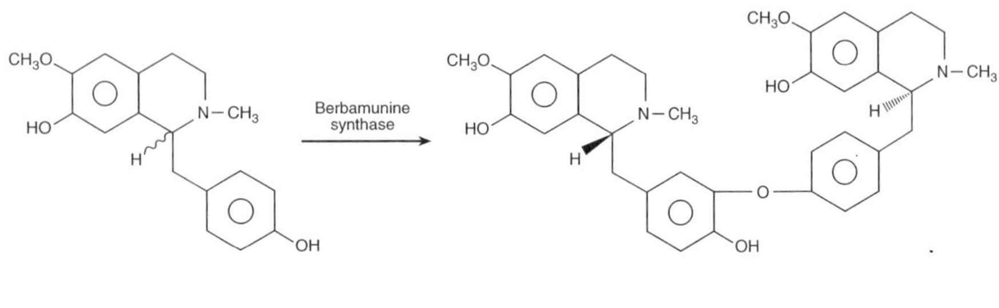

Figure 22. The reaction catalysed by berbamunine synthase.

putative substrate-recognition sites have been identified, each of which contains, relative to the rest of the nucleotide sequence, a high proportion of nucleotide substitutions leading to amino acid changes (132). It has been known for some years that minor amino-acid changes may lead to very substantial changes in substrate specificity; for example, substitution of Phe209 by Leu in the mouse enzyme "$P450_{coh}$" caused a change in substrate specificity from coumarin hydroxylation to steroid hydroxylation (133). In the case of berbamunine synthase, referred to above, the absence of the usual oxygenase activity has been ascribed to the substitution of a proline for an essential glycine residue in the oxygen-binding pocket (130).

Cytochrome P450 enzymes are amongst the most promising enzymes for site-directed mutagenesis and for the engineering of substrate specificity and exciting progress can be expected in the use of these enzymes in biotransformations within the next decade.

4. Conclusions and Outlook

We hope to have shown in this Chapter that plant cells and enzymes are a useful addition to the chemist's toolbox. It is difficult, however, to reach a generalised conclusion. The application of cell-free biocatalysis to chemistry has but a short history and most of the applications of enzymic preparations to date have involved well-characterised, stable enzymes (mainly esterases, lipases, peptidases, peroxidases, lipoxygenases and certain oxidoreductases) available in relatively large quantities by extraction from the native - usually microbial or animal - source. As indicated at the outset, the distinct appeal of plant enzymes is that many of them catalyse the biosynthesis or biotransformation of compounds which are specific to plants. The "berberine bridge" enzyme (134, 135) of the benzophenanthridine biosynthetic pathway is but one example of an enzyme catalysing an apparently unique reaction. Further work will reveal others. Although there is in principle no problem in generating large quantities of such proteins by heterologous gene expression, making widespread use of them in biotransformations is still some way in the future. An enzyme that functions as part of a highly-ordered biosynthetic pathway within the cell may not invariably function effectively outside it. A suitable substrate provided conveniently by the preceding enzyme of the pathway may be provided only with great difficulty by the organic chemist. The best use of such enzymes at present remains within the plant itself!

Against this background, it is not surprising that the range of plant-based biotransformations remains relatively small.Yet the potential of plant-cell and plant-enzyme biotransformations is enormous. Many selective, unique chemical reactions take place inside the plant cell and it is to be hoped that one day the biocatalytic power of plant cells can be utilised by man on a larger scale. Despite the potential of the plant-based biotransformations so far demonstrated, there is much progress yet to be made. Real advances will depend upon the combination of molecular genetics and protein chemistry to engineer enzymes with modified or tailored substrate specificity, stability or catalytic function, so that the most interesting and useful enzymes - whatever their source - can acquire new catalytic applications both inside and outside their cells of origin.

5. References

1. H. Sandermann, *Pharmacogenetics* **4** (1994) 225.
2. K.A. Marrs, *Annu. Rev. Plant Physiol. Plant Mol. Biol.* **47** (1996) 127.
3. P. M. Doran (ed.) *Hairy Roots* (1997) Harwood Academic Publishers, Amsterdam.
4. A.J. Parr, *J. Biotechnol.* **10** (1989) 1.
5. M.H. Zenk, in *Recent Advances in Phytochemistry: Plant Nitrogen Metabolism*, eds. J.E. Poulton, J.T. Romeo and E.F. Conn **23** (1988) 429.
6. Y. Fujita, in *Plant Tissue Culture: Applications and Limitations*, ed. S.S. Bhoujarani, (Elsevier, Amsterdam, 1990) pp. 259-275.
7. A.W. Alfermann and M. Petersen, *Plant Cell Tiss. Org. Cult.* **43** (1995) 199.
8. R.J.H. Clark, C.J. Cooksey, M.A.M. Daniels and R. Withnall, *Endeavour* **17** (1993) 191.
9. H.M. Tan and C.M. Cheong, *Biochem. Biophys. Res. Commun.* **204** (1994) 912.

10. B.D. Ensley, *Chimia* **48** (1994) 491.
11. J. March, *Advanced Organic Chemistry: reactions, mechanisms and structure*, 3rd edition (Wiley-Interscience, New York, NY, 1985) p. 1346.
12. T.R. Cech, *Science* **236** (1987) 1532.
13. R.A. Lerner, S.J. Benkovic and P.G. Schultz, *Science* **252** (1991) 659.
14. K. Faber, *Biotransformations in Organic Chemistry*, 3rd edition (Springer-Verlag, Berlin, 1997).
15. A.M. Klibanov, *Trends Biochem. Sci.* **14** (1989) 141.
16. H.G. Davies, R.H. Green, D.R. Kelly and S.M. Roberts, *Biotransformations in Preparative Organic Chemistry*, (Academic Press, London, 1989).
17. L. Poppe and L. Novák, *Selective Biocatalysis*, (VCH, Weinheim, 1992).
18. A.M. Klibanov, *Acc. Chem. Res.* **23** (1990) 114.
19. K. Faber and M.C.R. Franssen, *TIBTECH* **11** (1993) 461.
20. M.J. Wanner and G.J. Koomen, *J. Org. Chem.* **60** (1995) 5634.
21. G.J. Koomen and M.J. Wanner, *Proc. Kon. Ned. Acad. v. Wetensch.* **97** (1994) 423.
22. M.M. Yeoman, in *Cell culture and somatic cell genetics*, ed. F. Constabel and I.K. Vasil (Academic Press Inc., San Diego, CA, USA, 1987) p 197.
23. M.S.M. Pais and J.M.S. Cabral, in *Plant aging: basic and applied approaches.*, eds. R. Rodriguez *et al.* (Plenum Press, New York, NY, 1990) p 97.
24. J. Archambault, B. Volesky and W.G.W. Kurz, *Biotechnol. Bioeng.* **33** (1989) 293.
25. S.J. Gutcho, *Immobilized enzymes. Preparation and engineering techniques* (Noyes Data Corp., Park Rifge, NJ, 1974).
26. O. Zaborsky, *Immobilized enzymes* (CRC Press, Cleveland, OH, 1974).
27. A.J. Lambie, *Proc. Phytochem. Soc. Eur.* **30** (1990) 265.
28. A.C. Hulst and J. Tramper, *Enzyme Microb. Technol.* **11** (1989) 546.
29. A.C. Hulst, J. Tramper, K. van 't Riet and J.M.M. Westerbeek, *Biotechnol. Bioeng.* **27** (1985) 829.
30. M.L. Shuler, J.W. Pyne and G.A. Hallsby, *J. Am. Oil Chem. Soc.* **61** (1984) 1724.
31. J.E. Prenosil and H. Pedersen, *Enzyme Microb. Technol.* **5** (1983) 323.
32. F.J. Kim, H.N. Chang and J.R. Liu, *Biotechnol. Techn.* **3** (1989) 139.
33. R.J. Robins and M.J.C. Rhodes, *Appl. Microbiol. Biotechnol.* **24** (1986) 35.
34. S.M. Zhang and E. Drioli, *Separ. Sci. Technol.* **8** (1994) 1.
35. A.J. Parr, R.J. Robins and M.J.C. Rhodes, *Plant Cell Rep.* **3** (1984) 262.
36. J.D. Hamill and M.J.C. Rhodes in *Biosynthesis and Manipulation of Plant Products*, ed. D. Grierson (Blackie, London, 1993).
37. M.W. Signs and H.E. Flores, *Bioessays* **12** (1990) 7.
38. R.J. Robins, A.B. Hanley, S.R. Richards, G.R. Fenwick and M.J.C. Rhodes, *Plant Cell Tiss. Org. Cult.* **9** (1987) 49.
39. W. Kreis and E. Reinhard, *J. Biotechnol.* **26** (1992) 257.
40. S.J. Stohs and E.J. Staba, *J. Pharm. Sci.* **54** (1964) 56.
41. P. Obitz, S. Endreß and J. Stöckigt, *Phytochemistry* **40** (1995) 1407.
42. T. Furuya, M. Nakano and T. Yoshikawa, *Phytochemistry* **17** (1978) 891.
43. T. Furuya, T. Yoshikawa and M. Taira, *Phytochemistry* **23** (1984) 999.

44. H.F. Stakelin and A. von Wartburg, *Prog Drug Res.* **33** (1989) 169.
45. J.A. Broomhead, A.M.M. Rahman, P.M. Dewick, D.E. Jackson and J.A. Lucas, *Phytochemistry* **30** (1991) 1489.
46. J.P. Kutney, Y.P. Chen, S. Gao, G.M. Hewitt, F. Kuri-Brena, R.K. Milanova and N.M. Stoynov, *Heterocycles* **36** (1993) 13.
47. J.P. Kutney, *Pure & Appl. Chem.* **68** (1996) 2073.
48. B. Botta, G.D. Monache, P. Ricciardi, A. Vitali, V. Vinciguerra, D. Misiti, J.P. Kutney and N. Stoynov, *Heterocycles* **43** (1996) 2443.
49. J.P. Kutney, *Nat. Prod. Rep.* **7** (1990) 85.
50. J.P. Kutney, *Heterocycles* **27** (1988) 1845.
51. J.P. Kutney, *Pure & Appl. Chem.* **69** (1997) 431.
52. J.P. Kutney, K. Han, F. Kuri-Brena, R.K. Milanova and M. Roberts, *Heterocycles* **44** (1997) 95.
53. N.J. Walton, R.J. Robins and M.J.C. Rhodes, *Plant Sci.* **54** (1988) 125.
54. L.F. Fecker, C. Rügenhagen and J. Berlin, *Plant Mol. Biol.* **23** (1993) 11.
55. S. Herminghaus, D. Tholl, C. Rügenhagen, L.F. Fecker, C. Leuschner and J. Berlin, *Transgenic Research* **5** (1996) 193.
56. H.D. Boswell, A.B. Watson, N.J. Walton and D.J. Robins, *Phytochemistry* **34** (1993) 153.
57. T. Suga and T. Hirata, *Phytochemistry* **29** (1990), 2393.
58. Y. Orihara and T. Furuya, *Phytochemistry* **36** (1994) 55.
59. R. Whetten and R. Sederoff, *The Plant Cell* **7** (1995) 1001.
60. T.A. Holton and E.C. Cornish, *The Plant Cell* **7** (1995) 1071.
61. C. Schroeder, R. Lutterbach and J. Stöckigt, *Tetrahedron* **52** (1996) 925.
62. H. Mizukami, T. Terao, A. Amano and H. Ohashi, *Plant Cell Physiol.* **27** (1986) 645.
63. Y. Orihara, T. Noguchi and T. Furuya, *Phytochemistry* **35** (1994) 941.
64. Y. Orihara and T. Furuya, *Phytochemistry* **35** (1994) 641.
65. T. Kometani, H. Tanimoto, T. Nishimura, I. Kanbara and S. Okada, *Biosci. Biotech. Biochem.* **57** (1993) 2192.
66. C. Shroeder, J. Sommer, E. Humpfer and J. Stöckigt, *Tetrahedron* **53** (1997) 927.
67. G. Gil, P. Ferreira dos Santos and C. Bullard, *Phytochemistry* **38** (1995) 629.
68. H. Hamada, *Bull. Chem. Soc. Jpn.* **61** (1988) 869.
69. T. Suga, H. Hamada, T. Hirata and S. Izumi, *Chem. Letters* (1987) 903.
70. K. Nakamura, H. Miyoshi, T. Sugiyama and H. Hamada, *Phytochemistry* **40** (1995) 1419.
71. T. Hirata, H. Hamada, T. Aoki and T. Suga, *Phytochemistry* **21** (1982) 2212.
72. K. Kawagichi, M. Hirotani and T. Furuya, *Phytochemistry* **27** (1988) 3475.
73. P. Pawlowicz and A. Siewinski, *Phytochemistry* **26** (1987) 1001.
74. S. Budavari, *The Merck Index*, 11th edition (Merck & Co, Inc., Rahway, N.J., 1989).
75. C. Shin, M. Seki and N. Takahashi, *Chem. Lett.* (1990) 2089.
76. M. Capellas, P. Serra, M.D. Benaignes, G. Caminal, G. Gonzalez and J. Lopez-Santin, *Biocatalysis* **11** (1994) 273.
77. J. Brussee, E.C. Roos and A. van der Gen, *Tetrahedron Lett.* **29** (1988) 4485.

78. J. Brussee, W.T. Loos, C.G. Kruse and A. van der Gen, *Tetrahedron* **46** (1990) 979.
79. U. Niedermeyer and M.-R. Kula, *Angew. Chem. Int. Ed. Eng.* **29** (1990) 386.
80. N. Klempier, H. Griengl and M. Hayn, *Tetrahedron Lett.* **34** (1993) 4769.
81. W.R. Jackson, H.A. Jacobs, B.R. Matthews, G.S. Jayatilake and K.G. Watson, *Tetrahedron Lett.* **31** (1990) 1447.
82. J. Brussee, R.A.T.M. van Benthem, C.G. Kruse and A. van der Gen, *Tetrahedron: Asymmetry* **1** (1990) 163.
83. V.I. Ognyanov, V.K. Datcheva and K.S. Kyler, *J. Am. Chem. Soc.* **113** (1991) 6992.
84. T.A. van Beek, P. Maas, B.M. King, E. Leclercq, A.G.J. Voragen and Æ. de Groot, *J. Agric. Food Chem.* **38** (1990) 1035.
85. E. Leclercq, *Sesquiterpene Lactones and Inulin from Chicory Roots: Extraction, Identification, Enzymatic Release and Sensory Analysis*. PhD thesis, Wageningen (1992) .
86. D.P. Piet, R. Schrijvers, M.C.R. Franssen and Æ. de Groot, *Tetrahedron* **51** (1995) 6303.
87. D.P. Piet, A.J. Minnaard, K.A. van der Heyden, M.C.R. Franssen, J.B.P.A. Wijnberg and Æ. de Groot, *Tetrahedron* **51** (1995) 243.
88. D.P. Piet, M.C.R. Franssen and Æ. de Groot, *Tetrahedron* **50** (1994) 3169.
89. D.P. Piet, M.C.R. Franssen and Æ. de Groot, *Tetrahedron* **52** (1996) 11273 .
90. E.N. Vulfson and B.A. Law, in *Proceedings of "Chiral-90" symposium*, (Spring Innovation Ltd., Manchester, UK, 1990) 56.
91. A.R. Parente, M. Höchli and B.R. Lentz, *Biochim. Biophys. Acta* **812** (1985) 493.
92. N. Mitsuo, H. Takeichi and T. Satoh, *Chem. Pharm. Bull.* **32** (1984) 1183.
93. G. Ljunger, P. Adlercreutz and B. Mattiasson, *Enzyme Microb. Technol.* **16** (1994) 751.
94. E.N. Vulfson, R. Patel and B.A. Law, *Biotechnol. Lett.* **12** (1990) 397.
95. E.N. Vulfson, R. Patel, J.E. Beecher, A.T. Andrews and B.A. Law, *Enzyme Microb. Technol.* **12** (1990) 950.
96. Z. Chalid, D. Montet, M. Pina, F. Bonnot and J. Graille, *Biotechnol. Lett.* **16** (1994) 795.
97. M.C.R. Franssen, *Biocatalysis* **10** (1994) 87.
98. M.C.R. Franssen and H.C. van der Plas, *Adv. Appl. Microbiol.* **37** (1992) 41.
99. R. Wever, E. de Boer, H. Plat and B.E. Krenn, *FEBS Lett.* **216** (1987) 1.
100. P.M. Gschwend, J.K. MacFarlane and K.A. Newman, *Science* **227** (1985) 1033.
101. R. Wever, in *Microbial Production and Consumption of Greenhouse Gases: Methane, Nitrogen Oxides and Halomethanes.*, ed. J.E. Rogers and W.B. Whitman (Am. Soc. Microbiol., Washington, DC, 1991).
102. R. Wever, M.G.M. Tromp, B.E. Krenn, A. Marjani and M. van Tol, *Environ. Sci. Technol.* **25** (1991) 446.
103. R. Wever, *Nature* **335** (1991) 501.
104. P.R. Ortiz de Montellano, *Annu. Rev. Pharmacol. Toxicol.* **32** (1992) 89.
105. R.D. Libby, A.L. Shedd, A.K. Phipps, T.M. Beachy and S.M. Gerstberger, *J. Biol. Chem.* **267** (1992) 1769.

106. M.C.R. Franssen, J.D. Jansma, H.C. van der Plas, E. de Boer and R. Wever, *Bioorg. Chem.* **16** (1988) 352.
107. E. de Boer and R. Wever, *J. Biol. Chem.* **263** (1988) 12326.
108. M.C.R. Franssen, H.G. van Boven and H.C. van der Plas, *J. Heterocyclic Chem.* **24** (1987) 1313.
109. N. Itoh, A.K.M.Q. Hasan, Y. Izumi and H. Yamada, *Eur. J. Biochem.* **172** (1988) 477.
110. H. Yamada, N. Itoh, S. Murakami and Y. Izumi, *Agric. Biol. Chem.* **49** (1985) 2961.
111. W.D. Hewson and L.P. Hager, *Porphyrins* **7** (1979) 295.
112. S. Kobayashi, M. Nakano, T. Kimura and A.P. Schaap, *Biochemistry* **26** (1987) 5019.
113. S. Colonna, N. Gaggero, A. Manfredi, L. Casella, M. Gullotti, G. Carrea and P. Pasta, *Biochemistry* **29** (1990) 10465.
114. S.-I. Ozaki and P.R. Ortiz de Montellano, *J. Am. Chem. Soc.* **116** (1994) 4487.
115. C. Bolm, *Angew. Chem. Int. Ed. Eng.* **30** (1991) 403.
116. E.J. Allain, L.P. Hager, L. Deng and E.N. Jacobsen, *J. Am. Chem. Soc.* **115** (1993) 4415.
117. P.R. Ortiz de Montellano, Y.S. Choe, G. DePillis and C.E. Catalano, *J. Biol. Chem.* **262** (1987) 11641.
118. S.-I. Ozaki and P.R. Ortiz de Montellano, *J. Am. Chem. Soc.* **117** (1995) 7056.
119. G. Iacazio, G. Langrand, J. Baratti, G. Buono and C. Triantaphyllides, *J. Org. Chem.* **55** (1990) 1690.
120. V.K. Datcheva, K. Kiss, L. Solomon and K.S. Kyler, *J. Am. Chem. Soc.* **113** (1991) 270.
121. P.H. Markus, A.L.J. Peters and R. Roos, *EP 542,348* (1993); *Chem Abstr.* **119** (1993) 93,732.
122. B.A. Halkier, *Phytochemistry* **43** (1996) 1.
123. G.P. Bolwell, K. Bozak and A. Zimmerlin, *Phytochemistry* **37** (1994) 1491.
124. M.A. Schuler, *Critical Reviews in Plant Sciences* **15** (1996) 235.
125. A.G. Prescott and P. John, *Annu. Rev. Plant Physiol. Plant Mol. Biol.* **47** (1996) 245.
126. C.A.Mihaliak, F. Karp and R. Croteau, *Methods Plant Biochem.* **9** (1993) 261.
127. Z. Pan, F. Durst, D. Werck-Reichhart, H.W. Gardner, B. Camara, K. Cornish and R.A. Backhaus, *J. Biol. Chem.* **270** (1995) 8487.
128. M. Petersen and H.U. Seitz, *Biochem J.* **252** (1988) 537.
129. M. Petersen, A.W. Alfermann, E. Reinhard and H.U. Seitz, *Plant Cell Rep.* **6** (1987) 200.
130. P.F.X. Kraus and T.M. Kutchan, *Proc. Natl. Acad. Sci. USA* **92** (1995) 2071.
131. M. Hotze, G. Schroder and J. Schroder, *FEBS Lett.* **374** (1995) 345.
132. O. Gotoh, *J. Biol. Chem.* **267** (1992) 83.
133. R.P.L. Lindberg and M. Negishi, *Nature* **339** (1989) 632.
134. T.M. Kutchan and M.H.Zenk, *J. Plant Res.* **3** (1993) 165.
135. T.M. Kutchan, A. Boch and H. Dittrich, *Phytochemistry* **35** (1994) 353.

Chapter 8

PRODUCTION OF BIOLOGICALLY-ACTIVE PROTEINS IN PLANTS

GEORGE P. LOMONOSSOFF

1. Introduction

The development over the past 15 years of systems which allow the expression of heterologous genes in plants has led to an upsurge of interest in their potential use as bioreactors (1-3). Plants have the advantage over other expression systems, such as bacteria, yeast, fungi and insect cells, in that they can "look after themselves", requiring little more than sunlight, water and appropriate soil; they do not require the use of high-tech fermenters and sterile culture conditions. Enormous biomass can be produced by simply growing crops in the field and thus plants could be used as renewable and low cost sources of proteins of pharmaceutical and industrial interest. Although many heterologous proteins have been expressed in plants, this chapter will be limited to discussing situations where the expressed peptide or protein (or its activity) is of interest for an *ex situ* application. It will not cover situations in which the expression of the heterologous protein is designed to have its effect *in situ* e.g. by altering a metabolic pathway or conferring a desirable phenotype, such as resistance to insect attack, upon the plant. Some of these topics are covered in other chapters in this book.

2. Methods for the Expression of Foreign Proteins in Plants

Two systems have been developed for the introduction of foreign genes into plants: stable genetic transformation and the use of autonomously replicating viral vectors. The principles of the two methods are shown Figure 1.

Stable genetic transformation involves the integration of heterologous genes into the chromosomes of individual plant cells using either *Agrobacterium*-mediated transformation or particle bombardment. Cells harbouring the introduced sequence are selected, usually on the basis of antibiotic resistance conferred by a cointegrated gene, and plants are regenerated from these cells (for a review of the process, see (4)). The regenerated plants can then be used as a source of seed for subsequent

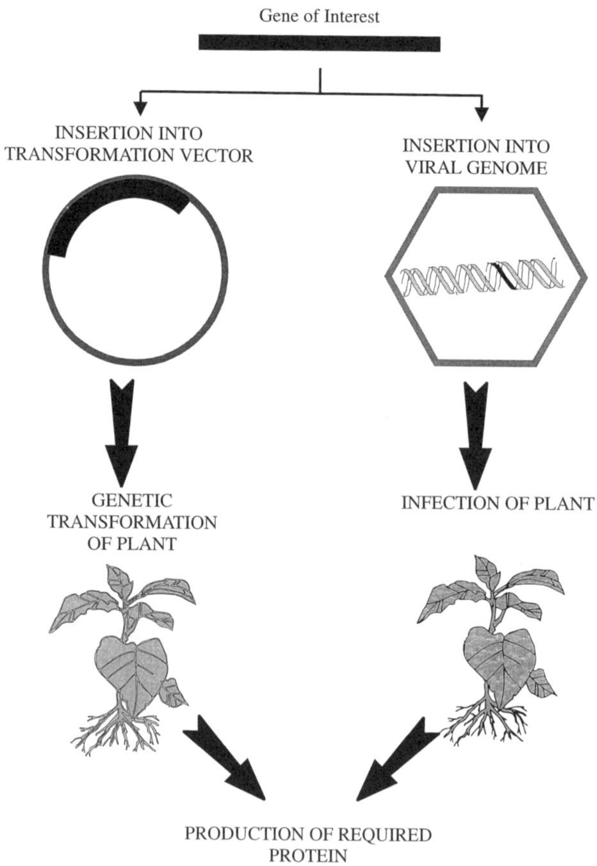

Figure 1. Representation of the two principal methods for expression of heterologous proteins in plants. In the route shown on the left, the gene of interest is inserted into a plasmid downstream of a plant-specific promoter. The construct is then genetically transformed into an appropriate plant species. In the route on the right, the gene of interest is inserted into a plant viral genome which is then used to infect a susceptible host plant.

generations of transformed plants. Expression of the integrated transgene is driven by a plant-specific promoter positioned immediately upstream of the coding sequence. This approach has been used successfully to introduce a wide variety of genes into an increasing number of plant species.

The alternative to stable genetic transformation, the use of plant virus-based vectors, has had a somewhat chequered history (5,6). Early attempts to use DNA plant viruses were only moderately successful (7,8) and it is only within the past 5 years that practically useful vectors, based on positive-strand RNA plant viruses, have been developed. The comparatively late development of such vectors was due to the difficulty in genetically manipulating RNA. However, the demonstration that cloned cDNA copies of RNA viral genomes can be transcribed *in vitro* to give infectious RNA (9,10) dramatically changed the situation and the genetic manipulation of RNA plant viruses is now routine. Indeed it is now possible to infect plants directly with a DNA copy of a RNA viral genome provided the viral sequence is cloned directly downstream of a plant promoter (11). In the RNA virus-based systems which have been successfully used to date, the heterologous sequence is introduced as an addition to the viral genome rather than substituting for a viral gene. Expression of the foreign gene in the target plant can be achieved by inoculating seedlings with modified virus, a procedure which can be as simple as rubbing the viral inoculum on a leaf in the presence of an abrasive.

The two alternative systems have their relative strengths and weaknesses. Stable genetic transformation and the subsequent regeneration of transformed plants can be a long and tedious business, often requiring several months or even years. The process is rather uncontrolled leading to the introduction of a varying number of copies of the gene of interest. This can lead to unexpected and unwanted effects such as transgene silencing and cosuppression (12,13). In addition, expression levels are often quite low though this is not invariably the case. On the other hand, stable transformation has certain advantages: the inserted gene is heritable and it is possible to cross transgenic lines to produce plants expressing more than one polypeptide chain simultaneously. To date, stable genetic transformation has been the main route for the expression of heterologous proteins in plants.

By contrast to the transformation/regeneration approach, the introduction of a foreign sequence using viral vectors is comparatively straightforward and quick. Many plant viruses, especially those which have been developed as vectors, grow to very high titres in infected plants with the consequent expression of large amounts of the heterologous protein. However, because the germline of the plant is not modified, the introduced gene is not heritable (unless a seed-transmissable virus were to be used) and it is necessary to inoculate each plant individually. Though this does not pose a problem when plants are grown on a laboratory scale it could be difficult when industrial-scale production is undertaken. There may also be limits to the size of gene which can be inserted into a viral genome without affecting the viability of the virus.

3. Choice of Plant

In many of the early studies on the expression of heterologous genes, the choice of plant was largely governed by the ease with which it could be transformed and regenerated. Thus many of the examples in this chapter involve the expression of genes in tobacco (*Nicotiana tabacum*). Though tobacco may be an appropriate species for some applications, it is far from ideal for many. This is especially true in situations where the expressed protein is intended for pharmaceutical use since tobacco plants produce an array of toxic alkaloids which must be removed from the final product. Likewise, tobacco is not edible so it is clearly unsuitable for applications which involve the direct feeding of plant material to animals. However, the number of plant species which can be efficiently transformed and regenerated is increasing all the time (see, for example, (14)) and it should shortly be possible to transform whichever species is most appropriate for a particular purpose. Likewise, viral vectors with a variety of host-ranges will undoubtably be developed.

A feasibility study on the potential of alfalfa for the large scale production of enzymes for industrial use has recently been carried out (15). Alfalfa has the advantage that it is a perennial, hardy crop that can be harvested 3 or more times a year. Being a legume it does not require the application of nitrogenous fertilisers and it also does not require irrigation. The target enzymes for expression were Mn^{2+}-dependent lignin peroxidase (isozyme H4) from *Phanerochaete chryosporium*, an enzyme which is of interest to the paper industry, and the α-amylase from *Bacillus licheniformis*, an enzyme which had previously been expressed in transgenic tobacco (see below). The efficiency of the transformation/regeneration process was initially monitored by examining the expression of the marker gene, β-glucuronidase (GUS). As a prelude to the large-scale production of plants expressing the target enzymes, methods for their extraction were evaluated by following the recovery of three endogenous enzyme activities. The information gained from these pilot studies should be of help in evaluating the economics of the large scale production of proteins in plants.

Another plant which has been investigated as a potential means of producing large amounts of heterologous proteins is the rubber tree, *Hevea brasiliensis*. This species is widely cultivated in south-east Asia and its latex is the source of natural rubber Latex is harvested by tapping and contains approximately 10 mg/ml soluble protein. If heterologous proteins could be expressed in the latex, they could be harvested in a non-destructive and continuous manner. Since a mature rubber tree can be tapped every other day and produces 70-150 ml of latex per tapping, transgenic trees could be sources of large amounts of proteins (16). Preliminary experiments have shown that it is possible to transform *Hevea* with the GUS gene (17) and expression of GUS could be detected in the latex of regenerated plants (unpublished results quoted in (16)).

4. Examples of the Production of Heterologous Proteins in Transformed Plants

4.1. Pharmaceuticals

4.1.1. Peptides

There are a large number of biologically-active peptides which are of pharmaceutical interest. In many cases only tiny amounts are required for biological activity and hence only small amounts are produced in their natural source. This makes isolation extremely expensive and a number of heterologous systems have been investigated as a way of producing relatively large quantities of such peptides cheaply. The first attempt to produce a pharmocologically-active peptide in plants involved the expression of the pentapeptide, Leu-enkephalin, fused to the 2S albumin from *Arabidopsis* (18). The 2S albumins are major seed storage proteins in many species and their purification is facilitated by their small size and specific solubility in low salt solutions. The proteins are synthesised as prepro-polypeptides which processed to give two subunits of 9 and 3-4kDa linked by disulphide bridges. The Leu-enkephalin sequence, flanked by lysine residues, was substituted for amino acids 35-41 of the large subunit and the modified sequence transformed into both *Arabidopsis* and *Brassica napus* (18). Following purification, the peptide was released from the albumin by a combination of trypsin and carboxypeptidase B digestion and purified by HPLC. In the case of *Arabidopsis* the final yield of purified peptide was 206nmol (113μg) enkephalin per gram of seed while for *B. napus* the maximum yield obtained was 50nmol per gram. The latter figure translates to 75 grams of peptide per hectare of rape, assuming an average productivity of 3000 kg seed per hectare.

Though in theory the above system could be used to express a variety of different peptides, the degree to which seed storage proteins can tolerate the insertion of heterologous sequences is not clear. Any insertion which interferes with post-translational processing or correct folding is liable to produce a protein which will probably be degraded quickly. Such an effect has been observed during an attempt to increase the methionine content of the 7S globulin, β-phaseolin, through the insertion of a synthetic 15 amino acid sequence (19).

4.1.2. Blood proteins

To investigate whether plants could express and correctly process a human protein, Sijmons *et al.* (20) transformed potato plants with a sequence encoding human serum albumin (HSA). In humans, HSA is synthesised as a prepro-protein in the liver. This is subsequently processed, by the sequential removal of the N-terminal pre- and pro-sequences (18 and 6 amino acids, respectively), to give a mature protein of 585 amino acids. Expression was driven by the cauliflower mosaic virus (CaMV) 35S promoter and translation maximised through the use of the leader sequence alfalfa mosaic virus (AlMV) RNA-4, a well-characterised translational enhancer (21). The sequence encoding the mature HSA protein was fused either to its homologous prepro-sequence or the leader peptide from the secreted pathogenesis-related (PR) protein S from tobacco. The maximum expression level achieved was 0.02% of soluble protein in leaf extracts. When purified, the plant-expressed protein appeared to have identical properties to natural HSA. N-terminal sequence analysis showed that while the protein fused to the PR-S leader was correctly processed, that expressed behind the human prepro-sequence still had the 6 amino acid pro-sequence attached. Thus in order to achieve correct N-terminal processing of a heterologous protein it is probably advantageous to use a plant-derived leader peptide. It was estimated that if expression could be maximised and targetted to the tubers, it might be possible to produce as much as 12kg of HSA per hectare of potatoes.

To investigate the potential of plants as a possible source of haemoglobin-based blood substitutes, the α and β globin chains of human haemoglobin (HbA) have been co-expressed in tobacco (22). The globin chains were individually fused to a 57 amino acid chloroplastic transit peptide and expression was driven by the CaMV 35S promoter. Expression of HbA reached 0.05% of total extracted protein in the seeds of the transgenic lines. The plant-expressed globin chains were the same size as their human equivalents, indicating that the signal peptide had been correctly cleaved. Flash photolysis studies on HbA extracted from the seeds indicated that the molecules were functional.

4.1.3. Cytokines and hormones

Interferons have attracted considerable attention as therapeutic agents over the past two decades and there have been numerous attempts to produce large quantities of material in a variety of expression systems. It has proved possible to produce detectable quantities of both human α- and β-interferons in transgenic tobacco (23,24). When rice was transformed with a sequence encoding human α-interferon under the control of the P' portion of the dual promoter from *Agrobacterium tumefaciens*, interferon activity could be detected in leaf extracts (25). The levels obtained (512-2048 units/gram of tissue) were similar to those previously achieved with tobacco (23).

Erythropoietin (Epo) is a cytokine which regulates the production of erythrocytes in mammals and is used therapeutically in the treatment of anaemia. Mature human Epo consists of 166 amino acids and is heavily glycosylated. The glycosylation, while not essential for activity in *in vitro* assays, plays an important role *in vivo* by prolonging the half-life of the protein in the bloodstream. Attempts to express human Epo in cultured tobacco cells without first fusing it to a signal peptide resulted in the accumulation of very low levels of protein (26). To overcome this, the sequence was fused to its homologous leader peptide, placed under the control of the CaMV promoter and transformed into cultured tobacco cells (27). One of the resulting calli was found to produce detectable levels of Epo and was subsequently maintained in suspension culture. Localisation studies indicated that although the protein was secreted across the plasma membrane, it was retained by the cell walls and was not released into the culture medium. Epo could be purified from cell extracts by immunoaffinity chromatography, the yield of purified protein being 13.1 μg from 6.3 kg of tobacco cells. N-terminal sequence analysis showed that the leader sequence had been correctly removed. An *in vitro* biological assay of the ability of Epo from tobacco to stimulate the formation of erythroid colonies showed that the protein was at least as active as Epo produced in mammalian cell culture. However, the tobacco protein was unable to stimulate erythrocyte formation when injected into rats. This lack of *in vivo* activity was attributed to instability of tobacco-expressed protein in the bloodstream due to the absence of sialic acid residues.

In an experiment to determine whether heterologous proteins are correctly folded when expressed in plants, the rainbow trout growth hormone (tGH-II) was transformed into tobacco (28). When passage of the protein into the secretory pathway was mediated by fusing it to the signal sequence from PR-1b, it was found to contain the correct disulphide bridges and to be partially glycosylated.

Unfortunately the amounts of protein obtained were too low to permit purification of the protein for an assessment of its biological activity. When tGH-II was expressed without the signal peptide, far lower levels of protein could be detected suggesting that, when expressed in the cytoplasm, the protein is rapidly degraded. No tGH-II could be detected in the seeds when *Arabidopsis* was transformed with constructs in which expression was under the control of a seed-specific promoter.

4.1.4. Immunogens

Since their first introduction approximately 200 years ago, vaccines have made an enormous contribution to both human and animal health. Traditionally vaccines are produced from attenuated or inactivated forms of the disease-causing organism. However there are many diseases where this approach is not feasible either because it has proved difficult or impossible to grow the pathogen in sufficient quantities in culture or because it is has not been possible to produce vaccines from such pathogens of sufficient guaranteed safety. Because of these constraints, there has been much interest in producing non-infectious immunogenic material in heterologous systems.

The first attempt to produce immunogenic material in plants involved transforming tobacco with a gene encoding the 24kDa hepatitis B surface antigen (HBsAg) (29). This protein can form virus-like particles (VLPs) of 17 nm diameter when expressed in yeast; injection of such particles into healthy subjects confers protective immunity. To express the protein in tobacco the gene was inserted downstream on an enhanced CaMV promoter and a translational enhancer derived from the 5' non-coding region of tobacco etch virus (TEV). Levels of expression of approximately 0.01% of soluble protein were achieved and VLPs with an average diameter of 22 nm could be immunoaffinity purified from leaf tissue (Figure 2). The plant-derived particles had a similar buoyant density to those expressed in yeast and when a partially purified preparation was injected into mice the production of anti-hepatitis B antibodies was similar to that obtained with yeast-expressed HBsAg (30).

To investigate the possibility of using plants to produce oral vaccines, Haq *et al.* (31) transformed potato and tobacco with the binding subunit of the *E.coli* heat-labile entertoxin (LT-B), a highly potent oral immunogen. As with the HBsAg example above, expression was driven by the CaMV 35S promoter and the TEV leader sequence. It was found that protein accumulation was enhanced when a hexapeptide microsome-retention sequence was fused to the C-terminus of LT-B although levels still only represented about 0.01% of total protein. When a crude extract of tobacco-derived LT-B was administered to mice by gastric intubation, it stimulated humoral and mucosal immune responses at a level equivalent to that achieved by the oral administration of bacterially expressed LT-B. The ability of plant-expressed LT-B to induce oral immunity was further verified by feeding slices of raw tubers from transgenic potatoes to mice. After 4 feedings of 5 g of transformed tuber, mucosal and serum antibodies against LT-B could be recovered. Similar success was obtained with tubers from potatoes transgenic for, and expressing, the capsid protein of Norwalk virus (32).

To develop oral vaccines for human use, it will be necessary to express the immunogen in a plant which is palatable when eaten raw. With this in mind, tomatoes have been transformed with either HBsAg (33) or the G-protein of rabies virus (34). The latter protein, which coats the outer surface of rabies virus, had previously

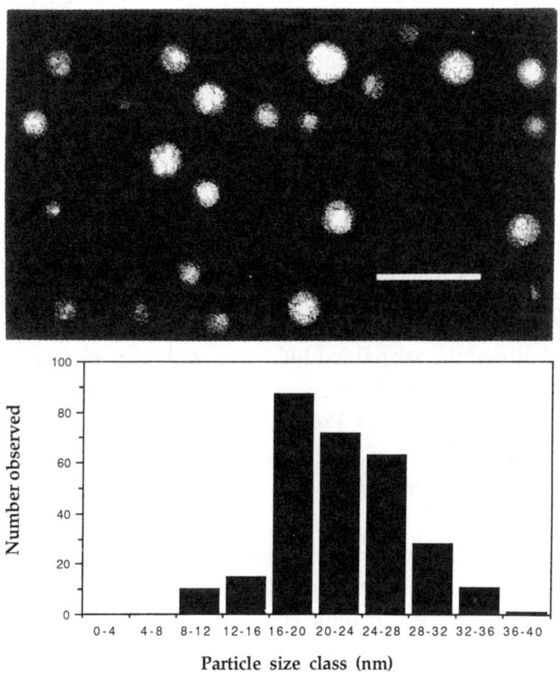

Figure 2. Upper Panel: Electron micrograph of hepatitis B virus surface antigen (HBsAg) particles immunoaffinity purified from a transgenic tobacco plant (Bar = 100nm). Lower Panel: Histogram of the distribution of the diameters of the plant-expressed HBsAg particles. Reproduced from (29) with permission.

been shown to be able to confer protective immunity in raccoons when administered orally. Expression of the G-protein was driven by the CaMV promoter but no leader peptide was fused to the protein. The recombinant G-protein could be detected at low levels (1-10 ng/mg soluble protein) in the leaves and fruit of transformed tomatoes and appeared to have the anticipated antigenic properties. To further extend the possibilities of producing oral vaccines for human use, a banana transformation/ regeneration system has recently been developed (35) and it is anticipated that this crop could be of particular use in tropical regions (36).

4.1.5. Antibodies

Antibodies are a vital component of the immune system which bind antigens with a very high specificity. They are widely used as diagnostic reagents, therapeutics and as reagents for the purification of proteins. The first example of the production of an antibody in plants concerned the expression of an intact mouse IgG in transgenic tobacco plants (37). This antibody consists of two identical heavy and two identical light chains covalently linked by disulphide bridges. Sequences encoding the light and heavy chains (plus their respective leader sequences) were separately transformed into tobacco plants behind the CaMV 35S promoter. Lines of plants expressing the individual chains were identified and sexually crossed to produce a F1 generation which would express both chains. The chains were shown to be correctly assembled into full-length antibodies which had the same binding properties as the original mouse antibody (37,38). Interestingly, the levels of intact antibody which accumulated in the F1 plants (up to 1% soluble protein) were considerably higher than the levels of the individual chains in the parental plants. This presumably reflects the enhanced stability of the assembled IgG. The antibodies produced in plants were glycosylated though there were differences in the glycan moiety between the plant- and murine-derived antibodies (39). Full-length antibodies have also been produced in plants either through the use of double-transformation (40) or by introducing the two chains simultaneously on a single *Agrobacterium* T-DNA (41,42).

Using the crossing approach for the generation of assembled protein complexes, Ma *et al.* (43) were able to express secretory IgA (SIgA) in tobacco plants. Like IgG, SIgA is composed of heavy and light chains but the full-length antibody is then dimerised by the J chain polypeptide and complexed with a fourth polypeptide, secretory component (SC). These modifications increase the avidity of binding of the antibodies and confer a degree of resistance to proteolysis. To produce an intact SIgA, genes encoding the heavy and light chains of a murine monoclonal antibody known as Guy's 13, a murine J chain and a rabbit SC were separately transformed into tobacco under the control of the CaMV 35S promoter (44,45). Through the series of crosses shown in Figure 3, plants expressing assembled, functional SIgA were obtained. Since Guy's 13 is a monoclonal antibody which recognises a cell-surface protein from the causative agent of dental caries, *Streptococcus mutans*, and topical application of the murine antibody had been shown to protect teeth against colonisation by the bacterium (46), it is hoped that plant material expressing the SIgA might be useful in combatting tooth decay (45).

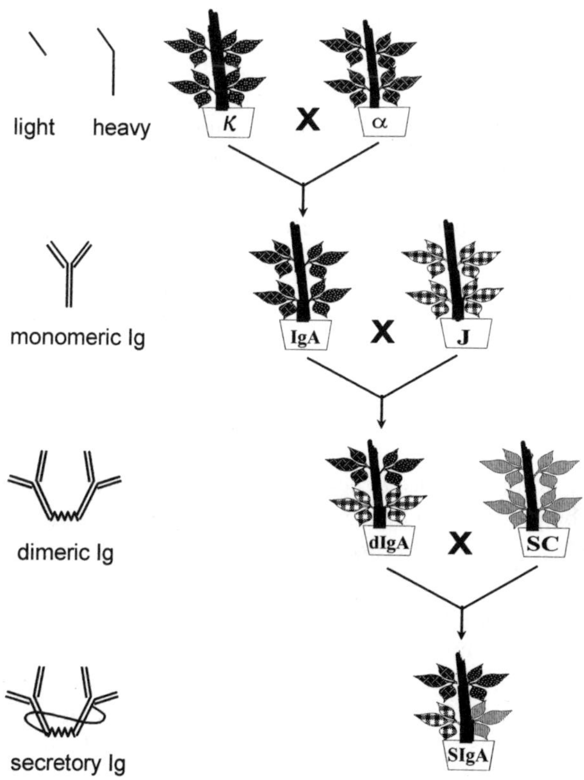

Figure 3. Assembly of a complete molecule of secretory IgA (SIgA) in plants. The genes encoding the four polypeptides of SIgA were individually introduced into transgenic tobacco plants which were subsequently cross-pollinated to generate a line of plants expressing the complete molecule (Ig = immunoglobulin; dIgA = dimeric IgA; SC = secretory component). Reproduced from (45) with permission.

As well as full-length antibodies, antibody fragments such as Fab, single-domain antibodies and single-chain Fv (scFv) antibodies have been expressed in plants (40,47,48). The latter consist of the variable (v) regions of the light and heavy chains joined by a synthetic polypeptide linker and are particularly suitable for the construction of bifunctional molecules such as immunotoxins. Because of their small size and lack of assembly requirements, it has proved possible to target scFv expression to particular compartments within the cell by linking them to appropriate signal sequences (49,50). By fusing the sequence of a scFv to the leader sequence from the seed storage protein, legumin B4, and driving expression from a seed-specific promoter from *Vicia faba*, Fiedler and Conrad (50) were able to direct expression of the protein to the seeds of transgenic tobacco plants. Functionally-active scFv accumulated to up to 0.67% of total seed protein and there was no loss of protein or antigen-binding material when mature seed was stored for one year at room temperature. To date, however, such antibody fragments have mainly been expressed for *in situ* applications such as perturbing the growth of plants (48,51) or conferring resistance to a pathogen (52). However, Owen *et al.* (48) did show that an anti-phytochrome scFv could be purified from transgenic plants by affinity chromatography.

4.2. Industrial Enzymes

4.2.1. α-Amylase

The α-amylase from *Bacillus licheniformis* was the first enzyme of industrial importance to be produced in transgenic plants (53). This enzyme is the one most commonly used in starch liquefaction, the first step in starch processing, because of its heat stability and activity over a wide pH range. To investigate whether this enzyme could be usefully expressed in plants, the *B.licheniformis* α-amylase gene was placed under the control of a modified CaMV 35S promoter, fused to either its homologous signal peptide or the signal peptide from the tobacco PR-S protein and transformed into tobacco. The maximum level of expression obtained in the leaves of transgenic plants was approximately 0.3% of total soluble protein when either signal peptide was used. The activity of the enzyme per mg of soluble protein was approximately 30 times higher in the extracellular fluid compared with total leaf extract implying that the enzyme was secreted. The presence of significant quantities of bacterial α-amylase did not appear to affect the phenotype of the plants. When seeds from the transgenic plants were milled and tested in a liquefaction assay using corn and potato starch, the patterns of degradation were similar to those obtained with the authentic bacterial enzyme. Furthermore, the plant-expressed enzyme retained its thermostability.

4.2.2. β-Glucanases

The degradation of β-glucans, the major constituent of the cell walls of the starchy endosperm of barley, is a vital part of the brewing process. The degradation takes place during malting through the action of (1-3, 1-4)-β-glucanases. These enzymes are normally heat-inactivated during kiln drying of malt and the subsequent mashing process and there has been a considerable interest in expressing thermostable versions in barley (54).

The first investigation of the properties of a thermostable β-glucanase expressed in barley predated the development of a system for the stable transformation of this plant and therefore relied on transient expression. Philipson (55) constructed a hybrid gene in which the 5' half was derived from the gene encoding the glucanase of *Bacillus amyloliquifaciens* and the 3' half from the corresponding gene from *B. macerans*. This was done so that the hybrid enzyme would have the low pH optimum of the former combined with the thermostability of the latter. When the hybrid sequence was transfected into barley aleurone protoplasts, activity could be detected when expression was driven by the constitutive pEMU promoter but not when the aleurone-specific α-amylase promoter was used. When no signal peptide was present, most of the β-glucanase activity was intracellular; however when the hybrid gene was inserted behind a sequence encoding an α-amylase signal peptide, 90% of the activity was extracellular.

The first stable transformation experiments involved the expression of the heat-stable β-glucanase from the fungus *Trichoderma reesei* in barley suspension culture cells (56). The sequence encoding the enzyme and its signal sequence was placed under the control of the CaMV 35S promoter and introduced into the cells by particle bombardment. The enzyme could be detected in the culture filtrate at levels of up to 0.5mg per litre of medium. The enzyme was active at 50°C and was able to degrade β-glucans during mashing. Attempts to regenerate transformed tobacco expressing the gene proved unsuccessful and it was observed that tobacco protoplasts transiently expressing the enzyme did not regenerate cell walls. This was interpreted as indicating that the expression of the β-glucanase interfered with cell wall synthesis, thereby preventing regeneration. This interpretation is questionable since dicots, such as tobacco, do not contain (1-3, 1-4)-β-glucans in their cell walls. Indeed, it has subsequently proved possible to express the β-glucanase domain (XYLD-C) encoded by the XynD gene of *Ruminococcus flavefaciens* in transgenic tobacco without a detrimental effect on the plant (57). Though the reason for the different results obtained with the the β-glucanases is not certain, one possibility is that the enzyme from *Trichoderma reesei* contains additional activities and it is these which inhibit cell wall synthesis in tobacco.

4.2.3. Xylanases

Xylanases convert xylans, which are present in the hemicellulose fraction of plant cell walls, to xylose. This activity is of considerable interest in a number of industrial areas such as the production of animal feed from the wastes of the paper-making, the removal of xylans from cellulose pulps and the selective removal of xylans from plant fibres. The first instance of the expression of a xylanase gene involved transforming tobacco plants with a gene derived from the yeast *Cryptococcus albidus* (58). This gene contains seven introns and it was demonstrated that these were correctly spliced in the transgenic plants. However, as the study was essentially concerned with splicing, no attempt was made to measure the activity of the expressed xylanase. Subsequently, tobacco was transformed with a sequence (XynZ) encoding the N-terminal domain of the thermostable xylanase from the bacterium *Clostridium thermocellum* (59). The CaMV 35S promoter was used to achieve constitutive expression and the protein was targeted to the apoplast by fusing the gene to the sequence encoding the signal peptide of proteinase inhibitor II. It was anticipated that the expression of high levels of xylanase in the plants would not interfere with their growth since the hemicellulose of dicotyledons, in contrast to that of monocotyledons, is composed mainly of xyloglucans (which are not susceptible to xylanase digestion) rather than xylans. Significant quantities of the XynZ protein could be detected in total leaf extracts of transformed plants and when the intercellular fluids were examined, XynZ made up more than 50% of the total protein. Total leaf extracts were shown to contain xylanase activity with a temperature optimum of 70°C. Treatment of leaf extracts at 60°C resulted in a 14- to 31-fold enrichment of enzymatic activity since at this temperature more than 90% of normal plant proteins are denatured. Expression of the xylanase domain of the xynD gene from *Ruminoccus flavefaciens* in tobacco also resulted in the synthesis of detectable xylanase activity but at a lower level (57).

4.2.4. Phytase

Phytate (myo-inositol hexaphosphate) is the principal storage form of phosphorus in plant seeds which are used in animal feeds. In plants, the stored phosphorus can be released through the action of phytases; monogastric animals, however, lack such enzymes and cannot utilise phytate. As a result, inorganic phosphate often has to be added as a feed supplement. An effective alternative is the addition of a phytase from

Aspergillus niger to the feed of pigs and poultry. This enzyme has a low pH optimum and is therefore active in the stomach. To exploit this observation, tobacco has been transformed with the gene encoding the *A. niger* phytase (60). To direct the enzyme to the apoplast, the coding sequence was fused in-frame to the signal peptide from PR-S and expression was driven by the CaMV 35S promoter. Phytase could be detected in the resulting plants, levels eventually reaching as much as 14.4% of total soluble leaf protein in mature plants (61). These levels of protein correspond to approximately 0.3% of total leaf dry weight and the ability of the protein to accumulate to this extent indicates that it is very stable. This confirmed the earlier observation that phytase activity in seeds of transgenic tobacco plants did not decline during storage for a year (60). Milled seeds from transgenic, but not control, plants were able to liberate phosphate from animal feed under conditions which mimicked those in the stomach (60). When milled seeds from transgenic plants were used to supplement the diet of chickens, the effect on growth was similar effect to that achieved by the addition of either inorganic phosphate or the phytase obtained from *A. niger*.

5. Examples of the Expression of Proteins Using Virus-Based Vectors

5.1. Expression of Whole Proteins

The double-stranded DNA virus, CaMV, was the first plant virus to attract attention as a candidate vector for the expression of foreign genes in plants. The genome of CaMV is circular and approximately 8kb in size (for a more detailed description of the genetic organisation of CaMV, see (5)). Gene II, which encodes a protein responsible for aphid transmissibility of the virus, was identified as being dispensable for virus replication and was the only region of the genome into which substantial lengths of foreign DNA could be inserted without abolishing virus infectivity (62). When a 200bp sequence encoding a mammalian metallothionein (MTII) was inserted into the gene II region, levels of MTII reached 0.5% of soluble protein in systemically-infected turnip leaves (63). De Zoeten *et al.* (64) subsequently replaced gene II with a gene encoding human interferon αD (IFN αD) and demonstrated that the modified virus could systemically infect turnip plants and direct the synthesis of biologically-active IFN αD. The maximum level of expression achieved corresponded to 196,830 international units of interferon per ml of plant sap. It was estimated that this corresponded to a yield of 2μg of IFN αD per gram of tissue fresh weight. However, no attempt was made to purify the protein.

Tobacco mosaic virus (TMV) is a rod-shaped, positive-strand RNA virus which reaches enormous titres in plants (65). Virus particles contain a single RNA of 6.4kb surrounded by more than 2000 copies of the 158 amino acid coat protein arranged with helical symmetry. The wild-type viral genome contains 3 open reading frames (Figure 4) but can be modified to express additional genes under the control of their own subgenomic promoters (66, 67).

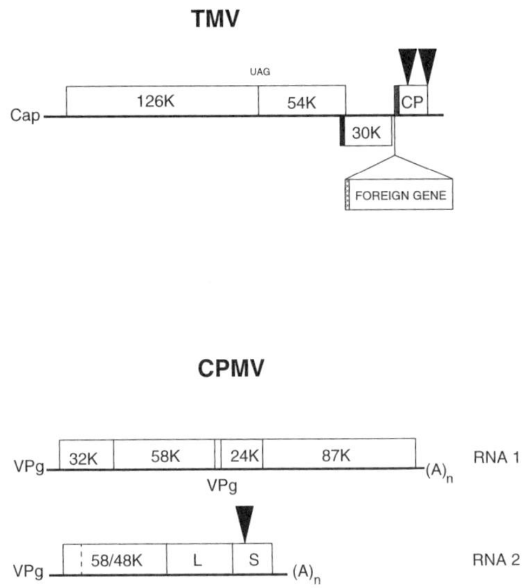

Figure 4. Genome organisations of two positive-strand RNA plant viruses which have been used as vectors. The open boxes represent the open reading frames on the viral RNAs and the protein products are labelled (e.g. 126K). In the case of TMV, the position of insertion of a foreign gene, complete with its own subgenomic promoter, is indicated below the wild-type genome structure. For both TMV and CPMV, the sites of insertion of peptides into the viral coat proteins are indicated by the filled arrows. Cap = ^{7}meGpppG, VPg = viral protein, genome-linked, $(A)_n$ = 3' poly(A) tail, CP = the coat protein of TMV, L and S = large and small coat proteins, respectively, of CPMV.

A TMV-based vector has been used to express α-trichosanthin (68). This is a plant-derived 27kDa ribosome-inactivating protein which can inhibit the replication of human immunodeficiency virus (HIV) in infected cells. The sequence encoding α-trichosanthin linked to a leader peptide was inserted into the viral genome between the regions encoding the the 30K movement protein and the viral coat protein (Figure

4). When the modified virus was inoculated on to *Nicotiana benthamiana*, it spread systemically throughout the plants and gave yields of virions similar to that obtained by infection with wild-type virus. α-Trichosanthin accumulated in the upper, systemically-infected leaves at levels corresponding to at least 2% of total soluble protein (68). The purified protein was shown to have been correctly processed at its N-terminus and to possess the ribosome-inactivating properties associated with natural α-trichosanthin. These results provided the first evidence that a RNA plant virus can be used to express high levels of a pharmacologically interesting protein.

5.2. Expression of Pharmacologically-Active Peptides

5.2.1. Pharmacologically active peptides

In an attempt to make use of the potentially high productivity of a TMV-based system, a pentapeptide corresponding to Leu-enkephalin was fused to the exposed C-terminus of the viral coat protein (69; Figure 4). A methionine residue was placed between the last amino acid of the coat protein and the enkephalin sequence so that active enkephalin could be released from the fusion protein by treatment with cyanogen bromide. When inoculated on to tobacco plants, the modified virus gave only local lesions and failed to spread systemically. Experiments in protoplasts demonstrated that the coat protein-enkephalin fusion protein was synthesised in substantial quantities but that the modified coat protein was unable to assemble into virus particles. This lack of assembly was presumably responsible for the lack of systemic spread of the virus in plants. Pure Leu-enkephalin could be obtained following treatment of protoplast extracts with cyanogen bromide (69). However, the necessity of using a plant protoplast system for the propagation of the recombinant virus clearly limited the potential for this particular system.

To overcome the problem of particle formation, Hamamoto *et al.* (70) designed a new vector which permitted the synthesis of both fused and unfused forms of the coat protein from the same RNA. This was achieved by incorporating a sequence which promotes readthrough of amber stop codons between the end of the coat protein gene and an in-phase sequence encoding the 12 amino acid peptide, angiotensin-I-converting enzyme inhibitor (ACEI). In addition an arginine codon was inserted immediately upstream of the ACEI coding sequence so that any ACEI produced could be cleaved from the coat protein by digestion with trypsin. The efficiency of readthrough could be varied between approximately 0.5% and 5% so that the ratio of modified: unmodified coat protein could be varied between 1:200 and 1:20. The presence of unmodified coat protein enabled virus assembly to occur normally and infections could spread systemically. Somewhat unexpectly, the modified subunits were also incorporated into the virions. Levels of ACEI-modified coat protein reached 100 μg/g fresh tissue in the leaves of either tobacco or tomato plants and tomato fruit was found to contain up to 10 μg/g of the modified coat protein. As yet no data have been presented regarding the pharmacological activity of the ACEI produced in plants though it was suggested that tomatoes expressing ACEI could act as dietary antihypertensives (70).

5.2.2. Vaccines

The last few decades have seen many attempts to produce novel stable and well-defined vaccines through the use of synthetic oligopeptides. Although it has been demonstrated that such oligopeptides are capable of stimulating the production of antibodies against the protein from which they are derived, the peptides themselves have generally been found to be insufficiently immunogenic to serve as useful vaccines. However their immunogenicity can generally be improved if they are coupled to large carrier molecules. There has therefore been considerable interest in developing systems (referred to as "epitope-presentation systems") in which the peptide sequence can be biochemically fused to a suitable carrier molecule. Many of the epitope-presentation systems which have been developed involve linking the epitope of interest to proteins which can self-assemble into macromolecular structures (71,72). The high yields and capsid stability of many RNA plant viruses make their coat proteins particularly attractive candidates for use as such carriers.

The first plant virus to be investigated as a potential epitope-presentation system was cowpea mosaic virus (CPMV). The genome of this virus consists of two molecules of positive-strand RNA which are separately encapsidated in icosahedral particles. Each RNA contains a single open reading frame and the virus uses a polyprotein processing strategy for the production of its mature proteins (Figure 4). The virus particles contain 60 copies each of a large (L) and small (S) coat protein which are arranged with pseudo T=3 (P=3) symmetry (73). Both coat proteins are encoded by the smaller genomic RNA, RNA 2. The initial attempt to produce particles displaying a foreign peptide involved the insertion of the immunodominant epitope from foot-and-mouth disease virus (FMDV) into the S coat protein (74). The precise position for the insertion (the βB-βC loop) was chosen on the basis of the known 3-dimensional structure of the virions (73). When the modified RNA 2 was inoculated on to cowpea plants in the presence of RNA 1, particles were produced that had the antigenic properties of the FMDV insert. However, the initial construct did not grow particularly well and proved to be genetically unstable, reverting to wild-type on serial passaging (75). By refining the site of insertion it has proved possible to produce genetically stable constructs expressing epitopes from a number of human and animal pathogens (75,76) which in many cases grew as well as wild-type CPMV. Immunological analysis of particles expressing epitopes from gp41 of HIV-1, VP1 of human rhinovirus 14 (HRV-14) and VP2 of the parvovirus mink enteritis virus (MEV) have shown that the particles possess the antigenic properties of the insert (75-77; Figure 5). The chimaeras have also been shown to be immunogenic: particles containing the HIV-1 epitope were able to raise antibodies in mice which are capable of neutralising an HIV-1 infection of laboratory T-cells (78,79) while particles expressing the MEV epitope have recently been shown to capable of protecting mink from a lethal challenge with MEV (76). The latter observation was the first demonstration of plant-derived material acting as an effective vaccine.

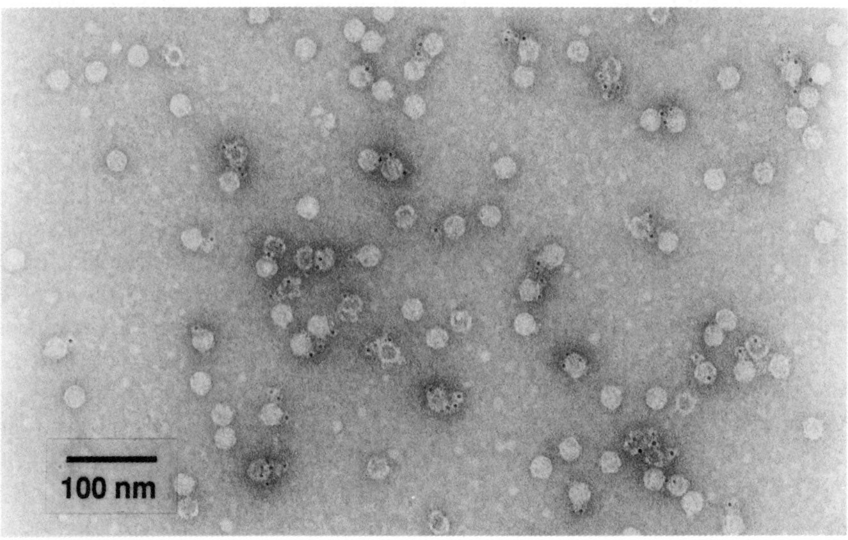

Figure 5. Particles of cowpea mosaic virus which have been modified to express an epitope from gp41 of HIV-1. The presence of the epitope on the surface of the virus particles is shown by immunogold-labelling of the virions using a monoclonal antibody specific for the inserted sequence.

A number of other RNA plant viruses have been investigated as potentional epitope-presentation systems. Principal among these is TMV, since detailed information is available regarding the structure of its coat protein (80). Using the vector employed for the expression of ACEI (70), TMV particles have been produced in which approximately 5% of the coat protein subunits have epitopes derived from either influenza virus haemagglutinin (HA) or HIV-1 gp120 fused to their exposed C-terminus (81). The modified subunits were incorporated into particles and precipitin reactions with the appropriate antipeptide antisera supported the conclusion that the inserted epitopes were located on the surface of the virions. By a similar approach, a dimer of an epitope derived from the circumsporozoite (CS) protein of *Plasmodium yoelii* has been expressed at the C-terminus of the TMV CP (82). In a separate construct, three copies of an epitope from the *P. vivax* CS protein were substituted for two amino acids in a surface loop of the coat protein (82). When either construct was inoculated on to tobacco, mosaic symptoms were induced on upper leaves. Virus particles could be recovered from systemically-infected tissue in yields ranging from 0.4 to 1.2 mg/g fresh weight. Expression of the respective epitopes was confirmed by immunoblot analysis of the recombinant coat proteins using monoclonal antibodies specific for sporozoites.

In the case of the construct expressing the epitope downstream of the leaky terminator at the C-terminus of the CP, approximately 5% of the subunits were modified; all subunits were modified in the case of the construct containing the repeated epitope at the surface loop. Subsequently, by slightly modifying the site of insertion it has proved possible to produce particles in which all the subunits carry epitopes near the C-terminus of the coat protein (83). A construct in which a 13 amino acid murine epitope was inserted after residue 154 of the coat protein produced particles which could raise antibodies specific for the insert when injected into mice.

Another RNA virus which has been explored as a possible epitope-presentation system is the monopartite virus, tomato bushy stunt virus (TBSV). Its capsids are composed of 180 copies of a single coat protein of 386 amino acids arranged with T=3 symmetry. The coat protein contains three distinct domains: the random (R), shell (S) and protruding (P) domains (84). The P domain is formed from the C-terminal portion of the coat protein and, as its name suggests, protrudes from the surface of the virus particles. An epitope consisting of 13 amino acids from the V3 loop of gp120 from HIV-1 has been successfully expressed as a fusion to the C-terminus of the TBSV coat protein (85). The epitope could be detected on the surface of the purified virions and the chimaeric particles could bind anti-HIV-1 antibodies in sera from HIV-1-positive patients.

Potato virus X (PVX) has been used to express foreign sequences on its virion surface. PVX is a monopartite RNA virus whose genomic RNA is encapsidated in flexuous rod-shaped particles made up of multiple copies of a single coat protein of 236 amino acids. Though there is no detailed information available regarding the structure of the subunits, the N-terminal portion of the coat is believed to be exposed on the surface of the virions. Utilising this fact, sequences have been fused to the N-terminus via a sequence encoding the FMDV 2A peptide (86). This 16 amino acid peptide mediates processing of the FMDV-encoded polyprotein (87) and is functional in a heterologous situation (88). Its presence allows the simultaneous production of native and fused forms of the coat protein. A construct with the 27kDa green fluorescent protein (GFP) from the jellyfish *Aequoria victoria* fused to the coat protein was infectious when inoculated on to *N. benthamiana* or tobacco. Electron microscopy of infected tissue showed the presence of virus particles which were approximately twice as wide as normal PVX virions. Subsequent experiments have shown that it is possible to express a variety of proteins ranging in size between 8.5 and 31kDa on the surface of PVX virions (unpublished data quoted in (86)). As yet, no data have been published regarding the immunogenicity of polypeptides expressed on the surface of PVX.

6. Purification of Plant-Expressed Proteins and Peptides

For certain purposes, such as the development of oral vaccines (31,32) or the incorporation of an enzymatic activity in feed (60), purification of the expressed protein is not envisaged and hence does not present a problem. For many other cases some level of extraction, enrichment or purification of the product is necessary for its subsequent use. The degree of purification required will depend on whether protein is intended for industrial or pharmacological use. Unlike proteins destined for medicinal use, industrial enzymes must be available in large quantities at a low price. However, the degree of purity of the product is less of an issue and it will probably be possible to use fairly crude extracts for the desired biocatalytic reactions. In cases where a high degree of purity is required, the plant-derived proteins can probably be purified by standard biochemical techniques although these may need to be appropriately adapted. Unfortunately obtaining a high enough level of purity by conventional methods might prove very expensive and a number of approaches to simplifying the process have been investigated.

One solution to the problem of purification is to fuse the peptide or protein of interest to an easily purifiable carrier molecule. Examples of this approach can be seen in the case of expression of peptides on the surface of plant virus particles (e.g. 70,74) or as part of the 2S seed storage protein (18). In some cases, such as the production of vaccines based on modified plant virus particles, it is envisaged that the whole molecule (peptide + carrier) will be the final product. However, in cases where release of the peptide from the carrier is necessary (18,69,70), additional purification will be required.

An elegant example to the carrier approach has involved fusing heterologous proteins to oleosins (89). Oleosins are highly lipophilic proteins associated with the surface of oil bodies. They accumulate to high levels in oil seeds (such as those from rape) and their main function appears to be to limit the tendency of oil bodies to coalesce. When seeds are extracted with aqueous buffers, the oil bodies, with their associated oleosins, remain intact and can be readily separated from other cellular components by their tendency to float during centrifugation. To determine whether oleosins could be used as carriers, the GUS gene was fused via a thrombin cleavage site to the C-terminus of the oleosin from *Arabidopsis*. Expression of the fusion protein was controlled by the *Arabidopsis* oleosin promoter and the construct was used to transform *B. napus*. GUS activity could detected only in the seeds of transgenic plants, confirming the seed-specificity of the oleosin promoter. After fractionation of the seeds, approximately 80% of GUS activity was found to be associated with oil bodies. This represented a 13-fold enrichment of GUS activity after a single flotation/centrifugation step. The GUS moiety could be released from the fusion protein by treatment with thrombin, though Rooijen and Moloney (89) also proposed that certain uses, such as the development of immobilised enzymes, would not require such a release.

7. Regulatory Considerations

Although plants have been used as sources of food, clothing, building materials and medicinal compounds since time immemorial, it is only comparatively recently that they have been proposed as bioreactors for the expression of heterologous proteins. Though the use of plant-expressed proteins in industrial processes, such as paper production, will probably depend solely on the cost-effectiveness of production process, any application which involves administering recombinant proteins to animals will have to receive regulatory approval. Though many of the regulatory requirements (e.g. concerning purity) will be no different from that required for the pharmaceutical use of any protein, there are some particular concerns which apply to plant-derived proteins (90). One of the most important of these is how comparable plant-derived proteins are to their conventionally-produced equivalents. One of the main differences between plant- and animal-derived proteins is likely to be their glycosylation state. Generally it has been found that animal-derived proteins expressed in plants, though glycosylated at the correct sites, have smaller carbohydrate moieties attached (27,34,91). In addition, the nature of the terminal sugars differs between plants and animals, with plant glycoproteins in particular lacking sialic acid residues (92). This difference has a number of possible consequences for the stability and antigenicity of plant-expressed proteins. For example, the lack of sialic acid residues has been implicated in the failure of plant-derived erythropoietin to be active in *in vivo* assays (27). An unfamiliar glycosylation pattern may also lead to a plant-expressed protein inducing an undesirable immune reaction.

Another consideration concerns variation in levels of expression obtained with plants. This is likely to be of particular concern when plant material is intended to be used directly, as in the case of edible vaccines. Clearly the dosage will have to be carefully regulated and it will therefore be important that consistent levels of expression of the heterologous protein are obtained. This should be relatively straightforward to achieve with transgenic plants where true-breeding lines can be developed. It may require more careful monitoring when virus-based vectors are used as there can be considerable plant-to-plant variations in virus titres. If plants are to be developed as a reliable source of proteins, especially those for pharmaceutical use, attention will also have to be paid to the genetic stability of the expression systems (90). In the case of transgenic plants, seed banks of appropriate lines will need to established and the stability of transgene expression over a number of generations will need to be investigated. Virus-based expression systems will need to be monitored closely because of the tendency of virus populations to evolve rapidly by genetic drift. At one time, the problem of such genetic drift was thought be so severe as to preclude the use of virus-based vectors for gene expression (93). However, subsequent analysis of the rate of drift of sequences expressed in TMV (94) indicated that fears were somewhat exaggerated. Nonetheless, it will be important to establish primary virus stocks and to limit the number of passages of a virus infection.

A further consideration concerns the potential environmental impact of the field release of the genetically modified plants. With transgenic plants the potential problems include the spread of the transgene to other related species by pollen transfer and the possibility that the transgenic plants may incidentally have acquired a selective advantage over non-transformed plants. With virus-based systems it should be borne in mind that the vector is itself a pathogen which is normally able to spread rapidly and cause disease. At present, physical containment is used to prevent such spread but in may be advisable in the future to develop systems based on disabled viruses.

8. Future Prospects

At the time of writing this chapter, the first genetically modified plants (e.g. Flavr Savr™ tomatoes and insect-resistant maize) have already reached the market. Though the examples to date have been aimed at conferring a desirable property on the target plant rather than its use as a bioreactor, they do suggest that the commercialisation of plants expressing heterologous sequences for *ex situ* uses may not be that far behind. At this stage it seems likely that the first examples of such plants to come on market will be those designed to express proteins for industrial uses where the regulatory issues are less complex. However, the recent demonstration that modified plant virus particles can protect target animals against an economically important disease (76) has brought the prospect of the use of plants for the production of pharmaceuticals that much nearer. Provided sufficient attention is paid to any problems associated with their use, the development of plants as bioreactors for the production of biologically-active proteins, whether for industrial or pharmacological use, should be of significant benefit in the future.

9. References

1. R.B. Flavell, *Trends Biotechnol.* **13** (1995) 313.
2. O.J.M. Goddijn and J. Pen, *Trends Biotechnol.* **13** (1995) 379.
3. D.N. Duvick, *Bitechnol. and Develop. Monitor* **27** (1996) 15.
4. J. Draper and R. Scott, in *Plant Genetic Engineering*, ed. D. Grierson (Blackie, Glasgow, London, 1991) 38.
5. C. Porta, and G.P. Lomonossoff, G.P. *Molecular Biotechnology* **5** (1996) 209.
6. H.B. Scholtof, K-B.G. Scholtof and A.O Jackson, *Annu. Rev. Phytopathol.* **34** (1996) 229.
7. S.N. Covey and R. Hull, in *Genetic Engineering with Plant Viruses*, eds. T.M.A. Wilson and J.W. Davies (CRC Press, Boca Raton, FL, 1992) 217.
8. J. Stanley, *Current Opinion in Genetics and Development* **3** (1993) 91.
9. P. Ahlquist and M. Janda, *Mol. Cell. Biol.* **4** (1984) 2876.
10. P. Ahlquist, R. French, M. Janda and S. Loesch-Fries, *Proc. Natl. Acad. Sci. USA* **81** (1984) 7066.
11. M. Mori, K. Mise, K. Kobayashi, T. Okuno and I. Furusawa, *J. Gen. Virol.* **72** (1991) 243.

12. R.B. Flavell, *Proc. Natl. Acad. Sci USA* **91** (1994) 3490.
13. J. Finegan and D. McElroy, *Bio/Technology* **12** (1994) 883.
14. P. Christou, *AGRO food INDUSTRY hi-tech* **5** (1994) 17.
15. S. Austin, E.T. Bingham, R.G. Koegel, D.E. Matthews, M.N. Shahan, R.J. Straub and R.R. Burgess, *Ann. NY Acad. Sci.* **721** (1994) 234.
16. H.Y. Yeang, P. Arokiaraj, H. Jaafar, S. Hamzah, M.A. Siti Arija and H. Jones, in *Engineering Crops for Inustrial End Uses*, eds. P.R. Shewry, J.A. Napier and P. Davis (Portland Press, London, 1998) 55.
17. P. Arokiaraj, H. Jones, K.F. Cheong, S. Coomber and B.V. Charlwood, *Plant Cell Reports* **13** (1994) 425.
18. J. Vandekerckhove, J. van Damme, M. van Lijsebettens, J. Botterman, M. de Block, M. Vandewiele, A. De Clercq, J. Leemans, M. van Montagu and E. Krebbers, *Bio/Technology* **7** (1989) 929.
19. L.M. Hoffman, D.D. Donaldson and E.M. Herman, *Plant Mol. Biol.* **11** (1988) 717.
20. P.C. Sijmons, B.M.M. Dekker, B. Schrammeijer, T.C. Verwoerd, P.J.M. van den Elzen and A. Hoekema, *Bio/Technology* **8** (1990) 217.
21. S.A. Jobling and L. Gehrke, *Nature* **325** (1987) 622.
22. W. Dieryck, J. Pagnier, C. Poyart, M.C. Marden, V. Gruber, P. Bournat, S. Baudino, B. Mérot, *Nature* **386** (1997) 29.
23. J. Chen, Y-R. Sun, X-H. Li, W-Q. Zeng, Y-Y. Lee and Y-D. Hou, *Scientia Sinica (Series B)* **3** (1990) 253.
24. O. Edelbaum, D. Stein, N. Holland, Y. Gafni, O. Livneh, D. Novick, M. Rubinstein and I. Sela, *J. Interferon Res.* **12** (1992) 449.
25. Z. Zhu, K.W. Hughes, L. Huang, B. Sun, C. Liu, Y. Li, Y. Hou and Z. Li. *Plant Cell, Tissue and Organ Culture* **36** (1994) 197.
26. S. Matsumoto, A. Ishii, K. Ikura, M. Ueda and R. Sasaki, *Biosci. Biotech. Biochem.* **57** (1993) 1249.
27. S. Matsumoto, K. Ikura, M. Ueda and R. Sasaki, *Plant Mol. Biol.* **27** (1995) 1163.
28. D. Bosch, J. Smal and E. Krebbers, *Transgenic Res.* **3** (1994) 304.
29. H.S. Mason, D.M-K. Lam and C.J. Arntzen, *Proc. Natl. Acad. Sci. USA* **89** (1992) 11745.
30. Y. Thanavala, Y-F. Yang, P. Lyons, H.S. Mason and C. Arntzen, *Proc. Natl. Acad. Sci. USA* **92** (1995) 3358.
31. T.A. Haq, H.S. Mason, J.D. Clements and C.J. Arntzen, *Science* **268** (1995) 714.
32. H.S. Mason, J.M. Ball, J-J. Shi, X. Jiang, M.K. Estes and C.J. Arntzen, *Proc. Natl. Acad. Sci. USA* **93** (1996) 5335.
33. H.S. Mason, O.Y-Y. Lam, C.A. Froberg and D.M-K. Lam, *Adv. Sci. Res.* **1** (1994) 1.
34. P.B. McGarvey, J. Hammond, M.M. Dienelt, D.C. Hooper, Z.F. Fu, B. Dietzschold, H. Koprowski and F.H. Michaels, *Bio/Technology* **13** (1995) 1484.
35. G.D. May, R. Afza, H.S. Mason, A. Wieko, F.J. Novak and C.J. Arntzen, *Bio/Technology* **13** (1995) 486.

36. C.J. Arntzen and H.S. Mason, In *New Generation Vaccines* eds. M.M. Levine, G.C. Woodrow, J.B. Kaper and G.S. Cobon (Marcel Dekker, New York, 1997).
37. A. Hiatt, R. Cafferkey and K. Bowdish, *Nature* **342** (1989) 76.
38. A. Hiatt and K. Mostov, In *Transgenic Plants: Fundamentals and Applications* ed. A. Hiatt (Marcel Dekker, New York, 1993) 221.
39. M.B. Hein, Y. Tang, D.A. McLeod, K.D. Janda, A.C. Hiatt, *Biotechnol. Prog.* **7** (1991) 455.
40. M. De Neve, M. De Loose, A. Jacobs, H. van Houdt, B. Kaluza, U. Weidle and M. van Montagu, *Transgenic Res.* **2** (1993) 227.
41. K. Düring, S. Hippe, F. Kreuzaler and J. Schell, *Plant Mol. Biol.* **15** (1990) 281.
42. F.A. van Engelen, A. Schouten, J.W. Molthoff, J. Roosien, J. Salinas, W.G. Dirkse, A. Schots, J. Bakker, F.J. Gommers and M.A. Jongsma, *Plant Mol. Biol.* **26** (1994) 1701.
43. J.K-C. Ma, A. Hiatt, M. Hein, N.D. Vine, F. Wang, P. Stabila, C. van Dolleweerd, K. Mostov and T. Lehner, *Science* **268** (1995) 716.
44. J.K-C. Ma, T. Lehner, P. Stabila, C.I. Fux and A. Hiatt, *Eur. J. Immunol.* **24** (1994) 131
45. J.K-C. Ma and M.B. Hein, *Plant Physiol.* **109** (1995) 341.
46. J.K-C. Ma, R. Smith and T. Lehner, *Infect. Immun.* **55** (1987) 1274.
47. E. Benvenuto, R.J. Ordas, R. Tavazza, G. Ancora, S. Biocca, A. Cattaneo and P. Galeffi, *Plant Mol. Biol.* **17** (1991) 865.
48. M.R. Owen, A. Gandecha, W. Cockburn and G.C. Whitelam. *Bio/Technology* **10** (1992) 790.
49. S. Firek, J. Draper, M.R.L. Owen, A. Gandecha, W. Cockburn and G.C. Whitelam, *Plant Mol. Biol.* **23** (1993) 861.
50. U. Fiedler and U. Conrad, *Bio/Technology* **13** (1995) 1090.
51. O. Artsaenko, E.W. Weiler, K. Muntz, U. Conrad, *J. Plant Physiol.* **144** (1994) 427.
52. P. Tavladoraki, E. Benvenuto, S. Trinca, D.D. Martinis, A. Cattaneo and P. Galeffi, *Nature* **366** (1993) 469.
53. J. Pen, L. Molendijk, W.J. Quax, P.C. Sijmons, A.J.J. van Ooyen, P.J.M. van den Elzen, K. Rietveld and A. Hoekema, *Bio/Technology* **10** (1992) 292.
54. D. McElroy and J. Jacobson, *Bio/Technology* **13** (1995) 245.
55. B.A. Phillipson, *Plant Science* **91** (1993) 195.
56. K. Aspegren, L. Mannonen, A. Ritala, R. Puupponen-Pimia, U. Kurten, M. Salmenkallio-Marttila, V. Kaupinen and T.H. Teeri, *Mol. Breeding* **1** (1995) 91.
57. K. Herbers, H.J. Flint and U. Sonnewald. *Mol Breeding* **2** (1996) 81.
58. J-F. Laliberté, O. Nicolas, S. Durand and R. Morosoli, *Plant Mol. Biol.* **18** (1992) 447.
59. K. Herbers, I. Wilke and U. Sonnewald, *Bio/Technology* **13** (1995) 63.
60. J. Pen, T.C. Verwoerd, P.A. van Paridon, R.F. Beudeker, P.J.M. van den Elzen, K. Geerse, J.D. van der Klis, H.A.J. Versteegh, A.J.J. van Ooyen and A. Hoekema, *Bio/Technology* **11** (1993) 811.

61. T.C. Verwoerd, P.A. van Paridon, A.J.J. van Ooyen, J.W.M. van Lent, A. Hoekema and J. Pen, *Plant Physiol.* **109** (1995) 1199.
62. B. Gronenborn, R.C. Gardner, S. Schaefer and R.J. Shepherd, *Nature* **294** (1981) 773.
63. D.D. Lefebvre, B.L. Miki and J.F. Laliberté, *Bio/Technology* **5** (1987) 1053.
64. G.A. De Zoeten, J.R. Penswick, M.A. Horisberger, P. Ahl, M. Schultz and T. Hohn, *Virology* **172** (1989) 213.
65. R.J. Copeman, J.R. Hartman and J.C. Watterson, *Phytopathology* **59** (1969) 1012.
66. W.O. Dawson, D.J. Lewandowski, M.E. Hilf, P. Bubrick, A.J. Raffo, J,J, Shaw, G.L. Grantham and P.R. Desjardins, *Virology* **172** (1989) 285.
67. J. Donson, C.M. Kearney, M.E. Hilf and W.O. Dawson, *Proc. Natl. Acad. Sci. USA* **88** (1991) 7204.
68. M.H. Kumagai, T.H. Turpen, N. Weinzettl, G. Della-Cioppa, A.M. Turpen, J. Donson, M.E. Hilf, G.L. Grantham, W.O. Dawson, T.P. Chow, M. Piatak and L.K. Grill, *Proc. Natl. Acad. Sci. USA* **90** (1993) 427.
69. N. Takamatsu, Y. Watanabe, H. Yanagi, T. Meshi, T. Shiba and Y. Okada, *FEBS Lett.* **269** (1990) 73.
70. H. Hamamoto, Y. Sugiyama, N. Nakagawa, E. Hashida, Y. Matsunaga, S. Takemoto, Y. Watanabe and Y. Okada, *Bio/Technology* **11** (1993) 930.
71. G.P. Lomonossoff and J.E. Johnson, *Seminars in Virology* **6** (1995) 257.
72. G.P. Lomonossoff and J.E. Johnson, *Current Opinion in Structural Biology* **6**, (1996) 176.
73. G.P. Lomonossoff and J.E. Johnson, *Prog. Biophys. Molec. Biol.* **55** (1991).
74. R. Usha, J.B. Rohll, V.E. Spall, M. Shanks, A.J. Maule, J.E. Johnson and G.P. Lomonossoff, *Virology* **197** (1993) 366.
75. C. Porta, V.E. Sall, J. Loveland, J.E. Johnson, P.J. Barker, and G.P. Lomonossoff, *Virology* **202** (1994) 949.
76. K. Dalsgaard, Å. Uttenthal, T.D. Jones, F. Xu, A. Merryweather, W.D.O. Hamilton, P.J.M. Langeveld, R.S. Boshuizen, S. Kamstrup, G.P. Lomonossoff, C. Porta, C. Vela, J.I. Casal, R.H. Meloen, and P.B. Rodgers, *Nature Biotechnology* **15** (1997) 248.
77. C. Porta, V.E. Spall, T. Lin, J.E. Johnson and G.P. Lomonossoff, *Intervirology* **39** (1996) 79.
78. L. McLain, C. Porta, G.P. Lomonossoff, G.P., Z. Durrani and N.J. Dimmock, *Aids Research and Human Retroviruses* **11** (1995) 327.
79. L. McLain, Z. Durrani, L.A. Wisniewski, C. Porta, G.P. Lomonossoff and N.J. Dimmock, *Vaccine* **14** (1996) 799.
80. A.C. Bloomer and P.J.G. Butler, In *The Plant Viruses* **Vol. 2**, eds. M.H.V. van Regenmortel and H. Fraenkel-Conrat (Plenum Press, New York, 1986) 19.
81. Y. Sugiyama, H. Hamamoto, S. Takemoto, Y. Watanabe, Y. Okada, *FEBS Lett.* **359** (1995) 247.
82. T.H. Turpen, S.J. Reinl, Y. Charoenvit, S.L. Hoffman, V. Fallarme and L.K. Grill, *Bio/Technology* **13** (1995) 53.

83. J. Fitchen, R.N. Beachy and M.B. Hein, *Vaccine* **13** (1995) 1051.
84. A.J. Olsson, G. Bricogne and S.C. Harrison, *J. Mol. Biol.* **171** (1983) 61.
85. T. Joelson, L. Åkerblom, P. Oxelfelt, B. Strandberg, K. Tomenius and T.J. Morris, *J. Gen. Virol.*, **78** (1997) 1213.
86. S. Santa Cruz, S. Chapman, A.G. Roberts, I.M. Roberts, D.A.M. Prior and K.J. Oparka, *Proc. Natl. Acad. Sci. USA* **93** (1996) 6286.
87. M.D. Ryan, A.M.Q. King and G.P. Thomas, *J. Gen. Virol.* **72** (1991) 2727.
88. M.D. Ryan and J. Drew, *EMBO J.* **13** (1994) 928.
89. G.J.H. van Rooijen and M.M. Moloney, *Bio/Technology* **13** (1995) 72.
90. L. Miele, *Trends Biotechnol.* **15** (1997) 45.
91. D.W. Galbraith, C.A. Zeiher, K.R. Harkins and C.L. Afonso, *Planta* **186** (1992) 324.
92. A. Driouich, L. Faye and L.A. Staehelin, *Trends Biochem. Sci* **18** (1993) 210.
93. L. van Vloten-Doting, J.F. Bol and B. Cornelissen, *Plant Molecular Biology* **4** (1985) 323.
94. C.M. Kearney, J. Donson, G.E. Jones and W.O. Dawson, *Virology* **192** (1993) 11.

Chapter 9

BIOTECHNOLOGY AND PLANT SECONDARY PRODUCTS: THE FUTURE

VINCENZO DE LUCA

1. Introduction

Expanded use of recombinant DNA technology to manipulate plant secondary metabolism has recently become possible with the increasing elucidation of a number of secondary metabolic pathways and with the availability of a continually increasing number of plant (1-5) and non-plant genes. This chapter proposes possible directions, based on examples from the literature, that research should focus on to successfully produce desired secondary metabolites. The argument is made that greater attention should be paid to results previously obtained from putatively simple industrial organisms, where successful production strategies were devised for the production of fungal or microbial primary and secondary metabolites, in the design strategies for producing plant secondary metabolites.

2. Understanding the Basis for Directed Metabolism in Plants

It has been argued that the unique ability of plants to produce large amounts of phenylpropanoid compounds derived from phenylalanine to synthesize structural supporting structures such as lignin and quantitative defence compounds such as condensed tannins, contributed largely to the ability of plants to colonize the land (6). This ancient capability of plants to mobilize large amounts of carbon from primary metabolism through the shikimate and phenylpropanoid pathways is coupled to developmentally- and environmentally- controlled events which activate in a cell-specific manner the efficient flow of carbon into phenylpropanoid-related compounds rather than into products of primary metabolism. The differential activation of pathway genes can thus simultaneously cause the accumulation of structural lignin in vasculature and the production of defensive phenylpropanoids triggered by environmental stress in particular tissues.

Similarly, the evolution of divergent pathways to produce alternative alkaloid- or terpenoid- based defences has often occurred by sacrificing phenylpropanoid-based structural as well as soluble defence compounds and has been correlated with the switch from woody to herbaceous growth (7). Interestingly, the switch from woody to herbaceous growth has also been successfully correlated with the evolution of angiosperms towards more short-lived life cycles through the development of increasingly efficient reproductive structures and an increased dependence on seed-based survival strategies. This implies that plants have successfully evolved the necessary genetic and biochemical capabilities to direct metabolism towards the production of desired natural products. Our knowledge about how plants do this is increasing at a rapid rate, but much more effort is needed to understand the extent to which control architectures (Chapter 2) which have evolved to produce a particular class of secondary metabolite, can be modified to accumulate totally new and perhaps unrelated products.

3. Metabolic Engineering of Plant Metabolism

Since the first report describing the technology for stable plant transformation in 1981 (8), the first genetically engineered crops were already being released for commercial use in 1994 in the United States for a fibre-producing bromoxynil-resistant cotton and this was quickly followed by release of the first food product, the slow-softening Flavr-Savr® tomato. In addition, almost one million hectares of transgenic tobacco resistant to cucumber mosaic virus were already under cultivation in 1994 in China (9). Technical advances coupled to rapid application of this technology clearly have ensured that many new genetically-engineered crops with single gene traits conferring resistance to viruses, herbicides and insects will soon be growing in many countries around the world. These results are confirming the great promise of this revolutionary technology and offer unlimited new opportunities for the creation of novel crops, which were not possible to produce in the past.

3.1. Metabolic Engineering of Plant Secondary Metabolism

Although a number reviews have been published (10-12) which illustrate the potential of metabolic engineering to alter plant secondary metabolism or to create totally new secondary metabolites, the theoretical framework to provide predicable engineering results is largely lacking. In this context, experiments to alter secondary metabolism have produced many unanticipated responses, which although providing insights about the complexity of metabolic control, have failed to produce the desired product.

In order to exploit predictably the potential of plants to produce desired secondary metabolites, a thorough understanding of the interface between intermediary/primary metabolism and secondary metabolism is essential. Plants have evolved the capacity to produce the majority of their secondary metabolites from relatively few universally-present precursors such as acetyl-CoA, amino acids or shikimate. A major gap in our knowledge here involves the evolution of metabolic specificity in different families of plants for the supply of precursors involved in the production of classes of secondary metabolites which are characteristic of those plant families. It is intuitively sensible to think that plants which evolved to produce particular secondary metabolites did so because they also evolved the ability to supply necessary primary precursors in the quantities required.

The tools of biotechnology should be used to probe the plasticity of different plant families to supply commonly available or unusual precursors, respectively, in order to accommodate a related or a totally new pathway which does not normally occur in those plant families. In addition detailed metabolic studies should be performed to characterize if and how this novel pathway affects normal plant metabolism. These studies will help to delineate the flexibility of target plants to supply precursors for the production of desired secondary metabolites and to define the design requirements to create new precursor pools required for the production of totally new secondary products.

4. Generating Substrate Precursors

The best-studied systems for the production of various metabolites occur in microorganisms where metabolic engineering processes were improved by mutation and selection rather than by genetic engineering (13,14). In this way strains of microorganisms which accumulate amino acids, drugs and other commercially useful metabolites have been produced. Although very successful commercial production systems have been designed in microorganisms, these processes were far from being predictive, but instead, required a repetitive cycle of directed genetic change followed by an analyses of the consequences, in order to remove each new limitation in order to optimize a strain for producing a particular metabolite (13).

Metabolic control analysis theory (MCA)(Chapter 2) which has been advanced as a major predictive tool for rational metabolic engineering suggests that flux through pathways is usually influenced by the activities of several individual steps. The validity of this theory has recently been illustrated for increasing the production of aromatic amino acids in microorganisms (15,16) by pathway engineering.

4.1. Increasing Aromatic Amino Acid Production in Microorganisms

A tryptophan-producing strain of *Corynebacterium glutamicum* engineered to express a 3-deoxy-D-arabino-heptulosonate 7-phosphate synthase (DS) no longer sensitive to feedback inhibition to phenylalanine and tyrosine, increased carbon flow through the common branch of the shikimate pathway (15). This modified strain did not produce higher levels of tryptophan, but instead accumulated chorismate. These results suggested that removal of DS as a limiting step to flux down the shikimate pathway, was now replaced by a new limiting step which allowed chorismate to accumulate rather than to be converted to aromatic amino acids. A second round of metabolic engineering with a plasmid containing the gene for DS as well as for the enzymes of tryptophan biosynthesis resulted in strains which accumulated anthranilate, the first metabolite in the tryptophan branch of the pathway. A final round of experiments expressing DS and tryptophan biosynthetic enzymes which were no longer sensitive to feedback inhibition by tryptophan finally resulted in a strain of *Corynebacterium glutamicum* which produced 43 g of tryptophan per litre of bacteria, a 54% increase in yield over the parent strain.

Another remarkable empirical application of MCA involved metabolic engineering of the pathway from glucose to aromatic amino acids in *Escherichia coli* (16). The overall pathway to aromatic amino acids could be divided into 5 main activities (17): *i)* getting glucose into the bacterial cell; moving glucose-6-phosphate into *ii)* phosphoenolpyruvate (PEP) and *iii)* erythrose-4-phosphate (E4P) via glycolysis and the pentose phosphate pathway, respectively; *iv)* producing 3-deoxy-D-*arabino*-heptulosonate 7-phosphate (DAHP) from PEP and E4P and *v)* producing aromatic amino acids from DAHP. The transport of glucose into *E. coli* involves glucose phosphotransferase (PTS) which consumes PEP and mutants were selected which effectively transport glucose by a non-PTS uptake system in order to effectively double the PEP available for other reactions inside the bacterial cell. This mutation coupled to the overexpression of an allosterically-insensitive DS resulted in a strain which excreted DAHP into the medium, reflecting the inability of downstream enzymes to keep up with DAHP production. Further studies also revealed that if the supply of E4P was increased by expressing transketolase, an enzyme involved in E4P synthesis, even higher levels of DAHP accumulated. These results confirm that increasing the PEP supply can be directed by suitable metabolic engineering into the production of shikimate pathway intermediates or into aromatic amino acid products and that E4P does become limiting under these circumstances.

4.2. Can the Concepts Learned from Microbial Metabolic Engineering Experiments be Applied to Plants?

In view of the subtleties of metabolic control observed as well as the unexpected difficulties encountered to modify secondary metabolism in microbial systems which which have been said to be "extremely simple and well characterized compared to plants", it is not surprising that the first pioneering efforts in plants have met with mixed results (10-12). It has been proposed that plants are infinitely more complex than microbes and that this complexity would be reflected in control architectures which regulate metabolic networks. In spite of this prediction, is it possible that control architectures in plants for pathways producing precursors for secondary metabolite biosynthesis are quite similar to those involved in microbial metabolism?

Several recent studies do suggest that plant pathways which supply substrates for production of secondary metabolite precursors are more similar to those of microbial pathways than previously anticipated. It appears, for example, that although genes for all amino acid biosynthetic pathways so far investigated are encoded by the plant nucleus, they are compartmentalized in the prokaryote-like chloroplast. This compartmentalization appears to extend to the pathways for aromatic amino acid biosynthesis (18), although some debate continues over the existence of a second pathway in the cytosolic compartment. Although the shikimate pathway in plants appears to be under complex developmental, spatial, temporal and environmental regulation, the biochemical properties of many of the enzymes of this pathway are under similar allosteric control to those of bacteria and fungi. This suggests that metabolic engineering approaches to modify the level of shikimate pathway intermediates or aromatic amino acid supply in bacteria might also be valuable in plants. However, similarities and differences in control architectures between different species of plants, which favour the production of one intermediate over another, need to be understood in order to succeed in a predictable manner to produce the desired precursor.

4.3. Evidence for Cross-Pathway Regulation in Plants for Amino Acid Biosynthesis

The ability of eukaryotes to regulate multiple amino acid pathways in response to amino acid deficiency has been defined as cross-pathway regulation in *Neurospora crassa* (19) and as general control of amino acid biosynthesis in *Saccharomyces cerevisiae* (20). This phenomenon, first illustrated in *Neurospora*, showed that cells starved for histidine caused derepression of the last 5 steps in tryptophan biosynthesis as well as for histidine and arginine biosynthetic enzymes. Starvation of cells for tryptophan or arginine also caused derepression of the three amino acid biosynthetic pathways. These properties suggest that cross-pathway regulation is a mechanism for eukaryotic cells to cope efficiently with amino acid imbalances.

Recent studies performed with *Arabidopsis thaliana* (21) suggest that amino acid biosynthesis in plants is also subject to general control. When histidine biosynthesis was blocked with a specific inhibitor of imidazoleglycerol-phosphate dehydratase, induction of eight genes involved in the biosynthesis of aromatic amino acids, histidine, lysine and purines occurred. Addition of histidine to the media eliminated the inductions and demonstrated that the changes in gene expression occurred as a consequence of inhibition of the histidine pathway.

Other studies with transgenic potato (22) also suggest that amino acid biosynthesis in plants is subject to control by amino acids. The transformation of potato with a tryptophan decarboxylase (TDC) gene from *Catharanthus roseus* redirected tryptophan into tryptamine and dramatically altered the pools of tryptophan, phenylalanine and phenlylalanine-derived phenylpropanoids in wounded potato tubers. The creation of an artificial metabolic sink for tryptophan caused a dramatic decrease of the tryptophan pool which then caused a decrease in phenylalanine pools, a product produced from the other branch of the shikimate pathway. The decreased phenylalanine pool in turn greatly decreased the accumulation of soluble and wall-bound phenylpropanoids and decreased the lignification of cell walls. These results (21,22) illustrate that histidine and tryptophan play important roles similar to those in *Neurospora* (19) as sensors of amino acid imbalance in plants and that substrate limitation rather than enzymic limitation could act as primary controlling factors of secondary metabolite biosynthesis (reviewed for the case of phenylpropanoids in 23) in plants. It is important therefore, that the understanding gained in relatively well-characterized model systems such as *Neurospora* and *Saccharomyces* be considered in developing substrate delivery systems in plants for the production of secondary metabolites.

5. Lessons Learned from Mutants and Current Metabolic Engineering Studies

Many interesting experiments have been performed which outline both the promise of metabolic engineering and the problems associated with this technology. Invariably problems for productive engineering appear as a consequence of lack of understanding of metabolism and suggest a requirement for more thorough knowledge of plant metabolism and regulation.

5.1. Manipulation of Lysine Levels to Improve Nutritional Quality of Plants

The improvement of the nutritional quality of crops has long been associated with increasing the content of essential amino acids such as lysine, either as the amino acid or associated with high-lysine-containing proteins. The fact that the plant and bacterial pathway for lysine biosynthesis is controlled by feedback inhibition of dihydropicolinate synthase (DHPS) by lysine, prompted experiments to overexpress a bacterial feedback-insensitive form of this enzyme in plants. Expression of the bacterial enzyme in the chloroplasts of tobacco plants was accompanied by a significant increase of free lysine in tobacco leaves (24). Interestingly, increased lysine levels were also found in leaves of a tobacco mutant possessing a form of DHPS no longer sensitive to feedback inhibition by lysine (25). However, the level of lysine in developing seeds remained unchanged in transgenic (24) and mutant (25) plants. The similarity of these results confirm the important and potentially ubiquitous role played by DHPS in regulating the level of lysine in plant cells as well as the relevance to target this gene to increase lysine levels.

The inability to increase lysine levels in transgenic tobacco seeds prompted further studies (26) which showed that in developing seeds, lysine controls its accumulation by a further regulatory mechanism not found in tobacco leaves. The degradative pathway involving lysine-ketoglutarate reductase (LKR) is stimulated by the accumulation of free lysine in tobacco seeds. This novel mechanism of induction of the degradative pathway appears to involve a signaling cascade mediated by Ca^{++} and protein phosphorylation which regulates the activity of LKR. These results show that a substrate can regulate its concentration through feedback inhibition as well as by regulating its degradation by a complex signal cascade. This is perhaps due to the potential toxicity of lysine since mature plants (24, 25) accumulating the highest lysine levels had abnormal phenotypes such as partial loss of apical dominance, mosaic green colour of leaf tips and some flowers were sterile. This body of work shows the potential of metabolic engineering to create new tools to study the regulation of fundamental processes and that different organs will indeed behave differently in response to a particular modification. These problems must be resolved initially to produce nutritionally superior plants and in the longer term to manufacture commercially-useful natural products derived from various amino acid substrates.

5.2. Creation of Tryptophan Metabolic Sinks by Overexpression of the TDC Gene

The TDC gene from *Catharanthus roseus* (27), has been used extensively in metabolic engineering experiments (22, 28-31). In the commercially important oil crop, *Brassica napus* (Canola), constitutive expression of TDC resulted in plants which accumulate lowered levels of tryptophan-derived indoleglucosinolates (28). Of particular significance, the indoleglucosinolate content of mature seeds from TDC-expressing plants, was only 3% of the level found in nontransformed seeds. These results showed that the creation of a new sink for tryptophan could be used to successfully divert metabolite flow from the indoleglucosinolates which are normally produced in this crop. The commercial value of removing indoleglucosinolates from the seed would be to improve the nutritive value of the oilseed meals which are produced after extraction of oil from the seed.

Constitutive expression of TDC in tobacco resulted in transgenic plants which accumulated up to 1% of their dry weight in tryptamine (29) These high levels of accumulation could not be reproduced in TDC-expressing potato or Canola (30), where tryptamine levels were only 10 and 2%, respectively, of those occurring in transgenic tobacco. These results are consistent with the existence of species-specific control architectures which regulate the availability of tryptophan for decarboxylation and which, in combination with other factors, may contribute to the unpredictable results often encountered in these types of experiments. Additional surprising results were also obtained with TDC-expressing transgenic tobacco, which accumulated high levels of tyramine (31) rather than tryptamine, even though the enzyme purified from *Catharanthus roseus* (32) and the cloned gene expressed in *E. coli* (33) accepted tryptophan as a substrate but not tyrosine. This unusual result, which remains to be explained, once again illustrates the empirical nature of metabolic engineering and suggests an essential need for careful chemical studies to completely understand how the transgenic plant was modified.

The TDC gene was also expressed constitutively in *Peganum harmala* cell suspension cultures (34) which synthesize β-carboline alkaloids and serotonin (5-hydroxytryptamine) as major secondary metabolites. Cell lines with greatly enhanced TDC activities were obtained compared with untransformed control lines. While tryptamine levels were unchanged, serotonin contents in transformed lines were increased 10-fold over untransformed controls to reach almost 2% of the cell dry weight. These results corroborated previous studies (35) which suggested that TDC, rather than the subsequent hydroxylase, was the rate-limiting step for serotonin accumulation. These results illustrate that elimination of a rate-limiting step can be used to enhance the production of a secondary metabolite in plants by using similar approaches to those previously discussed for in microorganisms (13-17).

5.3. Accumulation of Intermediates at the Interface Between Primary and Secondary Metabolism

An unexpected result of recent studies to modify secondary metabolism is the apparent lack of control observed in regulating the accumulation of metabolites at the interface between primary and secondary metabolism. This observation is especially striking in studies involving the relatively well characterized phenylpropanoid pathways. Attempts to manipulate lignin quality by downregulating genes involved in lignin biosynthesis have shown that unlike the shikimate pathway which appears to be highly regulated, the pathway beyond phenylalanine is not subject to the same type of fine control (reviewed in 36, 37-40). Experiments to manipulate lignin content by down-regulating cinnamyl alcohol dehydrogenase (CAD) (37) or caffeic/ferulic acid O-methyltransferase (OMT) (38-40) resulted in transgenic plants which accumulated their immediate precursors that were subsequently incorporated into lignin or lignin-like molecules with modified guaiacyl to syringyl lignin ratios (39, 40). The lack of feedback inhibition observed in these examples is not totally unexpected since similar observations were made with brown midrib mutants of maize (41) and sorghum (42) which contained reduced OMT and/or CAD activities.

Crucifers such as *Arabidopsis* accumulate hydroxycinnamic acid esters such as sinapoyl malate and sinapoyl choline in leaves and seeds, respectively. An *Arabidopsis* mutant unable to hydroxylate ferulate to yield sinapate, no longer accumulated sinapoylmalate or sinapoylcholine and furthermore produced a lignin lacking sinapic acid-derived components (43). This mutation in ferulate-5-hydroxylase was recently shown to be encoded by a novel gene belonging to the cytochrome P450-dependent monooxygenase family (44). Another important feature of this *Arabidopsis* mutant was that while sinapoyl choline almost completely disappeared from seeds, free choline accumulated in place of the choline ester. Similar metabolite studies were done with low-nicotine tobacco mutants (45) whose mutation occurred in the putrescine-*N*-methyltransferase gene which catalyses the first committed step in nicotine biosynthesis (46). Remarkably, nicotine and nornicotine levels decreased by over 80% in this mutant, while free putrescine levels almost increased to those which would normally be incorporated into nicotine.

These results show that in addition to the post-phenylalanine branch of the lignin pathway, neither choline nor putrescine biosynthesis appear to be under particular feedback control in the tissues specialized for producing these secondary metabolites. A particularly important question would be to investigate how generally this lack of regulation applies to other secondary metabolic pathways. Identification of weakly-regulated control points could be very useful to predict where intermediates would accumulate to high levels. These accumulating metabolites could then be used as feedstocks for production of novel commercially desirable secondary metabolites after introduction of genes encoding relevant biosynthetic pathways.

5.4. Accumulation of Altered End-Products at the End of Secondary Pathways

It has been suggested that the terminal portions of secondary pathways could be modified in a more predictable manner than early or intermediate steps of biosynthesis. This is based on the fact that the product which would normally accumulate in substantial quantity would serve as an abundant substrate for the subsequent desired reaction which is introduced by genetic engineering. This was realistically demonstrated in transgenic *Atropa belladonna* expressing a novel hyoscyamine 6β-hydroxylase gene previously cloned from *Hyoscyamus niger* (47). Expression of this enzyme, which catalyses the hydroxylation and epoxidation of hyoscyamine to yield scopolamine, in *Atropa belladonna* resulted in plants which accumulated almost exclusively scopolamine rather than hyoscyamine which occurs naturally in this plant. Other examples of the value of this approach include the expression of dihydroflavonol reductases or of hydroxylases to produce novel flower colours in plants (reviewed in 48). Numerous other successful examples will however be required in order to predict the ease with which novel secondary products will be created by manipulating terminal parts of relevant pathways. That potential unpredicted problems could arise was recently shown in the failure to produce delphinidin-based blue roses (49), since it was discovered that several additional traits would have to be modified in addition to producing anthocyanins with the proper substitution pattern in order to produce the blue flower colour.

6. Unique Opportunities for Manipulation of Plant Secondary Metabolism

The power and attraction of rDNA and related technologies in producing desired phenotypic traits, offers a complementary and very powerful tool to traditional plant breeding to achieve similar goals to those of plant breeding which lead to crop improvement. For example, the use of plant breeding for improving crops has sometimes resulted in the inadvertent selective elimination of defensive secondary metabolites which may have reduced the fitness of the crop (50). Plant biotechnology could be used to eliminate toxic secondary metabolites from edible plant organs such as energy-rich seeds but not from other parts of the plant where they serve as useful defensive compounds. This technology could be used not only to maintain but also to increase the chemical diversity of plant secondary metabolites that defend the crop. This would allow the production of a crop that can produce several different natural pesticides. The rotation of a crop whose defensive chemistry changes periodically may decrease the probability that plant pathogens could adapt to those chemistries. The improved fitness of such crops would stabilize and increase food production, while decreasing the use of pesticides as well as other synthetic compounds. On a larger scale these improved crops could alleviate numerous social and ecological problems associated with modern agricultural practices.

Thus plants with totally new secondary metabolite chemistries are possible since this technology has created the ability to combine genes encoding secondary metabolic pathways from organisms that have evolved along separate phylogenies to produce genotypes that no previous technology could have produced. The production of new secondary metabolites by combining genes from separate organisms should be actively pursued since this would lead to the creation of new diversity in the secondary chemistry of plants. This may lead to development of new crops which produce commercially useful "hybrid secondary metabolites" which have horticultural, agronomic or medicinal value. An early review on the commercial promise of creating novel "hybrid antibiotics" in industrial microorganisms (51) has recently begun to be fulfilled when combinatorial gene approaches were used to synthesize novel polyketide (52 & references therein) and cephalosporin antibiotics (53) in different *Streptomyces* species as well as in the production of more potent and useful semisynthetic derivatives of morphine and codeine in *E. coli* (54).

The commercial potential is unlimited for creating totally new food crops with improved nutritional value and with altered aroma and flavour. As human populations become more concerned with health and well-being, there has been an explosive interest in using food supplements (for example ginkgolide containing *Ginkgo biloba* leaf powders) containing particular plant ingredients which contribute to disease prevention. These products and the greatly increased interest in using traditional medicines for treating disease, create unique opportunities for the use of plant biotechnology to contribute to the creation of safe and proven food supplements and natural medicines which do not suffer from the problems of toxic side effects often associated with their use

Concerted efforts to study the regulation of secondary metabolism should be encouraged given the increasing availability of genes encoding all parts of a number of secondary metabolic pathways. These studies will reveal how plants regulate cell-, tissue- organ- and environment-specific events which control the biosynthesis, accumulation/compartmentation and degradation of secondary metabolites and which will be necessary to produce and accumulate desired compounds. In this context genetic engineering of secondary metabolism and its regulation will complement already highly developed and sophisticated methods for growing plants as cell or organized organ cultures. This complementary approach will certainly fulfill the long-sought-after goal (55) of using cell cultures for the commercial production of valuable plant drugs, pigments, flavours and numerous other secondary metabolites. Such breakthroughs will clearly decrease the need to exploit natural sources for these metabolites and could contribute significantly in combination with other measures for the preservation of natural ecosystems which are presently being overexploited and in danger of extinction.

7. Safety Concerns and Educating the Public

The unprecedented rate of change occurring in our world today as a result of innovations in Science and technology has resulted in major improvements in the quality of life for many of the Earth's people. However, many who have to live with this type of progress find it overwhelming, and are prone to fear and to believe the worst case scenarios that can be caused by improper use of these technologies. This is certainly the case for using rDNA technologies to modify living organisms to accomplish biological processes. Scientists have realised its revolutionary nature and from the beginning, they encouraged extensive participation by the public through lobby groups and the news media in setting the agenda for the direction of this science.

The initial cautious approach used by the scientific community in encouraging the use of this technology for metabolic engineering was followed by the development of risk assessment guidelines based on extensive testing and the subsequent realisation that the risks of genetic engineering were similar to those posed by other novel organisms (56 & references therein) which were produced by more traditional breeding methods. In the case of plants, this led to some of the first commercial uses of transgenic organisms described earlier in this chapter.

7.1. Should Different Guidelines Apply to Transgenic Plants with Altered Secondary Metabolites?

It could be speculated that the introduction of novel genes which create new or modified secondary metabolites in plants may create unanticipated modifications in metabolite pools (22, 28), which may have unknown consequences for the nutrition and health of consumers of the new food. It is clear that sensitive and quantitative methods of metabolite analysis are required to answer these questions and to ensure the safety of the modified plant. Crops created by classical breeding methods have also necessitated similar types of analyses to ensure licensing of new varieties. For example, the widespread exploitation of wild, tuber bearing *Solanum* species in potato breeding programmes required the development of sensitive methods for the quantification of glycoalkaloids in new varieties before they could be licensed for commercial use. The need for this was illustrated in the 1960's when a cultivar called Lenape was withdrawn from commercial production because of its high alkaloid content (57). The erroneous certification of this variety which increased the awareness of plant breeders and regulators, contributed to some of the guidelines presently used for certification of potatoes and other crops.

The similarity of the goals to be achieved through rDNA technology and plant breeding are remarkable. The end-product will be a plant with slightly altered chemical profiles which confer a selective advantage to the plant, a commercial advantage to the producer and a safe superior-quality product for the consumer. In both cases, a new organism will have been introduced into the environment which will establish new biotic and abiotic interactions. The regulations which govern the introduction of products produced by these two approaches should therefore be similar. Transgenic organisms should be evaluated and regulated according to their biological properties (phenotypes), rather than according to the genetic techniques used to produce them (56). As in the breeding approach, the products generated by rDNA methods must be evaluated for intended as well as for unintended phenotypes which are produced. The unintended phenotype could then be removed by breeding methods which would serve to complement the other approach.

The question that can be raised here is whether increased regulation of rDNA products is warranted because many novel combinations of properties can be achieved only by molecular and cellular techniques. The question is especially relevant with respect to the generation of transgenic crops with totally new natural products. Solutions to this question will evolve as we decide how to use this technology for engineering plant secondary metabolism and from the practical experience gained from generating transgenic plants accumulating different novel secondary metabolites.

8. Summary

The arrival of new technologies helps to create products for which they were designed, but more importantly many more unexpected benefits often accrue. The first attempts by agrochemical companies to apply this technology have resulted in products which will benefit growth of those companies and which will also benefit people and the environment.The huge continuing effort to sequence whole genomes and to clone genes has resulted in an abundance of the tools available for modifying plant primary and secondary chemistries. These tools have begun to be used for such modifications and researchers have quickly realized that even with all the studies of the last century, our understanding of the biochemistry of primary and secondary metabolism is still quite rudimentary. With the enthusiasm for acquiring molecular genetic tools, many plant scientists have abandoned more traditional areas of plant science such as plant physiology and biochemistry which are now increasingly necessary to complement molecular biology in order to realize the potential of plant biotechnology.

Future research should focus on understanding the evolution of metabolic specificity which permitted different plants to supply different precursor substrates for the production of species-specific secondary metabolites. In the design of research strategies for plant studies, much greater use should be made of the lessons learned in the creation of industrial processes for producing secondary metabolites with microoganisms. Efforts should be made to take into consideration and to take advantage of the differences observed in plants between the types of enzyme regulation occurring between pathways which supply primary metabolites and pathways which use them to make secondary metabolites. Creative use should be made of intermediates which would normally accumulate if endogenous pathways were interdicted. The time is also ripe for studying the regulation of secondary metabolism in order to elucidate common and unique regulatory cascades which control the production of the major classes of secondary metabolites in plants. Finally, scientists should make greater efforts to familiarize the public with the benefits and potential pitfalls of rDNA technology and the regulatory framework in place to ensure the effectiveness and safety of the products being produced.

9. References

1. R. Whetten and R. Sederoff *The Plant Cell* **7** (1995) 1001.
2. D.J. McGarvey and R. Croteau *The Plant Cell* **7** (1995) 1015.
3. T.M. Kutchan *The Plant Cell* **7** (1995) 1059.
4. T.A. Holton and E. Cornish *The Plant Cell* **7** (1995) 1071.
5. R.A. Dixon and N. Paiva *The Plant Cell* **7** (1995) 1085.
6. T. Swain *Comp. Biochem* **29A** (1970) 125
7. O.R. Gottlieb *Phytochemistry* **29** (1990) 1715.
8. L. Otten, H. De Greve, J.P. Hernalsteens, M. Van Montagu, O. Schieder, J. Straub and J. Schell, *Mol. Gen. Genet.* **183** (1981) 209.
9. P.J. Dale, *Trends Biotechnol.* **13** (1995) 398.
10. V. De Luca, *AgBiotech. News and Information* **5** (1993) 225N.
11. C. L. Nessler *Transgenic Research* **3** (1994) 109.
12. O.J.M. Goddijn and J. Pen, *Trends Biotechnol.* **13** (1995) 379.
13. J.E. Bailey, *Science* **252** (1991) 1668.
14. G. Stephanopoulos and J.J. Vallino **252** (1991) 1675.
15. R. Katsumata and M. Ikeda *Bio/technology* **11** (1993) 921.
16. N. Flores, J. Xiao, A. Berry, F. Bolivar and F. Valle *Nature Biotechnology* **14** (1996) 620.
17. A.L. Demain *Nature Biotechnology* **14** (1996) 580.
18. J. Schmidt and N. Amrhein *Phytochemistry* **39** (1995) 737.
19. M. Carsiotis, R.F. Jones and A.C. Wessling *J. Bacteriology* **119** (1974) 893.
20. P. Niederberger, G. Mozzari and R. Hütter *Mol. & Cell Biol.* **1** (1981) 584.
21. D. Guyer, D. Patton and E. Ward *Proc. Nat. Acad. Sci.* **92** (1995) 4997.
22. K. Yao, V. De Luca and N. Brisson *The Plant Cell* **7** (1995) 1787.
23. U. Margna *Phytochemistry* **16** (1977) 419.
24. O. Shaul anf G. Galili *The Plant Journal* **2** (1992) 203
25. M. Ghislain, V. Frankard and M. Jacobs *The Plant Journal* **8** (1995) 733.
26. H. Karchi, D. Miron, S. Ben-Yaacov and G. Galili *The Plant Cell* **7** (1995) 1963.
27. V. De Luca, C. Marineau and N. Brisson *Proc. Nat. Acad. Sci.* **86** (1989) 2582.
28. S. Chavadej, N. Brisson, J. McNeil and V. De Luca *Proc. Nat. Acad. Sci.* **91** (1994) 2166.
29. D.D. Songstad, V. De Luca, N. Brisson, W.G.W. Kurz and C.L. Nessler *Plant Physiol.* **94** (1989) 1410.

30. R.K. Ibrahim, S. Chavadej amd V. De Luca , in *Recent Advances in Phytochemistry* (Plenum Press, New York and London, 1994) **28** 125.
31. D.D. Songstad, W.G.W. Kurz and C.L. Nessler *Phytochemistry* **30** (1991) 3245.
32. W. Noé, C. Mollenschott and J. Berlin *Pl. Mol. Biol.* **3** (1984) 281.
33. P. Facchini and V. De Luca *J. Biol. Chem.* **269** (1994) 26684.
34. J. Berlin, C. Rügenhagen, P. Dietze, L. Fecker, O.J.M. Goddijn and J.H.C. Hoge *Transgenic Research* **2** (1993) 336.
35. F. Sasse, L. Witte and J. Berlin *Planta Medica* **53** (1987) 354.
36. C.J. Douglas *Trends in Plant Sciences* **1** (1996) 171.
37. C. Halpin, M.E. Knight, G.A. Foxon, M.M. Campbell, A.M. Boudet, J.J. Boon, B. Chabbert, M.T. Tollier and W. Schuch *The Plant Journal* **6** (1994) 339.
38. N. Weiting, N. Paiva and R.A. Dixon *Transgenic Research* **3** (1994)
39. R. Altanassova, N. Favet, F. Martz, B. Chabbert, M.T. Tollier, B. Monties, B. Fritig and M. Legrand *The Plant Journal* **8** (1995) 465.
40. J. Van Doorsselaere *The Plant Journal* **8** (1995) 855.
41. C. Grand, P. Parmentier, A. Boudet and A.M. Boudet *Physiol. Végét.* **23** (1985) 905.
42. F. Vignols, J. Rigau, M.A. Torres, M. Cappellades and P. Puigdomènech *Plant Cell* **7** (1995) 407.
43. C.S. Chapple, T. Vogt, B. Ellis and C.R. Somerville *Plant Cell* **4** (1992) 1413.
44. K. Meyer, J.C. Cusumano, C. Somerville and C.C. Chapple *Proc. Nat. Acad. Sci.* **93** (1996) 6869
45. P.D. Legg and G.B. Collins *Can. J. Gen. Cytol.* **13** (1971) 287.
46. N. Hibi, S. Higashiguchi, T. Hashimoto and Y. Yamada *The Plant Cell* **6** (1994) 723.
47. D. Yun, T. Hashimoto and Y. Yamada *Proc. Nat. Acad. Sci.* **89** (1992) 11799.
48. J.N.M. Mol, T.A. Holton and R.E. Koes *Trends in Biotechnol.* **13** (1995) 350.
49. T.A. Holton and Y. Tanaka *Trends in Biotechnol.* **12** (1994) 40.
50. M. Wink *Theor. Appl. Genet.* **75** (1988) 225.
51. H.G. Floss *Tibtech.* **5** (1987) 111.
52. M.A. Alvarez, H. Fu, C. Khosla, D.A. Hopwood and J.E. Bailey *Nature Biotechnology* **14** (1996) 335.
53. L. Crawford, A.M. Stepan, P.C. Mcada, J.A. Rambosek, M.J. Conder,V.A. Vinci, C.D. Reeves *Bio-Technology* **13** (1995) 58.
54. C.E. French, A.M. Halles, D.A. Rathbone, M.T. Long, D.L. Willey and N.C. Bruce *Bio-Technology* **13** (1995) 674.
55. N.J. Walton *Chemistry in Britain* (1992) 525.
56. D. Boulter *Phytochemistry* **40** (1995) 1.

INDEX

A

Abies grandis, 36
abietane, 124
abortifacient, 21
abscisic acid, 12, 13, 39, 66, 75
Acacia, 157
ACC (1-aminocyclopropane-1-carboxylic acid) oxidase, 40
ACC (1-aminocyclopropane-1-carboxylic acid) synthase, 74
Accurel, 283
acetogenins, 190
acetone cyanohydrin, 306
acetosyringone, 75
acetycholine, 233
acetyl-CoA, 357
acetyl-CoA carboxylase, 46, 52, 53
acetyl-CoA:salutaridinol-7-*O*-acetyltransferase, 44, 45
acetylenic bond, 167
α-acids, 46, 55
Acokanthera oubaio, 17
aconitan A, 227
aconitine, 227, 228
Aconitum, 227, 228
acridone alkaloids, 38
acridone synthase, 46
activated charcoal, 151
Activator, 39
adc gene, 60
S-adenosyl-L-methionine: (*S*)-coclaurine-*N*-methyltransferase, 44
S-adenosylmethionine, 57
adenylate cylase, 238
adriamycin, 225
adsorbent traps, 111
adsorption, 283
adsorption chromatography, 120, 122
adulteration, 147
Aequoria victoria, 347
affinisine, 172, 173
affinity chromatography, 36, 338
agarose, 285
agmatine, 60
$AgNO_3$, 134
Agrobacteria, 30, 55, 75, 287, 288, 328, 333, 336
Agrobacterium rhizogenes, 30, 287
Agrobacterium tumefaciens, 287, 333
ajmalicine, 165
ajmaline, 289
ajoene, 229
akuammicine, 195, 196, 201
albumin, 331
2S albumins, 331
alcohols, 307, 310
aldehydes, 301, 305
aleurone, 339
Alexa leiopetala, 223
alexine, 223
alfalfa, 53, 54, 61, 65, 66, 269, 330
alfalfa mosaic virus (AMV), 332
algae, 22, 122, 310, 311, 313, 314
alginate, 284, 291
alkaline phosphatase, 70
alkaloids, 1, 5-8, 11, 28, 29-30, 44, 46, 51, 56, 60, 77, 79, 92, 96, 114, 118, 121, 134, 151, 156, 162, 163, 165, 171, 172, 227, 234, 237, 238, 242, 281, 282, 295, 318, 319, 330, 356
  aconitine, 227, 228
  Aconitum, 227, 228
  acridone, 38

alkaloids (*cont.*)
affinisine, 172, 173
ajmalicine, 165
ajmaline, 289, 290
akuammicine, 195, 196, 201
Alexa leiopetala, 223
alexine, 223
Amaryllidaceae, 7
anabasine, 30, 295, 296
anisine, 227
Anisodus, 233
anti-cancer activity, 242
antihypertensive action, 238
Apocynaceae, 236, 242
aporphine, 129
atidine, 227
atisine, 227, 228
Atropa belladonna, 233, 234, 364
atropine, 28, 57, 233, 234
benzophenanthridine, 77, 291
benzoylecgonine, 3
benzylisoquinoline, 41, 44, 49, 54, 291
berbamunine, 319
berbamunine synthase, 319, 320
berberine, 49, 238, 239
berberine-bridge enzyme, 291
betalains, 7, 8
bisbenzylisoquinoline, 319
bisindole, 164, 293
brucine, 1
β-carboline, 362
caffeine, 11, 104
Camptotheca acuminata, 234, 235
camptothecin, 219, 225, 234, 235
castanospermine, 159, 223
Castanospermum australe, 223
Casuarina equisetifolia, 223
casuarine, 223
catharanthine, 293, 294
catharanthine *N*-oxide, 294
Catharanthus roseus, 165, 236

alkaloids (*cont.*)
cathenamine, 42
Celiptium®, 242
α-chaconine, 253
chloromethylene-affinisinium, 172, 173
Cinchona, 156, 238
cinnamoylcocaine, 3
Cl-opioid receptor, 242
cocaine, 3, 30, 140, 169, 170
(*S*)-coclaurine, 42, 44
codeine, 242, 291, 365
codeinone, 291
colchicine, 7
coniine, 29
conoduramine, 164
conodurine, 164
Coptis, 238, 239
coryneine, 227
COSY, 196
CPT-11, 235
Datura, 165, 233
Datura stramonium , 165
deadly nightshade, *see Atropa*
demissine, 18
Derris elliptica, 223
dibenzazonine, 44
dihydrochanoajmaline, 289
2*R*,5*R*-dihydroxymethyl-3R,4R-dihydroxypyrrolidine, 223
diterpenoid, 7, 227
Dragendorff reagent , 163
dregamine, 192
Duboisia, 233
ecgonine, 3
ellipticine, 225, 242
epi-cathenamine, 42
N-ethyl (*S*)-nornicotine, 296
ferric chloride / perchloric acid, 164
flash chromatography, 172
glucoalkaloids, 41, 42
glucobrassicin, 122

alkaloids (*cont.*)
glucosinolates, 122
glycoalkaloids, 57, 252, 253, 367
higenamine, 227
HMBC (Heteronuclear Multiple Bond Correlation), 201
homoerythrina, 121
homoharringtonine, 225
10-hydroxycamptothecin, 234
3-hydroxyconoduramine, 164
3-hydroxyconodurine, 164
21-hydroxyraumacline, 289
3-hydroxyvoacamine, 164
hyoscine, 233
hyoscyamine, 29-32, 56, 165, 233, 364
Hyoscyamus, 233
hypaconitine, 227
hysognavine, 227
ignarine, 227
indicine-*N*-oxide, 225
indole, 7, 51, 77, 155, 164, 192, 211, 234, 235, 290, 293, 318
indolizidine, 159
iso-vallesiachotamine, 42
isoquinoline, 7 *(see also* benzylisoquinoline)
jesaconitine, 227, 228
kobusine, 227, 228
leukaemia, 234
littorine, 30, 31
luxandrine, 151
Lycopodium, 7
Madagascar periwinkle (*Catharanthus roseus*), 11, 236
malaria, 238
maytansine, 225
mercuric iodide, 151
mesaconitine, 227, 228
methylcoclaurine, 319

alkaloids (*cont.*)
N-methyl-9-hydroxy-ellipticinum acetate, 242
monoterpenoid, 7, 29, 289, 293 *(see also* indole)
morphinans, 44, 291
morphine, 54, 140, 170, 242, 281, 365
motion sickness, 233
nacycline, 42
neopelline, 227
nicotine, 28, 32, 37, 41, 56, 57, 61, 77, 295, 363, 364
nicotine biosynthesis, 296
nitraramine, 281, 282
norcoclaurine, 42, 44
norditerpenoid, 129
nornicotine, 364
Nyssaceae, 234
Ochrosia borbonica, 242
Ochrosia elliptica, 242
Ochrosia sandwicensis, 242
palmatine, 238
Papaver somniferum, 242
Papaveraceae, 242
peptides, 7
phaeantharine, 134
piperidine, 7, 22, 29, 30, 34
piperine, 104
polyhydroxy, 222, 223, 226
pyridine, 57
pyrrolidine, 7, 60
pyrrolizidine, 7, 22, 56, 252
quaternary, 151
quinine, 1, 29, 115, 156, 238
quinoline, 7, 29
quinolizidine, 7, 22
Ranunculaceae, 227, 238
raumacline, 289, 290
reticuline, 49, 291
Rubiaceae, 238
sanguinarine, 54

alkaloids (*cont.*)
- scopolamine, 29, 31, 56, 165, 233, 234, 296, 364
- *Scopolia*, 233
- scoulerine, 49, 291
- senecionine, 16, 17
- serpentine, 165
- sesquiterpenes, 7
- silica, 172
- Solanaceae, 233
- α-solanine, 253
- steroidal, 7
- strictosidine, 35, 41, 42, 51
- strychnine, 1, 2, 96, 97, 115, 187, 190
- *Swainsonia canescens*, 223
- swainsonine, 223
- tabernaemontanine, 192
- tabersonine, 51
- taxines, 171, 172, 173
- Taxol™, 32, 281
- *Taxus baccata*, 171
- testicular teratoma, 237
- thebaine, 44
- tomatine, 57, 254
- topotecan, 234
- tropane alkaloids, 7, 30-32, 37, 55, 57, 167, 233, 234, 296
- tropine, 28, 30
- tropinone, 29
- vallesiachotamines, 42
- vinblastine, 51, 236, 237, 293, 294
- *Vinca rosea*, 11, 236
- vincamine, 119
- vincine, 119
- vincoside, 42
- vincristine, 51, 236, 237
- vindoline, 51, 293, 294
- voacamidine, 164
- voacamine, 164

alkenes, 315
allelopathics, 169
allene oxide cyclase, 77
allene oxide synthase, 77
allene oxide synthetase, 318
Alliaceae, 229
allicin, 171
alliin, 171, 229
alliinase, 171
Allium, 19, 228-230, 257, 267
Allium cepa, 19
alloaromadendrane-7α,11α-diol, 120, 121
(+)-*S*-allyl-L-cysteine sulphoxide, 171
allylisothiocyanate (AITC), 268
almonds, 305, 309
alpha-acids, 57
Alternaria alternata, 19
Alternaria longipes, 19
alumina, 120, 121, 123, 125, 129, 162, 174
Amaryllidaceae, 7
American mandrake, 242
amines, 7
amino acids, 6, 163, 303
- analogues, 255
- biosynthesis, 360
- 3-deoxy-D-*arabino*-heptulosonate 7-phosphate synthase, 358
- *Arabidopsis thaliana*, 360
- arginine, 29, 60, 360
- aromatic, 358-360
- aspartic acid, 303
- chorismate, 358
- compartmentalization, 359
- *Corynebacterium glutamicum*, 358
- cross-pathway regulation, 360
- dihydropicolinate synthase, 361
- DS (DAHP synthase), 52, 358, 359
- *Escherichia coli*, 358
- glutamic acid, 303
- histidine, 360
- imbalances, 360
- lysine, 29, 30, 46, 282, 295, 296, 360, 361

amino acids (*cont.*)
- lysine:ketoglutarate reductase, 361
- metabolic engineering, 361
- *Neurospora*, 360
- non-protein, 7
- nutritional quality, 361
- ornithine, 29, 37, 46, 295
- β-*N*-oxalyl-L-2,3-diaminopropionic acid, 255
- phenylalanine, 8, 9, 32, 44, 52, 358, 360, 363, 364
- phenylpropanoids, 360
- potato, 360
- proline, 29
- putrescine, 295
- *Saccharomyces*, 360
- shikimate pathway, 355, 358-360, 363
- tryptophan, 32, 38, 59, 288, 358, 360, 362
- tryptophan decarboxylase (TDC), 360
- tyrosine, 32, 358

1-aminocyclopropane-1-carboxylic acid oxidase (ACC oxidase), 40
Ammi majus, 230
Ammifurin®, 230
ammonium sulphate, 36
amperometric detection, 159
Amphiroa zonata, 313
amygdalin, 305, 306
α-amylase, 330, 338, 339
α-amylase promoter, 319
An1, 67
An2, 68
anabasine, 30, 295, 296
anaerobiosis, 66
anaesthesia, 233
analgesic, 281
anandamide, 236
androgen, 225
Angelica, 139, 141, 231, 232
Angelica sinensis, 231
angiotensin-I-converting enzyme inhibitor (ACEI), 343, 344, 346
animal feeds, 340, 341
anion-exchange chromatography, 31, 36
anisaldehyde-sulphuric acid, 163
anisine, 227
Anisodus, 233
anorexia, 236
anthochlors, 9
anthocyanins, 1, 9, 10, 14, 28, 39, 44, 51, 55, 66-69, 71, 75, 76, 121, 129, 162, 259, 298, 364
anthranilate, 358
anthranilate synthase, 38, 52
anthraquinones, 9, 163
anthrone-sulphuric acid, 163
anti-allergenic activity, 230
anti-asthmatic activity, 231
antibacterial activity, 238
antibodies, 169, 336, 338
anti-cancer activity, 32, 218, 219, 222, 224, 225, 237, 242, 243, 244, 292
anti-diabetic activity, 219, 237
antifeedants (*see also* toxins), 3, 28, 75, 169, 175
- aminoacid analogues, 255
- anabasine, 30
- anti-nutritional factors, 256
- atropine, 28, 57
- azadirachtin, 3, 10, 17
- benzylisoquinoline alkaloids, 41
- brominated phenols, 311
- canavanine, 11
- cardiac glycosides, 16
- α-chaconine, 253
- cocaine, 30
- coniine, 29
- coumarin, 20
- cyanogenic glycosides, 7, 16, 51, 253, 255
- L-Dopa, 216

antifeedants (*cont.*)
- demissine, 18
- FCs, 254, 255
- fluoroacetate, 16, 17
- furanocoumarins, 11, 12, 16, 252, 254
- furanoterpenoids, 255
- glucobrassicin, 122
- glucosinolates, 7, 51, 122, 255
- glycoalkaloids, 57, 252, 253, 367
- halogenated monoterpenes, 311
- halometabolites, 310
- herbivory, 216
- hyoscyamine, 29, 56
- indoleglucosinolates, 362
- ipomeamarone, 255
- 4-ipomeanol, 255
- isoflavonoids, 20
- isothiocyanates, 52
- lactupicrin, 18
- linamarin, 18, 253, 255
- miserotoxin, 16, 17
- neem, 11
- nicotine, 28, 29, 37, 56, 57, 61, 77
- nitriles, 52
- β-ocimene, 18
- oenanthetoxin, 17
- ouabain, 17
- β-*N*-oxalyl-L-2,3-diaminopropionic acid, 255
- papyriferic acid, 17
- proteinase inhibitors , 256
- psoralens, 254
- pyridine alkaloids, 57
- pyrrolizidine alkaloids, 252
- raffinose oligosaccharides (RFOs), 256
- rotenone, 18
- salicin, 17
- salicylic acid, 11
- saponins, 256, 257

antifeedants (*cont.*)
- scopolamine, 29, 56
- senecionine, 16, 17
- sinigrin, 16, 17
- α-solanine, 253
- soyasapogenol, 257
- tannins, 18
- tomatine, 57, 254
- tropane alkaloids, 37, 55, 57
- xanthotoxin, 12, 16, 18

antifungal agents, 18
- arbutin, 19
- avenacin AI , 19
- avenacoside A, 19
- catechol, 19
- chrysin 7-glucoside, 19
- epoxylinoleic acid, 18
- isoflavonoids, 18
- luteone, 18, 19
- mycotoxins, 20
- parthenolide, 18, 19
- 5-pentadecylresorcinol, 19
- sakuranetin, 19
- sapogenin, 19
- saponins, 18
- sclareol, 19
- sinigrin, 19
- α-tomatine, 19
- tulipalin A , 19
- 1-tuliposide A, 19
- 6-tuliposide A, 19

antiherbivore agents, 16
antihypertensive action, 238
anti-inflammatory activity, 231, 238, 241
anti-leukaemic activity, 293
anti-malarial activity, 231, 233
antimicrobial activity, 28, 241
antimuscarinic agents, 233
anti-nutritional factors, 256
antioxidant activity, 236
antioxidants, 36, 229, 251, 257-264, 271
- anthocyanins, 259

antioxidants (*cont.*)
ascorbic acid, 259
β-carotene, 259
caffeic acid, 259, 261, 262, 264
carotenoids, 259, 262
catechin, 262, 264
chlorogenic acid, 259, 263
curcumin, 259
ferulic acid, 259, 262
flavanols, 259, 261
flavonoids, 259, 261-263
hydroxycinnamates, 259, 262, 263
indole-3-carbinol, 264
indolyl glucosinolate, 264
isoflavones, 271
lignans, 271
lutein, 259
lycopene, 259
phenylindan, 259
quercetin, 259, 261-264
tocopherol, 259, 264
vitamin C, 259, 262, 264
vitamin E, 259, 261-263
antiphlogistic activity, 241
anti-phytochrome scFv, 338
Antirrhinum, 39, 63, 66-69
antisense, 53, 70, 71
anti-spasmodic activity, 231, 241
anti-tumour activity, 225, 244, 257, 281
antitumour assays, 169
anti-viral activity, 221, 222, 257
apc gene, 264
aphids, 18
Apiaceae, 230, 231, 254
apigenin, 240, 241
apigenin glycosides, 241
Aplysia, 311
Apocynaceae, 34, 236, 242
apoptosis, 264
aporphine alkaloids, 129
apples, 21, 259
Arabidopsis thaliana, 39, 41, 52, 53, 63, 69, 72, 74, 76, 79, 80, 331, 334, 348, 360, 363
arabinose, 12
Araceae, 222
arachidonate, 77
arbutin, 19, 139
arginine, 29, 60, 360
arginine decarboxylase, 60, 61
Armillaria mycelia , 162
Armillaria tabescens, 161
armyworms, 18
aroma compounds, 147
aromatic amines, 159
aromatic amino acids, 358-360
arrow poisons, 227
Artabotrys monteiroae, 129
artefacts, 170, 172
affinisine, 172, 173
alkaloids, 171, 172
allicin, 171
alliin, 171
(+)-*S*-allyl-L-cysteine sulphoxide, 171
alumina, 174
bilobalide, 174
chamazulene, 171, 172
chamomile, 171
chamomile oil, 171
chloromethylene-affinisinium, 172, 173
decomposition, 170
degradation, 174
diallyl disulphide, 171
dichloromethane, 172
extraneous compounds, 174
garlic, 171
impurities, 174
light, 172
matricin, 171, 172
polymerisation, 174
preparative TLC, 174

artefacts (*cont.*)
    sesquiterpene, 174
    steam distillation, 171, 172
    taxanes, 171
    taxines, 171, 173
    Taxus baccata, 171
artelinic acid, 233
artemetin, 233
Artemisia annua, 231, 232, 233
Artemisia douglasiana, 233
artemisinic acid, 233
artemisinin, 143, 144, 159, 231-233
Arum, 11, 14, 15
Ascophyllum nodosum, 314
ascorbic acid, 159, 259, 262
Asparagus, 257
aspartic acid, 303
Aspergillus niger, 341
Asteraceae, 56, 236
asthma, 236, 240
Astragalus atropubescens, 17
Astronium graveolens, 18
atherogenesis, 263
atherosclerosis, 258
atidine, 227
atisine, 227, 228
ATP, 73
Atropa, 28, 56, 233
Atropa belladonna, 233, 234, 364
atropine, 28, 57, 233, 234
aurea, 76
aurones, 9
Automated Multiple Development
    (AMD), 135, 136
auxins, 12, 39, 51, 71, 75, 76
Avena sativa, 18, 19
avenacin AI, 19
avenacoside A, 19
avocado, 259
Azadirachta indica, 175-177, 281
azadirachtins, 3, 10, 17, 152, 162,
    175-177, 281

B

Bacillus amyloliquifaciens, 339
Bacillus licheniformis, 330, 338
Bacillus macerans, 339
Baccharis megapotamica, 22
baculovirus, 42
baculovirus-based expression system, 319
bananas, 336
barbituric acid, 314
barley, 339
batatasin I, 13
bats, 14, 15
beans (*see also Phaseolus*), 257
bees, 14
benzalacetone synthase, 46
benzaldehyde, 305, 306, 315
benzenoids, 14
benzoate, 79
benzoate 2-hydroxylase, 79
benzofurans, 9
benzophenanthridine alkaloids, 77, 291
benzophenone synthase, 46
benzoquinones, 9
benzoyl-CoA, 46
benzoylecgonine, 3
benzylisoquinoline alkaloids, 41, 44, 49,
    54, 151, 291, 318
berbamunine, 319
berbamunine synthase, 41, 319, 320
Berberidaceae, 242
berberine, 49, 238, 239
berberine bridge enzyme, 41, 49, 291,
    321
Berberis, 49, 319
Berberis stolonifera, 319
bergapten, 141, 231
betalains, 7, 8, 28
Betula resinifera, 17
betulin, 1
Bignoniaceae, 220
bilobalides, 56, 136, 137, 157, 174

bioassays, 168
biocatalysts, 277, 279, 280, 282, 287
biological detectors, 168, 169
biomimetic reactions, 29, 34
bioreactors, 291, 292
biotransformations, 35, 41, 57, 277, 279, 282, 287, 288, 289, 291-297, 299, 300, 302, 303, 318, 320
biphenols, 13
bis-indole alkaloids, 293
α-bisabolol, 240, 241
bisabolol oxide A, 240
bisabolol oxide B, 240
bisbenzylisoquinoline alkaloids, 319
Bishop's weed, 230
bisindole alkaloids, 164
bistramide A, 202
black bean, 269
blackcurrants, 259
blocking agents, 266, 267
blood proteins, 332
blood substitutes, 332
borneol, 297, 301
Botrytis cinerea, 19
bouvadrin, 225
bowel disease, 258
Boyle, 1
Brassica, 17, 267
Brassica napus, 111, 331, 348, 362
Brassica oleracea, 19
brassinolides, 12, 188, 189
brassinosteroids, 12, 13, 39, 75, 76
breast cancer, 270, 271
breeding, 367
brevitoxin A, 22
brine shrimp assay , 169
Bristol-Myers /Squibb, 244
broccoli, 259, 267
brominated phenols, 311
bromoperoxidase, 313, 314
bronchodilator, 230
bronze locus, 39
brown midrib mutants, 363
bruceantin, 225
bruchid beetles, 11, 18
brucine, 1
Brugmansia candida x *aurea*, 296
Brussels sprouts, 267, 268
Bryonia, 77
Buddleja, 137
bufadienolides, 4, 5
bumble bees, 15
n-butanol, 116
butterflies, 15
bZIP proteins, 67

C

^{13}C, 30
^{13}C-^{13}C INADEQUATE, 207
^{13}C 1D spectra, 207
^{13}C NMR, 194
c-fos, 264
c-jun, 264
C1 (transcription factor), 67, 68
Ca^{2+} ions, 76
cabbage, 267
cadaverine, 46, 295, 296
caffeic acid, 60, 259, 261, 262, 264
caffeic acid *O*-methyltransferase, 77
caffeic/ferulic acid *O*-methyltransferase, 363
caffeine, 11, 104
caffeoyl-CoA, 44
calcium antagonists, 231
Caldariomyces fumago, 314, 315
californium plasma desorption, 192
callus cultures, 300
cAMP, 238
camphor, 104, 106, 297, 301
(+)-camphor, 298, 299
camphor 5-exo-hydroxylase, 319
Camptotheca acuminata, 234, 235
camptothecin (CPT), 219, 225, 234, 235

CaMV 35S promoter, 332, 334, 336-341
Canavalia ensiformis, 11
canavanine, 11
cancer, 224, 234, 236, 242-245, 258, 259, 270
cancer chemotherapy, 236
Cannabinaceae, 235
cannabinoid receptor, 236
cannabinoids, 236
Cannabis sativa, 235, 236
canola, 362
Capillary Electrophoresis (CE), 137, 142, 153
  Angelica, 139, 141
  arbutin, 139
  bergapten, 141
  CE-MS, 160
  cocaine, 140
  columbianadin, 140, 141
  columbianetin, 141
  columbianetin acetate, 141
  coumarin, 141
  coumarins, 139-141
  electroosmotic flow, 138
  electrophoretic flow, 138
  flavonols, 139
  illicit drug substances, 140
  mescaline, 140
  micellar electrokinetic capillary chromatography (MEKC), 139, 140, 141
  morphine, 140
  osthol, 140, 141
  psilocybin, 140
  psoralen, 140, 141
  Δ9-THC, 140
  umbelliferone, 141
  Uvae ursi, 139
  xanthotoxin, 140, 141
capillary zone electrophoresis, 137, 140
Caprifoliaceae, 34
capsaicin, 57, 299, 300
Capsicum annuum, 57, 299
capsid protein, 335
β-carboline alkaloids, 362
carcinogenesis, 264, 266, 267, 270
carcinogens, 265-268
carcinomas, 224
cardenolide glycosides, 239
cardenolides, 4, 5, 163
cardiac glycosides, 16
cardioactive drugs, 288
Δ2-carene, 300, 301
Carica papaya, 303
β-carotene, 259
carotenoids, 5, 14, 15, 40, 55, 155, 259, 262
κ-carrageenan, 284, 285
carrots, 255, 259
Carthamus tinctorius, 236
carvomenthones, 301, 302
carvone, 302, 303
carvone oxide, 15
caryophyllene, 134
casbene synthase, 37
cassava, 253, 255
Cassia didymobotrya, 293
castanospermine, 159, 223
Castanospermum australe, 223
casticin, 233
Casuarina equisetifolia, 223
casuarine, 223
catalase, 79
cataracts, 258
Catasetum, 15
catechin, 262, 264
catechol, 19, 78
catecholamines, 159
catharanthine, 293, 294
catharanthine *N*-oxide, 294
Catharanthus roseus, 34, 39, 40, 51, 165, 219, 236, 237, 293, 302, 360, 362
cathenamine, 42
cation-exchange , 159
cauliflower, 267

cauliflower mosaic virus (CaMV) 35S
promoter, 332, 334, 336-341
Caventou, 1
cDNA, 54, 319
cDNA libraries, 38, 40
CE-MS, 160
celery, 255
Celiptium®, 242
Celite, 283, 304
cell cultures, 30, 35, 38, 39, 49, 57, 63, 65, 66, 77, 288, 291, 292, 293, 297-299, 301-303
cellulose, 120, 121
cellulose pulps, 340
Central Drug Research Institute, 238
centrifugal counter-current chromatography, 120
centrifugal layer chromatography, 136
centrifugal partition chromatography, 119
centrifugal TLC, 125
cereals, 270
α-chaconine, 253
chalcone reductase, 41, 46, 47
chalcone synthase, 35, 43, 46, 47, 54, 61, 63, 66, 68, 69, 71, 76, 77
chalcone-flavanone isomerase, 35, 63, 68
chalcones, 9
chamazulene, 171, 172
chamomile, 171, 240, 241
chamomile oil, 171
Chamomilla, 240
chaperonins, 73
chelation, 258, 259
chemical ionization (CI), 168, 192
chemical shifts (δ), 193
chemoselectivity, 280
chemotaxonomic information, 222, 224
chemotaxonomic investigations, 226
chemotherapy, 236
chick peas, 257, 269
chicory, 99, 306, 307
chilli peppers, 57, 299, 300
Chinese cabbage, 267
Chinese medicine, 227, 231
chiral chromatography, 120
chiral GC, 123
chiral LC, 123
chiral separations, 123, 142
chirality, 147
chives, 229
chloramine-trichloroacetic acid, 163
chlorogenic acid, 13, 60, 259, 263
chloromethane, 311
chloromethylene-affinisinium, 172, 173
chloroperoxidase, 314, 315
chloroplastic transit peptide, 332
chloroquine, 238
cholera, 229
cholesterol, 187, 230
choline esterase, 233
chorismate, 358
chorismate synthase (CS), 52
chromatofocusing, 36
chromatography, 116 *et seq.*
- abietane, 124
- adsorption chromatography, 120, 122
- adulteration, 147
- affinisine, 173
- affinity chromatography, 36, 338
- $AgNO_3$, 134
- alkaloids, 118, 134
- alloaromadendrane-7α,11α-diol, 120, 121
- alumina, 120, 121, 123, 129, 162
- anion-exchange, 36
- anthocyanins, 129, 162
- *Armillaria tabescens*, 161
- aroma compounds, 147
- *Artabotrys monteiroae*, 129
- artemisinin, 143, 144, 159
- automated multiple development (AMD), 135, 136
- *Azadirachta indica*, 176, 177

chromatography (*cont.*)
azadirachtin, 162, 176, 177
bilobalides, 136, 137, 157, 174
Buddleja, 137
caryophyllene, 134
castanospermine, 159
catecholamines, 159
cation-exchange , 159
cellulose, 120, 121
centrifugal counter-current chromatography, 120
centrifugal layer chromatography, 136
centrifugal partition chromatography (CPCC), 119
centrifugal TLC, 125
chiral chromatography, 120
chiral GC, 123
chiral LC, 123
chiral separations, 142
chirality, 147
chloromethylene-affinisinium, 173
chromatofocusing, 36
chromosorb®, 145
CO_2, 142, 143
columbianadin, 140
column chromatography, 123, 124, 125
complexation chromatography, 120, 123
coniferin, 119
counter-current chromatography, 117, 119
CPCC (centrifugal partition chromatography), 119
Craig, 117
Crataegus, 159
β-cyclocitral, 146
DCCC, 118, 119
9,10-dehydrofusaric acid, 126

chromatography (*cont.*)
Delphinium barbeyi, 129
derivatisation, 146, 147
dihydroanthracenones, 124
2-dimensional GC, 147
2-dimensional TLC, 134, 135
diode array detection (DAD), 129, 154, 155, 161
3-directional TLC, 135
droplet counter-current chromatography (DCCC), 118
ECD, 143, 145
eleutheroside B, 119
eleutheroside B1, 119
eluotropic series, 121
4-epi-γ-gurjunene, 130
essential oils, 144, 147
Extrelut®, 132, 133
α-farnesene, 112
farnesol, 120-122
flame ionization detector (FID), 143, 146, 165, 169
flash chromatography, 123-126, 150, 172, 175
flavonoid aglycones, 147
flavonoid glycosides, 118, 127
flavonoids, 129, 130
flavonols, 144
Florisil®, 120
fluorescence, 143
FT-IR, 143
fusaric acid, 126
Fusarium nygamai, 126
garlic, 127
gas chromatography (GC), 110-113, 122, 127, 142-144, 146-148, 150, 153, 165, 167-169, 191
Gasteria bicolor, 124
GC columns, 145
GC detectors, 143, 165
GC-AED, 168
GC-FID, 146

chromatography (*cont.*)
GC-FT-IR, 168
GC-FT-IR-MS, 167
GC-FTIR, 168
GC-MS, 147, 155, 160, 161, 167-169
gel filtration, 36, 120
geraniol, 121, 122
ginger, 149
ginger oils, 135, 148
Ginkgo biloba, 136, 157
ginkgolides, 106, 136, 157
glucobrassicin, 122
glucosinolates, 122
glycosides, 118, 125
gravity column chromatography, 123, 124
γ-gurjunene, 130
heptafluorobutyric anhydride, 147
high performance liquid chromatography (HPLC), 38, 119, 122, 126-129, 136,140,142-144, 146-148, 150, 153, 157, 161, 162, 165, 331
high performance TLC, 134
homoerythrina, 121
HPLC troubleshooting, 128
HPLC-DAD, 155
HPLC-ELSD, 158
HPLC-MS, 160
HPLC-NMR, 162
HPLC-RI, 158
HPLC-UV, 153, 158
HPTLC, 134
hydrophobic-interaction chromatography, 36
immunoaffinity chromatography, 37, 333
ion-exchange chromatography, 120, 122
ion-pair chromatography, 120
ion-pair HPLC, 129

chromatography (*cont.*)
ion-pair reagents, 122, 129
iridal triterpenoids, 144
iridoid glycosides, 125
iridoids, 126
lapachol methylether, 124
larkspur, 129
LC detectors, 143
LC-GC, 128
LC-MS, 161, 162
LC-MS interfaces, 160
linalool, 147, 148
liquid chromatography (LC), 120, 143, 148, 154, 162, 191
Lobar, 126
low pressure liquid chromatography (LPLC), 126, 127
mass spectrometry (MS), 143, 147, 168
medium pressure liquid chromatography (MPLC), 126, 127
2-methoxy-eudesma-1,4,6-trien-3-one, 126
monoterpene hydrocarbons, 135, 144
nuclear magnetic resonance (NMR), 143, 148
nordamnacantal, 124
norditerpenoid alkaloids, 129
normal phase HPLC, 129
osthol, 140
overpressured layer chromatography (OPLC), 136
oxygenated monoterpenes, 144
oxygenated sesquiterpenes, 144
paper chromatography, 134
partition chromatography, 120, 121
phaeantharine, 134
Plantago media, 125
polyamide, 120, 121
polygodial, 132, 133
Polygonum hydropiper, 132
preparative HPLC, 128

chromatography (*cont.*)
preparative LC, 148
preparative Thin Layer Chromatography (TLC), 124, 137, 138, 174
psoralen, 140
qinghaosu, 144
reversed phase (RP) chromatography, 115, 121
Roman chamomile, 118
Rosa, 129, 130
rotation locular counter current chromatography, 119
rotation planar chromatography, 136
RP, 126-128
RP-C18, 127
RP-HPLC, 132, 139, 159
saffron, 146
safranal, 146
sample clean-up, 131
saponins, 118
β-selinene, 134
separation mechanisms, 120
separatory funnel, 131, 132, 133
Sephadex® LH-20, 118, 120, 123, 124
Sephasorb®, 120
β-sesquiphellandrene, 148, 149
sesquiterpene, 120, 126, 174
sesquiterpene hydrocarbons, 129, 134, 135, 144, 149
SFC-MS, 160
SFC-NMR, 162
Siberian ginseng, 119
silica, 120, 121, 123, 126, 127, 129, 132-134, 136, 137, 150, 162, 172
silver, 129
silylation, 147
β-sitosterol, 144, 145

chromatography (*cont.*)
sniffing port, 169
solid phase extraction (SPE), 131, 133-133
SPE sample clean-up, 131
Spherisorb®, 129
supercritical fluid chromatography (SFC), 142-144, 146, 148, 153, 167
tannins, 118
terpenes, 146
terpenetrilactones, 136
tetrabutylammonium sulphate (TAS), 122
thin layer chromatography (TLC), 125, 133-136, 153, 162, 165
tolu balsam, 129
vacuum liquid chromatography (VLC), 124-126, 175
vincamine, 119
vincine, 119
vitamin A, 162
water-pepper, 132
XAD resins, 123
xanthotoxin, 140
Zanthoxylum budrunga, 125
α-zingiberene, 148, 149
chromolyn, 230
chromones, 9
chromoplasts, 49
Chromosorb®, 145
chrysin 7-glucoside, 19
chrysoplenetin, 233
chrysoplenol D, 233
chalcone synthase (CHS), 43, 68
*chs*15 gene, 69
Cicer arietinum, 269
cichoriin, 116
Cichorium intybus, 18, 306, 308
Cimifuga simplex, 14, 15
Cinchona, 156, 238, 288
1,8-cineole, 104, 106, 297, 299

cinnamaldehydes, 43
cinnamate, 44, 61, 74 79
cinnamate 4-hydroxylase (C4H), 43, 49, 61, 318
cinnamoyl-CoA reductase, 43, 44
cinnamoylcocaine, 3
cinnamyl alcohol dehydrogenase (CAD), 43, 363
cinnamyl alcohols, 43
cis-elements, 62, 63, 66, 67, 70
cisplatin, 244
citral, 229
Citrus, 19
4CL (4-coumarate:CoA ligase), 43, 44, 55, 66
Cl-opioid receptor, 242
Clevenger-type apparatus, 108
Clostridium thermocellum, 340
clover, 22
CNS disorders, 221
CO_2, 103-106, 142, 143
CoA thioesters, 57
coat proteins, 342-344, 346, 347
coca, 3
cocaine, 3, 30, 140, 169, 170
cocaine biosensor, 169
(*S*)-coclaurine, 42, 44
codeine, 242, 291, 365
codeinone, 291
Coffea arabica, 299
coffee, 11, 103, 259
colchicine, 7
Coleus forskohlii, 238
collagenase, 225
Colletotrichum circinans, 19
colon cancer, 264
Colorado beetle, 18
colubrinol, 225
columbianadin, 140, 141
columbianetin, 141
columbianetin acetate, 141
column chromatography, 123, 124, 125
comfrey, 252
Commelina communis, 14
compartmentalization, 359
complexation chromatography, 120, 123
Compositae, 5, 11, 231, 240
computational chemistry, 215, 227
Concentration Control Coefficients, 58
coniferin, 119
coniferyl alcohol, 270, 317
coniine, 29
Conium maculatum, 29
Connectivity Theorem, 58
conoduramine, 164
conodurine, 164
controlled-pore glass (CPG), 283
*cop*9, 76
Coptis, 49, 238, 239
Corallina officinalis, 313
Corallina pilulifera, 313
corn oil, 112
coronary heart disease, 229, 263
correlation maps, 194
Corynebacterium glutamicum, 358
coryneine, 227
cosuppression, 59, 60, 70, 71, 81, 329
COSY, 195-198, 204, 207-209
cotton, 356
4-coumarate: CoA ligase (4CL), 35, 43, 44, 55, 66
p-coumarate: CoA ligase, *see* 4-coumarate: CoA ligase
coumarate 3-hydroxylase (C3H), 43
4-coumaric acid, 9, 44, 61
p-coumaric acid, *see* 4-coumaric acid
coumarins, 9, 13, 20, 139-141, 231, 240, 318
4-coumaroyl-CoA, 35, 44, 46
p-coumaroyl-CoA, *see* 4-coumaroyl-CoA
coumestans, 269, 271
counter-current chromatographic methods, 119
counter-current chromatography, 117

covalent binding, 283
cowpea mosaic virus (CPMV), 342, 344, 345
cowpeas, 256
CPT-11, 234, 235
Craig, 117
Crataegus, 159
cremophor, 244
Crescentia cujete, 15
m-cresol, 9
p-cresol, 14, 15
cross-pathway regulation, 360
cross-talk, 79
Cruciferae, 2, 6, 20
Cryptococcus albidus, 340
crystallisation, 152
crystallography, 188, 189
cucumber mosaic virus, 356
Cucurbitaceae, 5
cucurbitacins, 5
curcumin, 259
cyanidin 3,5-diglucoside, 15
2-cyano-6-oxazolopiperidine, 34
cyanogenic glycosides, 6, 7, 16, 51, 253, 255, 318
cyanohydrins, 305, 306
cyclases, 36, 306, 307, 308
cyclic GMP, 76
β-cyclocitral, 146
1,6-cyclodecadienes, 307
cyclo-oxygenases, 263
cycloheximide, 72
Cyperus iria, 21
cysteine sulphoxides, 267
cytochrome, 266
cytochrome oxidase, 233
cytochrome P450, 37, 40, 49, 51, 263, 267, 289, 297, 317-319, 363
cytokines, 264, 333
cytokinins, 13, 75, 76
cytosol, 51
cytotoxicity, 225
CZE, 137, 139

D

2D NMR, 194, 195, 206, 300
2D spectra, 207
DAHP, 358, 359
DAHP synthase (DS), 52, 358, 359
daidzein, 269-271
Dalton's Law, 106
databases, 221, 227
Datura, 55, 165, 233
Datura stramonium, 165
DCCC, 118, 119
DDMP, 257
deadly nightshade, 233
DEAE-cellulose, 283
decaffeination, 103
decomposition, 170, 174
defence, 28, 52, 75, 79
degradation, 174
dehydroabietic acid, 294
dehydroalanine, 74
9,10-dehydrofusaric acid, 126
dehydroleucodin, 233
delayed COSY, 196
Delila, 67-69
delphinidin, 364
delphinidin 3,5-diglucoside, 15
Delphinium barbeyi, 129
Delphinium nelsonii, 15
4'-demethylepipodophyllotoxin, 292
demissine, 18
DENDRAL, 206
densitometer, 165
dental caries, 337
3-deoxy-D-*arabino*-heptulosonate 7-phosphate (DAHP), 358
3-deoxy-D-*arabino*-heptulosonate 7-phosphate (DAHP) synthase, 52, 358, 359
depsides, 22

depsidones, 22
DEPT or J-modulated spectra, 194
DEPT mode, 207
derivatisation, 146, 147
dermatitis, 229
dermatological diseases, 230
Derris elliptica, 18, 223
desiccation, 77
detection
    Acacia, 157
    acetylenic bond, 167
    ajmalicine, 165
    alkaloids, 156, 163, 165
    allelopathics, 169
    amperometric detection, 159
    anisaldehyde-sulphuric acid, 163
    anthocyanins, 162
    anthrone-sulphuric acid, 163
    antibodies, 169
    antifeedants, 169
    antitumour assays, 169
    Armillaria tabescens, 161
    aromatic amines , 159
    artemisinin, 159
    ascorbic acid, 159
    azadirachtin, 162
    bilobalide, 157
    biological detectors, 168, 169
    bisindole alkaloids , 164
    brine shrimp assay, 169
    capillary electrophoresis, 153
    cardenolides, 163
    carotenoids, 155
    castanospermine, 159
    catecholamines, 159
    Catharanthus roseus, 165
    cation-exchange, 159
    CE-MS, 160
    chloramine-trichloroacetic acid, 163
    chemical ionisation (CI), 168
    CO_2, 143
    cocaine, 169, 170
detection (*cont.*)
    cocaine biosensor, 169
    conoduramine, 164
    conodurine, 164
    Crataegus, 159
    Datura stramonium, 165
    densitometer, 165
    derivatise, 156
    2,3-dihydro-2,5-dihydroxy-6-methyl-4H-pyran-4-one (DDMP), 257
    dihydroindole, 155
    2*R*,5*R*-dihydroxymethyl-3*R*,4*R*-dihydroxypyrrolidone, 223
    diode array detection, 129, 161
    diode array detector (DAD), 129, 154, 155, 161
    double bonds, 154
    Dragendorff reagent, 163
    Ehrlich reagent , 163
    electrochemical detection, 138, 158
    electron capture detector (ECD), 143, 166, 167
    ELSD (evaporative light scattering detection), 143, 157, 158
    enzyme immunoassays, 169, 170
    essential oils, 154, 163, 168
    evaporative light scattering detection (ELSD), 143, 157, 158
    Fastblue B, 163
    ferric chloride / perchloric acid, 164
    flame ionization detector (FID), 143, 144, 165-167, 169
    fluorescence, 143, 163, 165
    fluorescent detection, 156
    fluorescent indicator, 162
    Fourier Transform infrared detector (FT-IR), 143, 144, 167
    GC, 143, 165 *et seq.*
    GC detectors, 143, 165
    GC-AED, 168
    GC-FT-IR, 168
    GC-FT-IR-MS, 167

detection (*cont.*)
GC-MS, 147, 155, 160, 161, 167-169
ginger, 149
ginger oil, 148
Ginkgo biloba, 156, 157
ginkgo terpenetrilactones, 156
ginkgolides, 157
glycosides, 162, 163
herbicides, 169
HPLC, 143 *et seq.*
HPLC-DAD, 155
HPLC-ELSD, 158
HPLC-MS, 160
HPLC-NMR, 162
HPLC-RI, 158
HPLC-UV, 153, 158
3-hydroxy-conoduramine, 164
3-hydroxy-conodurine, 164
3-hydroxy-voacamine, 164
hyoscyamine, 165
immunoassays, 169
indole, 155
indole alkaloids, 155, 164
indolizidine alkaloids, 159
ion-selective electrodes, 170
kairomone, 167
LC, 143, 162
LC detectors, 143
LC-MS, 161, 162
LC-MS interfaces, 160
LC-NMR, 162
mass spectrometry, 147, 153, 160, 168
morphine, 170
MS, 138, 143, 160, 162, 168
ninhydrin, 163
nitrogen phosphorus detector (NPD), 167
NMR, 143, 148, 149, 153, 162
non-protein amino acids , 157
off-line, 153, 168

detection (*cont.*)
on-line, 153
oviposition deterrents, 169
oviposition stimulants, 169
perillene, 167, 168
permanganate - soda, 163
peroxides, 159
phenolics, 163
phenols, 159
photodiode array detector, 130
potassium hydroxide, 163
quinine, 156
quinones, 159
refractive index detection (RID), 155, 156, 158
scopolamine, 165
selective ion monitoring, 161
serpentine, 165
β-sesquiphellandrene, 148, 149
sesquiterpene, 159
sesquiterpene hydrocarbons, 149
sesquiterpenes, 161, 162
SFC detector, 158
SFC-MS, 160
SFC-NMR, 162
sniffing port, 169
spraying reagents, 163, 165
sulphuric acid , 163
supercritical fluid chromatography (SFC) 142-144, 146, 148, 153, 167
Tabernaemontana, 164
terpene trilactones, 158
terpenes, 156, 163, 168
thermal conductivity detector (TCD), 166
Thin Layer Chromatography (TLC), 162-165
triacylglycerols, 158
tropane alkaloids , 167
UV, 143
UV detection, 144

detection (*cont.*)
  UV spectroscopy, 154
  vanillin-sulphuric acid, 163
  vitamin A, 162
  voacamidine, 164
  voacamine, 164
  α-zingiberene, 148, 149
Deuterophoma tracheiphila, 19
diabetes, 221, 236, 240
diallyl disulphide, 171, 267
1,4-diaminobutane, 29, 46, 295
1,5-diaminopentane, 46, 295
diastereoselectivity, 302
dibenzazonine alkaloids, 44
dibenzofuran, 20
dibenzylbutanolides, 292
Dichapetalum cymosum, 17
dichloromethane, 172
diferulate, 48
differential screening, 41
differentiation, 287
diffraction studies, 188
Digitalis, 288
Digitalis purpurea, 100, 151, 238, 239
digitoxin, 100, 239, 288
digitoxin 12β-hydroxylase, 288
digoxin, 288
dihydroanthracenones, 124
dihydrocarvone, 302
dihydrochanoajmaline, 289
dihydroconiferyl alcohol, 13
dihydroflavonol reductases, 364
dihydroflavonols, 9
dihydroindole, 155
dihydrophenanthrenes, 9
dihydropicolinate synthase (DHPS), 361
2*R*,5*R*-dihydroxymethyl-3*R*,4*R*-dihydroxypyrrolidone (DMDP), 223
4,8-dimethyl-1,3E,7-nonatriene, 111
dimethyl disulphide, 15
dimethylallyl pyrophosphate, 3
dimethylamine, 171
diode array detection , 129, 161
diode array detector (DAD), 129, 154, 155, 161
dioxygenase, 318
diterpene lactones, 13
diterpene triepoxides, 294
diterpenoids, 3, 5, 7, 32, 124, 238, 244
diterpenoid alkaloids, 227
2,4-dithiapentane, 15
DNA-chip technology, 80
DNA damage, 265
DNA footprinting, 63
DNA plant viruses, 329
DNA-affinity chromatography, 63
DNA plant viruses, 329
L-DOPA, 18, 216
double bonds, 154
doxorubicin, 244
Dragendorff reagent, 163
dregamine, 192
droplet counter-current chromatography (DCCC), 118
drugs and drug development
  aconitan A, 227
  aconitine, 227, 228
  Aconitum, 227, 228
  adriamycin, 225
  ajoene, 229
  alkaloids, 234, 237, 242
  Alliaceae, 229
  alliin, 229
  Allium, 228-230
  American mandrake, 242
  Ammi majus, 230, 231
  Ammifurin®, 230
  anaesthesia, 233
  analgesic, 281
  androgen, 225
  Angelica, 231, 232
  Angelica officinalis, 231
  Angelica sinensis, 231
  anisine, 227
  Anisodus, 233

drugs and drug development (*cont.*)
- anorexia, 236
- anti-allergenic activity, 230
- anti-asthmatic activity, 231
- antibacterial activity, 238
- anti-cancer activity, 218, 219, 222, 224, 225, 237, 242, 243, 244
- anti-diabetic activity, 219, 237
- anti-fungal activity, 221
- antihypertensive activity, 238
- anti-inflammatory activity, 231, 238, 241
- anti-leukaemic activity, 293
- anti-malarial activity, 231, 233
- antimicrobial activity, 241
- antimuscarinic agents, 233
- antioxidant, 229
- antiphlogistics, 241
- anti-spasmodic activity, 231, 241
- anti-tumour activity, 225, 244, 281
- anti-ulcerous activity, 233
- anti-viral activity, 221, 222
- Apiaceae, 230, 231
- apigenin, 240, 241
- apigenin glycosides, 241
- Apocynaceae, 236, 242
- Araceae, 222
- arrow poisons, 227
- artelinic acid, 233
- artemetin, 233
- *Artemisia*, 233
- *Artemisia annua*, 231, 232, 233
- *Artemisia douglasiana*, 233
- artemisinic acid, 233
- artemisinin, 231-233
- asthma, 236, 240
- atidine, 227
- atisine, 227, 228
- *Atropa*, 233
- *Atropa belladonna*, 233, 234

drugs and drug development (*cont.*)
- atropine, 28, 57, 233, 234
- Berberidaceae, 242
- berberine, 238, 239
- bergapten, 231
- Bignoniaceae, 220
- bisabolol oxide A, 240
- bisabolol oxide B, 240
- α-bisabolol, 240, 241
- Bishop's weed, 230
- bouvadrin, 225
- Bristol-Myers/Squibb, 244
- bronchodilator, 230
- bruceantin, 225
- calcium antagonists, 231
- *Camptotheca acuminata*, 234, 235
- camptothecin, 219, 225, 234, 235
- cancer, 224, 234, 236, 242-245
- cancer chemotherapy, 236
- Cannabinaceae, 235
- cannabinoid receptor, 236
- cannabis, 235, 236
- *Cannabis sativa*, 235, 236
- carcinoma, 224
- cardenolide glycosides, 239
- cardioactive drugs, 288
- casticin, 233
- *Catharanthus roseus*, 219, 236, 237
- Celiptium®, 242
- Central Drug Research Institute, 238
- chamomile, 240, 241
- *Chamomilla*, 240
- chemotaxonomic information, 222, 224
- chemotaxonomic investigations, 226
- chemotherapy, 236
- Chinese medicine, 227, 231
- chives, 229
- chloroquine, 238
- cholera, 229
- cholesterol, 230
- chromolyn, 230

drugs and drug development (*cont.*)
chrysoplenetin, 233
chrysoplenol D, 233
Cinchona, 238
cisplatin, 244
citral, 229
Cl-opioid receptor, 242
cocaine, 30
codeine, 242
Coleus forskohlii, 238
collagenase, 225
colubrinol, 225
Compositae, 231, 240
computational chemistry, 215, 227
Coptis, 238, 239
coronary heart disease, 229
coryneine, 227
coumarins, 231, 240
CPT-11, 234, 235
cytotoxicity, 225
databases, 217, 221, 227
Datura, 233
deadly nightshade, 233
dehydroleucodin, 233
dermatitis, 229
dermatological diseases, 230
diabetes, 236, 240
diabetes mellitus, 221
Digitalis purpurea, 238, 239
digitoxin, 239, 288
digoxin, 288
diterpenoid alkaloids, 227
diterpenoids, 238, 244
L-dopa, 216
doxorubicin, 244
drug discovery, 215
Duboisia, 233
Elder, 230
Eli Lilly, 236, 237
ellipticine, 225, 242
endocrine hormones, 225
etoposide, 292

drugs and drug development (*cont.*)
epilepsy, 236
ethnobotanical information, 218-221, 224, 226
etoposide, 242, 243
expectorant, 231
exploitation, 215
extraction, 224
false positives, 221, 225
fermentation, 224
flavones, 233
flavonoids, 229, 240
forskolin, 238
foxgloves (*Digitalis*), 238, 239
furanocoumarins, 230
garlic, 229, 230
geraniol, 229
German chamomile, 240
gerot, 230
Ginkgo biloba, 239, 240
Ginkgoaceae, 239
ginkgolide A, 240
ginkgolides, 240
Glaxo Wellcome, 225, 226
glucosidase inhibitors, 222
haemorrhoids, 229, 231, 238
hashish, 235
heart conditions, 229
heart disease, 238
hemp, 235
hepatitis, 231
herbal remedies, 240
herniarin, 240
herpes, 221
higenamine, 227
HIV, 224
homoharringtonine, 225
hormone receptors, 225
10-hydroxycamptothecin, 234
hyoscine, 233
hyoscyamine, 29-32, 56, 233
Hyoscyamus, 233
hypaconitine, 227

drugs and drug development (*cont.*)
hypertension, 229
hysognavine, 227
ignarine, 227
imperatorin, 231
immunosuppressive activity, 294
indicine-*N*-oxide, 225
indole alkaloids, 51, 234, 235
indomethacin, 241
isopimpinellin, 231
isosafrol, 231
Japanese medicine, 227
jesaconitine, 227, 228
khellin, 230
kobusine, 227, 228
labdane, 238
Lamiaceae, 238
Lastet®, 243
leeks (*Allium porrum*), 229
legumes, 222
leukaemia, 224, 234, 243
lignans, 242, 243
ligustilide, 231
linalool, 229
luteolin, 240
lymphoblastic leukaemia, 237
lymphomas, 243
Madagascar periwinkle (*Catharanthus roseus*), 219, 236
maidenhair tree, 239
malaria, 231, 238
Matricaria recutita, 240, 241
matricin, 240, 241
maytansine, 225
Medexport, 230
medicine men, 226
Meladinine® , 230
melanoma, 244
memory loss, 240
Memphis Chemical Co., 230
menstruation, 231

drugs and drug development (*cont.*)
mesaconitine, 227, 228
N-methyl-9-hydroxy-ellipticinum acetate, 242
microtubules, 244
mitotic activity, 225
morphine, 281
monkshood, 227
monoterpenoids, 229
Moraceae, 222
morphinan, 44
morphine, 54, 242
motion sickness, 233
Myrtaceae, 222
Nabilone®, 236
National Cancer Institute (NCI), 217-221, 224-226, 243, 244
National Pharmacopoeiae, 227
nematodes, 222
neopelline, 227
neuralgia, 227
nicotine, 28, 29, 41, 56, 57, 61, 77
Nyssaceae, 234
Ochrosia, 242
Ochrosia borbonica, 242
Ochrosia elliptica, 242
Ochrosia sandwicensis, 242
oestrogen, 225
onion, 229
Oxsoralen®, 230
paclitaxel, 243
paediatric leukaemia, 237
palmatine, 238
Papaver, 242
Papaver somniferum, 242
Papaveraceae, 242
parasitic diseases, 229
Parkinson's Disease, 216
pharmaceutical companies, 216, 218, 219
phenolics, 221, 224
phytosterols, 224

drugs and drug development (*cont.*)
plant collection, 218
plant screening, 218
Plasmodium, 238
Plasmodium falciparum, 231
platelet-aggregating factors, 240
podophyllin, 242
podophyllotoxin, 242
Podophyllum peltatum, 242, 243
polyhydroxyalkaloids, 222, 226
polysaccharides, 220, 221
post-harvest processing, 241
primaquine, 231
Provir®, 221
psoralens, 230, 231
psychotropic effects, 235
PUVA (psoralens + UV-A), 231
pyridine alkaloid, 57
qinghaosu, 231
quercetin, 240
Quinh Hao, 231
quinine, 29, 238
Ranunculaceae, 227, 238
receptor binding, 225
receptor-based assays, 225
receptor-based mechanisms, 225
Research Triangle Institute, 225
respiratory diseases, 230, 231
rheumatism, 227, 229
Rhone-Poulenc, 244
Royal Botanic Gardens, Kew, 222, 226
Rubiaceae, 238
salicylamide, 241
Sandoz, 242
saponins, 224, 229
sarcoma, 224, 243
scopolamine, 29, 31, 56, 233, 234
scopoletin, 231
Scopolia, 233
screening, 217, 221, 224, 225
Scrophulariaceae, 238
sesquiterpene, 231, 233, 240

drugs and drug development (*cont.*)
Solanaceae, 233
sweet Annie, 231
sweet wormwood, 231
tannins, 220, 221, 224
Taxaceae, 243
taxanes, 244
Taxol®, 32, 219, 225, 243, 244, 281
Taxotere®, 244
Taxus, 244
Taxus brevifolia, 243
teniposide, 243
terpenes, 240
testicular teratoma, 237
tetrahydrocannabinols (THCs), 140, 235, 236
thebaine, 44
toothpick plant, 230
topoisomerase I, 225, 234
topoisomerase II, 225
topotecan, 234
traditional healers, 226
triglycerides, 230
tripdiolide, 225
tropane alkaloids, 32, 37, 55, 57, 233, 234
tubulin, 244
tumours, 244
ulcer-protective activity, 241
umbelliferone, 240
University of Illinois, 225
vanillic acid, 231
Vepeside®, 243
vinblastine, 51, 219, 225, 236, 237
Vinca rosea (syn. Catharanthus roseus), 236
vincristine, 51, 219, 225, 236, 237
viral infections, 221
Virend®, 221
warts, 242
xanthotoxin, 231
yew, 243
DS (DAHP synthase), 52, 358, 359

Duboisia, 57, 233
dye-ligand matrices, 36

E

E.coli heat-labile entertoxin, 334
E-2-hexenal, 113
ECD, 143, 145
ecdysone, 21
ecgonine, 3
Ehrlich reagent, 163
EI (Electron Impact), 168, 192
Elasticity Coefficients, 58
Elder, 230
electrochemical detection, 138, 158, 159
electron capture detector (ECD), 166, 167
electron impact (EI), 168, 192
electroosmotic flow, 138
eleutheroside B, 119
eleutheroside B1, 119
Eli Lilly, 236, 237
elicitors, 52-54, 61, 63, 65, 66, 77
ellipticine, 225, 242
ELSD (evaporative light scattering detection), 143, 157, 158
eluotropic series, 121
emulgators, 307
enantioselectivity, 297, 302, 314
endocrine hormones, 225
endoplasmic reticulum, 51
enkephalin, 331, 343
enterodiol, 270
enterolactol, 270
enterolactone, 270
entrapment, 284, 285
environmental impact, 349
enzyme assays, 32 *et seq.*
enzyme immunoassays, 169, 170
enzyme inhibition, 263
enzyme isolation, 32
enzyme purification, 35, 42
enzyme turnover, 73
epi-cathenamine, 42
4-epi-γ-gurjunene, 130, 131
epilepsy, 236
epitope, 346
epitope-presentation systems, 344-346
epo (erythropoietin), 333, 349
epoxidation, 294, 314, 315, 364
epoxides, 294
epoxylinoleic acid, 18
12,13-epoxylinoleic acid, 19
equol, 270, 271
erucic acid, 2
erythropoietin, 333, 349
erythrose-4-phosphate, 358
essential oils, 36, 55, 69, 108, 144, 147, 150, 154, 163, 168
esters, 304
esterases, 303
estradiol (*syn.* oestradiol), 270
estriol (*syn.* oestriol), 21
ethnobotanical information, 218-221, 224, 226
ethnobotanical investigations, 226
ethyl acetate, 15
ethyl 2-methyl-3-oxobutanoate, 302
N-ethyl (*S*)-nornicotine, 296
ethylene, 12, 13, 40, 56, 74-77
N-ethylputrescine, 296
etioplasts, 49
etoposide, 242, 243, 292
Eucalyptus, 297
Eucalyptus perriniana, 297, 298
eudesmanes, 306, 307
eugenol, 14
euglossine, 15
evaporative light scattering detection (ELSD), 143, 157, 158
expectorants, 231
expert systems for structural elucidation, 205
expression (gene), 319, 327-329, 332, 341

extraction techniques, 95, 224
- (+)-*S*-allyl-L-cysteine sulphoxide, 171
- adsorbent traps, 111
- alkaloids, 96
- allicin, 171
- alliin, 171
- antifeedants, 175
- apigenin, 241
- apigenin glycosides, 241
- *Azadirachta indica*, 175-177
- azadirachtin, 175-177
- *Brassica napus*, 111
- caffeine, 104
- camphor, 104, 106
- chamazulene, 171, 172
- chamomile, 171
- chamomile oil , 171
- chicory, 99
- 1,8-cineole, 104, 106
- Clevenger-type apparatus, 108
- CO_2, 103-106
- coffee, 103
- corn oil, 112
- Dalton's Law, 106
- decaffeination, 103
- diallyl disulphide, 171
- *Digitalis purpurea*, 100
- digitoxin, 100
- 4,8-dimethyl-1,3*E*,7-nonatriene, 111
- essential oils, 108
- α-farnesene, 112
- flavonoid glycoside, 100
- freon, 104
- garlic, 171
- ginkgolides, 106
- glycosides, 100
- headspace (HS) sampling, 113
- headspace techniques, 109
- heptanal, 112
- hesperidin, 100
- E-2-hexenal, 113

extraction techniques (*cont.*)
- (Z)-3-hexenyl acetate, 112
- hexyl acetate, 112
- hops, 103
- hydrodistillation, 108
- infrared spectroscopy (NIR) , 95
- kairomone, 111
- kiwi fruit, 112
- kiwi fruit flowers, 112
- Likens and Nickerson-type, 108
- maceration, 101, 102
- matricin, 171, 172
- monoterpene, 107
- multiple headspace extraction, 110
- nonanal, 112
- orchids, 111
- pentadecane, 112
- percolation, 101, 102
- piperine, 104
- post-harvest processing, 241
- *Rosmarinus officinalis*, 105
- separatory funnels, 117
- sesquiterpene alcohols, 107
- sesquiterpene lactones, 99
- solid phase extraction (SPE), 114
- solid phase microextraction, 112
- solvent extraction, 99
- solvent microextraction, 113
- solvents, 96, 97
- Soxhlet extraction, 101, 102
- SPE, 114
- steam distillation, 106-108, 171, 172
- steam distillation extraction (SDE), 108, 109
- strawberries, 112
- strychnine, 97
- supercritical fluid extraction (SFE), 103, 105
- supercritical fluids, 103, 104
- tetranortriterpenes, 175
- 4,8,12-trimethyl-1,3*E*,7*E*,11-tridecatetraene, 111

extraneous compounds, 174

Extrelut®, 132, 133

F

FAB (fast atom bombardment), 191, 192
false positives, 225
farnesol, 21, 120-122
fast atom bombardment (FAB), 191, 192
Fastblue B, 163
fatty acids, 13, 315
Fed-1, 72
feeding deterrent, 281
feeding experiments, 29
fenchone, 297
Fenton reaction, 259, 261
fermentation, 224
ferns, 9
ferredoxin, 72
ferric chloride / perchloric acid, 164
ferritin, 259
ferulate, 363
ferulate 5-hydroxylase, 43, 363
ferulic acid, 259, 262
feruloyl-CoA, 44
feverfew, 18
Flame Ionisation (FI) detection, 143, 144, 146, 165-167, 169
flash chromatography, 123-126, 150, 172, 175
flatulence, 256
flavan-3,4-diols, 9
flavan-3-ols, 9
flavanols, 259
flavanones, 9
flavanone 3β-hydroxylase, 35, 40
flavanonols, 9
flavolans, 9
flavones, 9, 14, 19, 65, 233
flavonoid 3'-hydroxylase, 318
flavonoid 3' 5'-hydroxylase, 318
flavonoids, (*incl.* isoflavonoids) 9-11, 14, 35, 39, 46, 51, 53, 54, 56, 60, 66,
flavonoids (*cont.*) 68, 75, 114, 121, 129, 130, 197, 229, 240, 259, 261-263, 268, 318
aglycones, 147
Ammi majus, 230
anthocyanins, 9, 14, 28, 39, 44, 51, 55, 66-69, 75, 76, 121, 129, 162, 259, 298, 364
anti-malarial activity, 233
antioxidant, 229
Apiaceae, 230
apigenin, 240, 241
apigenin glycosides, 241
artemetin, 233
casticin, 233
catechin, 262, 264
chamomile, 241
Chamomilla, 240
chrysin 7-glucoside, 19
chrysoplenetin, 233
chrysoplenol D, 233
Compositae, 11, 240
coronary heart disease, 229
COSY, 197
coumestans, 269
cyanidin 3,5-diglucoside, 15
daidzein, 269, 270
delphinidin, 364
delphinidin 3,5-diglucoside, 15
dihydroflavonol reductases, 364
dihydroflavonols, 9
flavan-3,4-diols, 9
flavan-3-ols, 9
flavanols, 259
flavanones, 9
flavanone glycosides, 152
flavanonols, 9
flavolans, 9
flavones, 1, 9, 14, 233
flavonoid glycosides, 100, 118, 127
flavonoid 3'-hydroxylase, 318
flavonoid 3' 5'-hydroxylase, 318

flavonoids (*cont.*)
flavonols, 9, 14, 65, 139, 144, 259, 261
flower colour, 364
formononetin, 21
genistein, 269, 270
glyceollin, 41, 46
glyceollin I, 65
glycitein, 269
glycosides, 118, 127, 152
hesperidin, 100, 152
HOHAHA, 197, 198
isoflavones, 21, 269, 270
isoflavone synthase, 318
isoflavonoids, 9, 18, 20, 268, 269
Leguminosae, 9
luteolin, 240
luteone, 18, 19
Matricaria recutita, 240, 241
medicarpin, 54, 65
microbial degradation, 270
pelargonidin 3,5-diglucoside, 15
phaseollin, 65
phytoalexins, 20
pisatin, 65
pterocarpans, 61, 65
quercetin, 10, 240, 259, 262-264
relayed COSY, 197
rutin, 60, 198
flavonols, 9, 14, 65, 139, 144, 259, 261
flavoprotein oxidases, 259
Flavr-Savr® tomatoes, 350, 356
floral meristems, 69
floral pigments, 14
Florisil®, 120
flower colour, 364
fluoroacetate, 16
fluoroacetic acid, 17
fluorescence, 143, 163, 165
fluorescent detection, 156
fluorescent indicator, 162
flux control, 58
Flux Control Coefficients, 58, 59
FMDV 2A, 347
food flavouring, 230
foot-and-mouth disease virus (FMDV), 344, 347
foreign proteins, 327
formononetin, 21
forskolin, 238
Forsythia, 48
Fourier Transform infrared detector (FT-IR), 167
foxgloves (*Digitalis*), 238, 239
frachinoside, 116
fractional (vacuum) distillation, 149, 150
essential oils , 150
monoterpenes, 150
Pulicaria gnaphalodes, 150
sesquiterpenes, 150
frame-shift mutations, 72
Fraxinus chinensis, 115
free radicals, 258, 259, 261, 263, 264
French bean (*Phaseolus vulgaris*), 49, 53, 54, 61, 63, 65, 66, 70, 72, 269
freon, 104
FT-IR, 143, 144, 167
furanocoumarins, 11, 12, 16, 56, 65, 66, 230, 252, 254, 255
psoralens, 230
furanoterpenoids, 255
fusaric acid , 126
Fusarium, 22
Fusarium avenaceum, 19
Fusarium nygamai, 126
Fusarium oxysporum, 19
Fusarium solani, 19
fusca, 76

G

G-box, 63, 66
G-protein, 335, 336
Gaeumannomyces graminis, 19

galactinol, 256
galactose, 12, 256
α-galactosides, 256
α-1,6 galactosidase, 256
gallic acid, 12, 13
gamma effect rule, 203
Gardenia jasminoides, 298
garlic, 127, 171, 229, 230, 257
gas chromatography (GC), 106, 109-113, 122, 127, 131, 142-144, 146-148, 150, 153, 165, 167-169, 191
Gasteria bicolor, 124
GC columns, 145
GC detectors, 143
GC-AED, 168
GC-FID, 146
GC-FT-IR, 168
GC-FT-IR-MS, 167
GC-FTIR, 168
GC-MS, 147, 155, 160, 161, 167-169
GC/MS, 296
2-D gel electrophoresis, 37
gel filtration, 44
gel filtration chromatography, 36, 120
gel-retardation, 63
gelatin, 285
gene isolation, 38
gene silencing, 70, 71
genetic transformation, 327, 329
genistein, 269-271
Gentianaceae, 9
geraniol, 15, 121, 122, 229
geraniol / nerol 10-hydroxylase, 318
geraniol 10-hydroxylase, 51
geranyl diphosphate synthase, 49
geranylgeranyl diphosphate synthase, 49
germacranes, 306-308
1,5-germacrane-1,10-epoxides, 307
1,5-germacrane-4,5-epoxides, 307
1,5-germacranes, 307
germacrene B, 14
germacrone 4,5-epoxide, 307
German chamomile, 240
gerot, 230
giant hogweed, 255
gibberellins, 4, 5, 12, 13, 37, 39, 75, 76, 208-211
ginger, 149
ginger oils, 135, 148
ginkgo, 136, 158
Ginkgo biloba, 56, 151, 156, 157, 203, 239, 240, 365
ginkgo terpenetrilactones, 156
Ginkgoaceae, 239
ginkgolide A, 106, 136, 151, 157, 204, 240
ginkgolide B, 136, 151, 157
ginkgolide C, 136, 151, 157
ginkgolide J, 106, 136, 157
ginkgolides, 56, 106, 152, 157, 240, 365
GlaxoWellcome, 225, 226
glucans, 227
β-glucans, 339
glucanase, 339
β-glucanases, 339
(1-3, 1-4)-β-glucanases, 339
glucoalkaloids, 41, 42
glucobrassicanapin, 267
glucobrassicin, 122, 267
glucoiberin, 267
gluconapin, 267
gluconasturtiin, 267
glucose, 12
glucose esters, 79
glucose-6-phosphate dehydrogenase, 61
β-glucosidase, 19, 42, 309
β-D-glucosidase, 300
glucosidase inhibitors, 222
glucosidases, 299, 300
glucosides (*see also* glycosides), 48, 79, 298, 299, 300
  (+)-camphor, 298
  Eucalyptus perriniana, 298
glucosinolates, 6, 7, 39, 51, 122, 255, 267

glucosyl transferase, 51
β-glucuronidase (gus), 53, 63, 330
glutamic acid, 303
glutathione, 70
glutathione *S*-transferase, 266
glyceollin, 46
glyceollin I, 65
glycine, 302
Glycine max (soybean), 46, 65, 72, 257, 269, 302, 314, 315, 317
glycitein, 269
glycoalkaloids, 57, 252, 253, 367
glycoconjugates, 192
glycolipids, 307
glycosidases, 303, 307-309
glycosidation, 309, 310
glycosides, 6, 10, 51, 56, 65, 66, 100, 114, 116, 118, 125, 152, 162, 163, 239, 257, 298, 307-309
alcohols, 307, 310
alkaloids, 41, 42
apigenin glycosides, 241
arbutin, 19, 139
avenacin AI, 19
avenacoside A, 19
camphor 5-exo-hydroxylase, 319
(+)-camphor, 299
capsaicin, 299
cardenolides, 163, 239
α-chaconine, 253
chrysin 7-glucoside, 19
cichoriin, 116
1,8-cineole, 299
coniferin, 119
COSY, 197
cyanidin 3,5-diglucoside, 15
cyanogenic glycosides, 6, 7, 51, 253, 255
delphinidin 3,5- diglucoside, 15
demissine, 18
Digitalis purpurea, 238, 239
digitoxin, 100, 239
Eucalyptus perriniana, 299
glycosides (*cont.*)
eleutheroside B, 119
eleutheroside B1, 119
flavanone glycosides, 152
flavonoid glycosides, 100, 118, 127
foxgloves (*Digitalis*), 238, 239
frachinoside, 116
GAs, 255
glucobrassicanapin, 267
glucobrassicin, 122, 267
glucoiberin, 267
gluconapin, 267
gluconasturtiin, 267
glucosidase inhibitors, 222
glucosinolates, 7, 39, 51, 122, 255, 264, 267
glycoalkaloids, 57, 252, 253, 367
glycoconjugates, 192
glycosidases, 308, 309
glycosyl transferases, 308, 309
hesperidin, 100, 152
1-hexanol, 310
HOHAHA, 197, 198
HPLC, 162
indoleglucosinolates, 362
iridoids, 5, 34, 51, 125, 126
linamarin, 253, 255
Myrsine pellucida, 191
1-octanol, 310
1-octylglucoside, 308
pelargonidin 3,5-diglucoside, 15
phenethyl glucosinolate, 267
progoitrin, 267
quercetin, 10
relayed COSY, 197
rutin, 198
salicin, 17, 299
saponins, 18, 256, 257
saxifragolin B, 191
secologanin, 34, 35, 41, 42, 51
sinigrin, 16, 17, 19, 267, 268
α-solanine, 253
soyasapogenols, 257

glycosides (*cont.*)
steroidal saponins, 5
steroids, 192
strictosidine, 35, 41, 42, 51
tomatine, 19, 57
α-tomatine, 254
triterpenes, 192
triterpenoid saponins, 5
1-tuliposide A, 19
6-tuliposide A, 19
vanillin β-D-glucoside, 300
vincoside, 42
glycosyl transferases, 308, 309
gradient-enhanced NMR spectroscopy, 205
gradient RP-HPLC , 129
grafting experiments, 56, 78
gravity liquid chromatography, 123, 124
green fluorescent protein (GFP), 347
green pepper, 259
GroEL, 73
GroES, 73
growth inhibitors
batatasin I, 13
chlorogenic acid, 13
coumarins, 13
dihydroconiferyl alcohol, 13
diterpene lactones, 13
fatty acids, 13
gallic acid, 13
lactones, 13
oleic acid, 13
palmitic acid, 13
parasorbic acid, 13
phenolic acids, 13
phenylpropanoids, 13
podolactone E, 13
sesquiterpene lactones, 13
stilbenoids, 13
3,4,5-trihydroxybenzoic acid, 12
umbelliferone, 13
growth inhibitors (*cont.*)
xanthinin, 13
guaiane lactone, 306
guaianes, 306, 307
guayule, 318
gum-tree, 305
γ-gurjunene, 130, 131
gus, 53, 63, 330, 331, 348
gut membrane, 257
Guttiferae, 9
Guy's antibody, 336, 337
Gymnodium breve, 22

H

H-box, 63, 66-70
^1H-^1H COSY, 207
^1H NMR, 194, 206, 207, 209
H_2O_2 burst, 79
haem, 311, 314, 315, 318
haemoglobin, 332
haemorrhoids, 229, 231, 238
Hafnia alvei, 296
hairy root cultures, 30, 288, 295, 296
halogenated monoterpenes, 311
halogenation, 310, 312
halometabolites, 310
haloperoxidases, 292, 310, 311, 312, 313, 314
hashish, 235
HbA, 332
HbsAg (hepatitis B virus surface antigen), 334, 335
headspace (HS) sampling, 113
headspace techniques, 109
heart conditions, 229
heart disease, 238, 258, 259
heat shock proteins, 73
heavy atoms, 189, 190
Helminthosporium carbonum, 19
hemicellulose, 340

hemp, 235
hepatitis, 231
hepatitis B virus surface antigen, 334, 335
heptafluorobutyric anhydride, 147
heptanal, 112
2-heptanone, 14
Heracleum mantegazzianum, 255
herbal remedies, 240
herbicides, 169
herbivores, 16
herbivory, 11, 216
herniarin, 240
herpes, 221
hesperidin, 100, 152
heterologous expression, 319, 321
heteronuclear correlation, 198
heteronuclear multiple bond correlation (HMBC), 201,204, 205, 207-209
heteronuclear multiple quantum correlation (HMQC), 199, 204, 205, 207, 208
heteronuclear single quantum correlation (HSQC), 199, 208
heterotrimeric GTP-binding proteins, 76
Hevea brasiliensis, 305, 331
1-hexanol, 310
(Z)-3-hexenyl acetate, 112
hexyl acetate, 112
higenamine, 227
High Performance TLC, 134
High Pressure Liquid Chromatography (HPLC), 38, 119, 122, 126-129, 131, 136, 140, 142-144, 146-148, 150, 153, 157, 160-162, 165, 331
histidine, 360
HIV-1, 345, 346
HMG-CoA reductase, 53
HMQC, 199, 204, 205, 207, 208
HOHAHA, 197, 198, 204
homeobox genes, 69
homeodomain proteins, 67
homoerythrina, 121
homoharringtonine, 225
Homonuclear Hartmann Hahn (HOHAHA), 197, 198, 204
hops (*Humulus lupulus*), 9, 46, 55, 103, 268
hormone receptors, 225
hormones (*see also* plant growth substances), 333
horseradish peroxidase (HRP), 314, 315
HPLC troubleshooting, 128
HPLC-DAD, 155
HPLC-ELSD, 158
HPLC-MS, 160
HPLC-NMR, 162
HPLC-RI, 158
HPLC-UV, 153, 158
HPTLC, 134
HSA, 332
HSP70 (heat shock protein), 73
HSQC (heteronuclear single quantum correlation), 199, 208, 199, 208
HT29 cell line, 268
Human Immunodeficiency Virus (HIV), 224, 342
human interferon αD (IFN αD), 134
human rhinovirus 14 (HRV-14), 345
human serum albumin, 332
humming bird, 15
Humulus lupulus (hops), 9, 46, 55, 103, 268
hydrocarbons, 75, 310
hydrodistillation, 108
hydrogen cyanide, 305, 306
hydrogen donation, 261
hydrogen peroxide, 259, 264
hydrolysis, 302, 304, 305
hydroperoxides, 315, 318
hydrophobic-interaction chromatography, 36
3-hydroxy-3-methylglutaryl-CoA reductase(HMG-CoA reductase), 53
hydroxybenzoates, 10
2-hydroxybenzoic acid, 78

10-hydroxycamptothecin, 234
hydroxycinnamates, 259, 262, 263
p-hydroxycinnamic acid *see* 4-coumaric acid
hydroxycinnamoyl-CoA, 44
4-hydroxycinnamoyl-CoA substrates, 55
3-hydroxyconoduramine, 164
3-hydroxyconodurine, 164
20-hydroxyecdysone, 20, 21
10-hydroxygeraniol, 51
5-hydroxytryptamine, 362
3-hydroxyvoacamine, 164
hydroxyl radicals, 259
hydroxylases, 364
hydroxylation, 51, 296, 297, 299, 300, 318, 320, 364
β-hydroxymethylglutaryl-CoA, 49
21-hydroxyraumacline, 289
21-hydroxyraumacline reductase, 289
Hyocyamus niger, 364
hyoscine, 233
hyoscyamine, 29-32, 56, 165, 233, 364
hyoscyamine 6β-hydroxylase, 56, 364
Hyoscyamus, 56, 233
hypaconitine, 227
hypertension, 229
hypocholesterolemic activity, 257
hysognavine, 227

I

IAA oxidase, 13
IgG, 336
ignarine, 227
illicit drug substances, 140
imidazoleglycerol-phosphate dehydratase, 360
immobilisation, 282-287
immunoaffinity chromatography, 37, 333
immunoassays, 169
immunocytochemical techniques, 48, 56
immunogens, 334
immunogold labelling, 49
immunosuppressive activity, 294
immunotoxins, 338
imperatorin, 231
impurities, 174
in situ hybridisation, 68
in situ nucleic-acid hybridisation, 56
INADEQUATE, 202, 203, 208
indicine-*N*-oxide, 225
indigo, 278
Indigofera, 278
indoles, 7, 14, 15, 41, 155, 318
indole alkaloids, 51, 77, 155, 164, 192, 211, 234, 235, 289, 293
indole-3-acetic acid, 12, 13
indole-3-carbinol, 264
indoleglucosinolates, 362
indolizidine alkaloids, 159
indolyl glucosinolate, 264
indomethacin, 241
industrial enzymes, 338
influenza virus, 346
infrared spectroscopy (NIR) , 95
insect cell culture, 319
interferon αD (IFN αD), 341
interferons, 333, 341
α-interferons, 333
β-interferons, 333
ion-exchange chromatography, 63, 120, 122
ion-pair chromatography, 120
ion-pair HPLC, 129
ion-pair reagents, 122, 129
ion-selective electrodes, 170
ionic binding, 283
ionising radiation, 28
ipomeamarone, 255
4-ipomeanol, 255
Ipomoea batatis, 255
Ipomopsis aggregata, 15
iridal triterpenoids, 144
iridoid glucosides, 34, 51

iridoid glycosides, 125
iridoids, 4, 5, 126
  polygodial, 132, 133
  Polygonum hydropiper, 132
  RP-HPLC, 132
  SPE, 132
  water-pepper, 132
iron, 259, 261
Isatis tinctoria, 278
iso-vallesiachotamine, 42
isoborneol, 301
isobutyryl-CoA, 46
isocoumarins, 268
isodihydrocarvone, 302
isoenzymes, 52, 53-55, 70
isoeugenol, 14, 15
isoflavone synthase, 318
isoflavones, 21, 269-271
isoflavonoid phytoalexins, 46, 53
isoflavonoids, 9, 18, 20, 46, 53, 54, 65, 66, 268, 269
isopentenyl diphosphate, 3, 49
isopimpinellin, 231
isoprenoid biosynthesis, 33
isoprenoids, 3, 32, 33, 36, 49, 55, 80
  abietane, 124
  abscisic acid, 39, 75
  aconitine, 227, 228
  Aconitum, 228
  ajmaline, 289, 290
  alloaromadendrane-7α,11α-diol, 120, 121
  anisine, 227
  anti-cancer activity, 243, 244
  anti-malarial activity, 233
  anti-oxidant activity, 236
  anti-tumour activity, 244
  anti-ulcerous activity, 233
  antifeedants, 175
  Armillaria, 161, 162
  artelinic acid, 233
  Artemisia, 231, 233
  Artemisia annua, 231, 232, 233
isoprenoids (*cont.*)
  Artemisia douglasiana, 233
  artemisinic acid, 233
  artemisinin, 143, 144, 159, 231-233
  atidine, 227
  atisine, 227, 228
  Azadirachta indica (neem), 175-177
  azadirachtins, 3, 10, 17, 162, 175-177, 281
  betulin, 1
  bilobalides, 56, 136, 137, 157, 174
  bis-indole alkaloids, 293
  α-bisabolol, 240, 241
  bisabolol oxide A, 240
  bisabolol oxide B, 240
  borneol, 297, 301
  brassinolides, 188, 189
  brassinosteroids, 39, 75, 76
  bufadienolides, 5
  camphor, 104, 106, 297, 301
  (+)-camphor, 298, 299
  camphor 5-exo-hydroxylase, 319
  cancers, 244, 245
  cardenolide glycosides, 239
  cardenolides, 4, 5
  Δ2-carene, 300, 301
  β-carotene, 259
  carotenoids, 5, 14, 15, 40, 55, 155, 259, 262
  carvomenthones, 301, 302
  carvone, 302, 303
  carvone oxide, 15
  caryophyllene, 134
  casbene synthase, 37
  catharanthine, 293, 294
  catharanthine *N*-oxide, 294
  α-chaconine, 253
  chamazulene, 171, 172
  chamomile, 171, 240, 241
  chamomile oil, 171
  Chamomilla, 240
  cholesterol, 187

isoprenoids (*cont.*)
1,8-cineole, 104, 106, 297, 299
citral, 229
Coleus forskohlii, 238
Compositae, 231, 240
COSY, 209
cucurbitacins, 5
cyclases, 36, 306, 307, 308
1,6-cyclodecadienes, 307
dehydroleucodin, 233
Digitalis purpurea, 238, 239
digitoxin, 239, 288
digoxin, 288
dihydrocarvone, 302
dihydrochanoajmaline, 289
diterpene lactones, 13
diterpene triepoxides, 294
diterpenoid alkaloids, 227
diterpenoids, 3, 5, 32, 238, 244, 294
ecdysone, 21
4-epi-γ-gurjunene, 130, 131
essential oils, 144, 150, 163, 168
eudesmanes, 306, 307
farnesol, 21, 120-122
fenchone, 297
forskolin, 238
foxgloves (*Digitalis*), 238, 239
furanoterpenoid, 255
GAs (glycoalkaloids), 255
GC, 148
geraniol, 15, 121, 122, 229
geraniol / nerol 10-hydroxylase, 318
germacranes, 306, 307, 308
1,5-germacrane-1,10-epoxides, 307
1,5-germacrane-4,5-epoxides, 307
1,5-germacranes, 307
germacrone 4,5-epoxide, 307
German chamomile, 240
gibberellic acid, 12, 13, 208-211

isoprenoids (*cont.*)
gibberellins, 4, 5, 12, 37, 39, 75, 76
ginger, 149
ginger oils, 135, 148
Ginkgo biloba, 151, 156, 157, 203, 239, 240, 365
ginkgo terpenetrilactones, 156
Ginkgoaceae, 239
ginkgolides, 56, 106, 136, 151, 152, 157, 204, 240, 365
glycoalkaloids, 41, 42, 252, 253, 367
glycoconjugates, 192
guaiane lactones, 306
guaianes, 306, 307
γ-gurjunene, 130, 131
1H NMR, 209
halogenated monoterpenes, 311
HMBC (heteronuclear multiple bond correlation), 201, 204, 205, 207-209, 209
HPLC, 148
HSQC (heteronuclear single quantum correlation), 199, 208, 208
20-hydroxyecdysone, 20, 21
10-hydroxygeraniol, 51
21-hydroxyraumacline, 289
hypaconitine, 227
hysognavine, 227
ignarine, 227
indole alkaloids, 290, 293
ipomeamarone, 255
4-ipomeanol, 255
iridal triterpenoids, 144
iridoid glucosides, 34, 51, 125
iridoids, 4, 5, 126
isoborneol, 301
isodihydrocarvone, 302
isoeugenol, 15
jesaconitine, 227, 228
kairomone, 167
kobusine, 227, 228

isoprenoids (*cont.*)
- labdane, 238
- lactucin, 306, 307
- Lamiaceae, 238
- limonoids, 4, 5
- linalool, 147, 148, 229
- LSD (Logic for Structure Determination), 207-211
- lutein, 259
- lycopene, 259
- maidenhair tree, 239
- malaria, 231
- marrubiin, 32
- *Matricaria recutita*, 240, 241
- matricin, 171, 172, 240, 241
- melanoma, 244
- menthol, 202, 297
- mesaconitine, 227, 228
- 2-methoxy-eudesma-1,4,6-trien-3-one, 126
- monoterpene cyclases, 36
- monoterpene hydrocarbons, 135, 144
- monoterpenoid indole alkaloids, 293
- monoterpenoids, 3, 5, 14, 17, 29, 55, 107, 150, 229
- *Myrsine pellucida*, 191
- neocarvomenthols, 301, 302
- neopelline, 227
- neoprocurcumenol, 307
- NMR, 149
- NOE, 204
- norditerpenoid alkaloids, 129
- nortriterpenoids, 4, 5
- oestradiol (*syn.* etradiol), 270
- oxygenated monoterpenes, 144
- oxygenated sesquiterpenes, 144
- Paclitaxel®, 243
- papyriferic acid, 16
- parthenolides, 18, 19, 92
- patuletin, 15
- perillene, 167, 168

isoprenoids (*cont.*)
- phytosterols, 5, 224
- *Plasmodium falciparum*, 231
- podolactone E, 13
- prenyltransferases, 36
- *Pulicaria gnaphalodes*, 150
- qinghaosu, 144, 231
- quassinoids, 4, 5
- Quinh Hao, 231
- raumacline, 289, 290
- rubber, 55
- rubber-particle protein, 318
- *Ruptiliocarpon caracolito*, 188
- saffron, 146
- safranal, 146
- sapogenins, 19
- saponins, 4, 18, 118, 190, 224, 229, 256, 257
- saxifragolin B, 191
- secologanin, 34, 35, 41, 42, 51
- secoterpenoids, 188
- β-selinene, 134
- β-sesquiphellandrene, 148, 149
- sesquiterpene alcohols, 107
- sesquiterpene hydrocarbons, 129, 134, 135, 144, 149
- sesquiterpene lactones, 5, 13, 99
- sesquiterpenes, 3, 5, 14, 49, 120, 126, 150, 159, 161, 162, 174, 231, 233, 240, 306
- silica, 136
- β-sitosterol, 144, 145
- α-solanine, 253
- soyasapogenols, 257
- spirocaracolitones, 188, 189, 190
- steroid saponins, 5
- steroidal hormones, 271
- steroids, 192, 253, 268
- strictosidine, 35, 41, 42, 51
- sweet Annie, 231
- sweet wormwood, 231
- Taxaceae, 243
- taxanes, 244

isoprenoids (*cont.*)
- Taxol®, 32, 225, 243, 244, 281
- Taxotere®, 244
- *Taxus*, 243, 244
- *Taxus brevifolia,* 243
- terpene cyclases, 49
- terpene trilactones, 158
- terpenes, 146, 156, 163, 168, 177, 211, 240, 268
- terpenoid biosynthesis, 46
- terpenoid esters, 302
- α-terpineol, 297
- β-terpineol, 297
- α-terpinyl acetate, 297
- tetranortriterpenes, 175
- tetraterpenoids, 3
- tocopherols, 236
- α-tocopherol, 264
- tolu balsam, 129
- α-tomatine, 254
- α-trichosanthin, 342
- tripdiolide, 225, 294
- triterpenes, 192, 257
- triterpenoid saponins, 5
- triterpenoids, 4
- tumours, 244
- vinblastine, 293, 294
- vincoside, 42
- vindoline, 293, 294
- vitamin E, 261-263
- xanthinin, 13
- yew, 243
- α-zingiberene, 149
- isoquinoline (*see also* benzylisoquinoiline), 7

isosafrol, 231
isothiocyanates, 52, 267, 268
isotriptophenolide, 294
isovaleryl-CoA, 46

J

J chain, 336
J-modulated spectra, 207
*Jaf*13, 67
Japanese medicine, 227
jasmonate, 75, 77-79
jasmonate esters, 56
jasmonates, 75, 77
jasmonic acid, 75
jesaconitine, 227, 228

K

kairomone, 111, 167
Kalanchoe blossfeldiana, 13
kale, 259
KAP-1, 63, 69
KAP-2, 63, 69
β-ketoacylsynthases, 46
ketones, 301
khellin, 230
kiwi fruit, 112
kobusine, 227, 228
Kunitz trypsin inhibitor, 72

L

labdane, 238
labelled precursors, 29, 32, 35
labelling, 30
lactones, 13, 240
lactucin, 306, 307
lactupicrin, 18
lambing, 21
Lamiaceae, 55, 69, 238
lapachol methylether, 124
larkspur, 129
Lastet®, 243
latex, 331

Lathyrus sativus, 255
laticifers, 55
Laurencia, 311
lavanone glycosides, 152
LC, 68, 69, 143, 148, 162, 191
LC detectors, 143
LC-GC, 128
LC-MS, 161, 162
LC-MS interfaces, 160
LC-NMR, 162
LDL (low-density lipoprotein), 263
lead(II) ions, 151
leader peptides, 332, 333, 336, 342
leader sequences, 332, 336, 338
leader sequence, alfalfa mosaic virus (AlMV) RNA-4, 332
leaf-cutting ants, 18
leeks, 229
legumes, 222, 255-257, 269
legumin B4, 338
Leguminosae, 9, 20, 65, 269
lentils, 257
lettuce, 13
leucoplasts, 296
leu-enkephalin, 331, 343
leukaemia, 224, 234, 243
lichens, 9, 22
lignans, 9, 48, 242, 243, 268, 270, 271, 291, 292
lignin, 43, 44, 48, 52, 55, 69, 291, 298, 355, 363, 364
lignin peroxidase, 330
ligustilide, 231
limonoids, 4, 5
linalool, 147, 148, 229
linamarin, 18, 253, 255
linoleic acid, 236, 315, 317
linolenate, 77
linseed, 270
lipases, 303
lipid hydroperoxides, 261
lipid peroxidation, 261
lipid peroxyl radicals, 261
lipoproteins, 263
5-lipoxygenase, 263
lipoxygenases, 77, 263, 315, 317
Lithospermum erythrorhizon, 285, 286
littorine, 30, 31
Lobar system, 126
Loblolly pine, 53
locusts, 18
Lonicera, 34
Lotus corniculatus, 18
low-density lipoprotein (LDL), 263
low pressure liquid chromatography (LPLC), 126, 127
LSD (Logic for Structure Determination), 207-211
LT-B, 334, 335
lupin, 22, 256
Lupinus albus, 18, 19
lutein, 259
luteolin, 240
luteone, 18, 19
luxandrine, 151
lycopene, 259
Lycopersicon esculentum, 19
Lycopodium, 7
lymphoblastic leukaemia, 237
lymphomas, 243
lysine, 29, 30, 46, 282, 295, 296, 360, 361
lysine decarboxylase, 295
lysine α-ketoglutarate reductase (LKR), 361

M

maceration, 101, 102
Madagascar periwinkle (*Catharanthus roseus*), 11, 51, 219, 236, 293
MADS-box proteins, 67
maidenhair tree, 239
maize, 39, 66-69, 350, 363
malaria, 231, 238
malonic acid, 269

malonyl transferases, 51
malonyl-CoA, 9, 35, 46, 52, 53
malting, 339
maltol, 257
Malus sieboldii, 19
mandelonitrile, 305, 306
Mangifera indica, 19
MAPKKKs, 76
Marchantia polymorpha, 302
marrubiin, 32, 33
Marrubium vulgare, 32
mass spectrometry (MS), 147, 153, 160, 168, 190-192
- acetogenins, 190
- alkaloids, 162
- *Armillaria tabescens*, 161
- californium plasma desorption, 192
- CE-MS, 160
- chemical ionization (CI), 168, 192
- cocaine, 30
- diode array detection, 161
- dregamine, 192
- electron impact (EI), 168, 192
- essential oils, 168
- fast atom bombardment (FAB), 191, 192
- GC, 191
- GC-AED, 168
- GC-FT-IR-MS, 167
- GC-FTIR, 168
- GC-MS, 160, 161, 167-169
- glycoconjugates, 192
- glycosides, 162
- HPLC, 160
- HPLC-MS, 160
- indole alkaloids, 192
- LC, 191
- LC-MS, 161
- LC-MS interfaces, 160
- *Myrsine pellucida*, 191
- palytoxin, 192

mass spectrometry (MS) (*cont.*)
- saponins, 190, 192
- saxifragolin B, 191
- selective ion monitoring, 161
- sequencing, 190
- sesquiterpenes, 161, 162
- SFC-MS, 160
- squamocin, 190, 191
- stereochemistry, 192
- steroids, 192
- tabernaemontanine, 192
- terpenes, 168
- triterpenes, 192
- tropane alkaloids, 30

matairesinol, 270
Matricaria recutita, 240, 241
matricin, 171, 172, 240, 241
Mayer's reagent, 151
maytansine, 225
MCA, 57-59, 358
MECC, 139
Medexport, 230
Medicago sativa, 53, 54, 302
medicarpin, 54, 65
medicine men, 226
medium pressure liquid chromatography, 126
MEKC, 139, 140
Meladinine®, 230
melanoma, 244
membrane bioreactors, 285
membrane reactors, 285
membranes, 233
memory loss, 240
Memphis Chemical Company, 230
Mentha, 36, 55, 69, 302
menthol, 202, 297
mercuric iodide, 151
mesaconitine, 227, 228
mescaline, 140
metabolic channelling, 42, 44, 46, 48, 58
metabolic compartmentation, 76

metabolic control analysis (MCA), 57, 69, 358
metabolic engineering, 356, 358, 359, 361
metabolic grids, 44
metabolic sinks, 362
metallothionein (MTII), 341
2-methoxy-eudesma-1,4,6-trien-3-one, 126
methyl anthranilate, 14, 15
methyl benzoate, 14
N-methyl-9-hydroxy-ellipticinum acetate, 242
methyl sulphides, 14
N-methylanthraniloyl-CoA, 46
methylcoclaurine, 319
α-methylenebutyrolactones, 92
4-(methylnitrosamino)-1-(3-pyridyl)-1 butanone (NNK), 268
δ-*N*-methylornithine, 32
N-methylputrescine, 295
mevalonate, 3, 32, 33, 49, 53, 80
mevalonate pathway, 49
micellar electrokinetic capillary chromatography, 139-141
microbial degradation, 270
microtubules, 244
millet, 305
mink enteritis virus (MEV), 345
mint, 14
miroestrol, 21
miserotoxin, 16, 17
mitogen-activated protein kinase (MAPK), 77
mitogen-activated protein kinase kinase kinases (MAPKKK), 76
mitotic activity, 225
mobility-shift determinations, 63
molecular chaperones, 73
molluscan, 18
monkshood, 227
monoterpenoids, 229
monolignols, 43, 44, 48
monooxygenase, 363
monophenols, 13
monoterpene cyclases, 36
monoterpene hydrocarbons, 135, 144
monoterpenoid indole alkaloids, 29, 293
monoterpenoids, 3, 5, 14, 17, 55, 150, 289, 297
Moraceae, 222
morphinans, 44, 291
morphine, 54, 140, 170, 242, 281, 365
moth larvae, 16
moths, 15, 18
motion sickness, 233
moulting hormones (insect), 20
mouse IgG, 336
MPLC, 126, 127
mRNA degradation, 72
mRNA processing, 72
mRNA stability, 71, 72
MS, 30, 138, 143, 160, 167, 168
MTII (metallothionein II), 341
Mucuna, 18
multienzyme complexes, 46
multigene families, 53, 54
multiple headspace extraction, 110
mutants, 38, 39, 75, 76, 79, 314, 361, 363
Myb 305, 63, 67-69, 305, 308, 340
Myb-related transcription factors, 67-69
Myc-related transcription factors, 67
Mycotheceum, 22
mycotoxins, 20
myo-inositol, 12
myrosinase, 267
Myrtaceae, 222
Myrtillocactus geometrizans, 300

N

^{15}N, 30
N-terminal processing, 332
Nabilone®, 236
nacycline, 42

nahG, 78
naphthoquinones, 9
naringenin, 35
National Cancer Institute (NCI), 217-221, 224-226, 243, 244
National Pharmacopoeiae, 227
Nectria cinnabarina, 19
neem, 3, 11, 281
nematodes, 222
neocarvomenthols, 301, 302
neopelline, 227
neoprocurcumenol, 307
Nernst equation, 262
neuralgia, 227
neurolathyrism, 255
Neurospora crassa, 360
neutron diffraction, 190
Nicotiana (see also tobacco), 56, 57, 60, 295, 296
  N. benthamiana, 343, 347
  N. glutinosa, 19
  N. plumbaginifolia, 300
  N. sylvestris, 293
  N. tabacum, 41, 60, 61, 297, 300-302, 330
nicotine, 28, 29, 32, 37, 41, 56, 57, 61, 77, 295, 296, 363, 364
nicotine biosynthesis, 296
nicotinic acid, 30
ninhydrin, 163
nitraramine, 281, 282
Nitraria schoberi, 282
nitriles, 52
nitrogen phosphorus detector (NPD), 167
NNK (4-(methylnitrosamino)-1-(3-pyridyl)-1 butanone), 268
nobiletin, 19
nodulation, 75
nodulation-inducers, 56
nOe, 203, 204
NOESY, 203
non-mevalonate pathway, 32, 33
non-protein amino acids, 7, 157
nonanal, 112
(*S*)-norcoclaurine, 42, 44
nordamnacantal, 124
norditerpenoid alkaloids, 129
norharman, 288
normal phase HPLC, 129
nornicotine, 364
nortriterpenoids, 4, 5
Norwalk virus, 335
nuclear magnetic resonance (NMR), 30, 32, 143, 148, 149, 162, 190, 193, 204
  akuammicine, 195, 196, 201
  bistramide A, 202
  13C-13C INADEQUATE, 207
  13C 1D spectra, 207
  13C NMR, 194
  chemical shifts (δ), 193
  chromophores, 206
  correlation maps, 194
  COSY, 195-198, 204, 207-209
  1D 1H NMR, 194
  1D NMR, 206
  2D NMR, 194, 195, 206
  2D spectra, 207
  delayed COSY, 196
  DENDRAL, 206
  DEPT mode, 207
  DEPT or J-modulated spectra, 194
  expert systems for structural elucidation, 205
  flavonoids, 197
  gamma effect rule, 203
  gibberellic acid, 208-211
  Ginkgo biloba, 203
  ginkgolide A, 204
  gradient-enhanced NMR spectroscopy, 205
  1H 1D spectra, 207
  1H-1H COSY, 207
  1H NMR, 194, 209

nuclear magnetic resonance (NMR) (*cont.*)
- heteronuclear multiple bond correlation (HMBC), 201, 204, 205, 207-209
- heteronuclear multiple quantum correlation (HMQC), 199, 204, 205, 207, 208
- heteronuclear single quantum correlation (HSQC), 199, 208
- Homonuclear Hartmann Hahn (HOHAHA), 197, 198, 204
- HPLC-NMR, 162
- improvements, 205
- INADEQUATE, 202, 203, 208
- indole alkaloids, 211
- J-modulated spectra, 207
- LC-NMR, 162
- LSD (Logic for Structure Determination), 207-211
- marrubiin, 32
- menthol, 202
- nicotine, 32
- NMR chromophores, 206
- nOe, 203, 204
- NOESY, 203
- nuclear Overhauser effect, 203
- palytoxin, 212
- relayed COSY, 197, 198
- ROESY, 203
- rutin, 198
- SFC-NMR, 162
- spin-spin coupling measurements (J), 193
- stereochemistry, 203
- strategy, 204
- terpenes, 211
- Total Correlation Spectroscopy (TOCSY), 198
- tropane alkaloids, 30, 32

nuclear magnetic resonance (NMR) (*cont.*)
- vitamin A, 162
- W couplings, 196
- X-H COSY, 199
- nuclear Overhauser effect, 203

nuclear run-on transcription studies, 63, 72
nutritional quality, 361
Nyssaceae, 234

O

oats, 18
Ochrosia, 242
- *O. borbonica*, 242
- *O. elliptica*, 242
- *O. sandwicensis*, 242

β-ocimene, 18
1-octanol, 310
1-octylglucoside, 308
β-ODAP, 255
odours, 14
Oenanthe crocata, 17
oenanthetoxin, 17
oestradiol (*syn.* estradiol), 270
oestriol (*syn.* estriol), 21
oestrogen receptors, 268, 270, 271
oestrogens, 21, 225, 268
oestrone, 21
oil bodies, 348
oil seeds, 348, 362
olefinic double bonds, 129
olefins, 163
oleic acid, 13
oleosins, 348
oncogene, 268
one-pot synthesis, 294
onions, 229, 257, 259
Ophrys lutea, 14, 15
opium poppy (*Papaver somniferum*), 54, 242, 281, 291
OPLC, 136
oral immunity, 335

oral vaccines, 334-336, 347
oranges, 259
orchids, 14, 111
organ cultures, 30, 35, 39, 55, 286, 288
ornithine, 29, 37, 46, 295
Orobanche, 22
Oryza sativa, 19
osmotic stress, 77
osmotin, 77
osthol, 140, 141
ouabain, 17
Overhauser effect, 203
overpressured layer chromatography (OPLC), 136
oviposition deterrents, 169
oviposition stimulants, 169
β-*N*-oxalyl-L-2,3-diaminopropionic acid, 255
β-oxidation, 77
oxidative coupling reactions, 48
N-oxides, 56
2-oxoglutarate-dependent dioxygenases, 40, 41, 51, 56
Oxsoralen®, 230
oxygenated monoterpenes, 144
oxygenated sesquiterpenes, 144
oxynitrilase, 305

P

P450cam, 319
p53 gene, 264
Paclitaxel®, 243
paediatric leukaemia, 237
PAL (phenylalanine ammonia lyase), 43, 44, 49, 53, 54, 60, 61, 63, 66, 67, 74, 77
palmatine, 238
palmitic acid, 13
palytoxin, 192, 212
papain, 303, 304
Papaver, 242
Papaver somniferum (opium poppy), 54, 242, 281, 291
Papaveraceae, 242
papaya, 303
paper chromatography, 134
paper-making, 340, 348
papyriferic acid, 16, 17
parasitic diseases, 229
parasorbic acid, 13
Parkinson's disease, 216
parsley (*Petroselinum crispum*), 35, 53, 54, 56, 63, 65, 66, 255
parsnip (*Pastinaca sativa*), 11, 18, 255
Parthenium, 126, 318
Parthenium argentatum, 318
parthenolide, 3, 18, 19, 92
particle bombardment, 328, 339
partition chromatography, 120, 121
partition coefficients, 114, 258, 262, 263
Pastinaca sativa (parsnip), 11, 18, 255
pathogen, 53, 77, 79
pathogenesis-related (PR) proteins, 77, 78, 332
patuletin, 15
Pb(II) ions, 151
PCR® (Polymerase Chain Reaction®), 40, 62
peanuts, 259
peas, 65, 256, 257
Pedicularis, 22
Peganum harmala, 362
pelargonidin 3,5-diglucoside, 15
Pelletier, 1
pentadecane, 14, 112
5-pentadecylresorcinol, 19
pentose phosphate pathway, 61
peptidases, 303
peptide bonds, 304
peptides, 7, 304, 331, 342, 343, 347, 348
percolation, 101, 102
pericycle, 56
perillene, 167, 168

permanganate - soda, 163
Peronospora parasitica, 19
peroxidase, 49, 291, 293, 314
peroxidases, 10, 293, 314
peroxidative reactions, 57
peroxides, 159
pertraction, 287
pervaporation, 287
Petroselinum crispum (parsley), 35, 53, 54, 56, 63, 65, 66, 255
Petunia, 35, 39, 54, 66-68, 71
Petunia hybrida, 35, 63, 71
phaeantharine, 134
Phanerochaete chryosporium, 330
pharmaceutical companies, 216, 218, 219
phase I enzymes, 266, 267
phase II enzymes, 266, 267
phaseollin, 65
β-phaseolin, 332
Phaseolus vulgaris (French bean), 49, 53, 54, 61, 63, 65, 66, 70, 72, 269
phenethyl glucosinolate, 267
phenethyl isothiocyanate (PEITC), 266, 267, 268
phenol, 10
phenolases, 10
phenolic acids, 9, 13
phenolic ketones, 9
phenolic metabolites (*see also* flavonoids), 8, 51, 56, 163, 221, 224
  acetosyringone, 75
  American mandrake, 242
  Ammi, 230, 231
  Ammifurin®, 230
  Angelica, 139, 141
  anthochlors, 9
  anthocyanins *see* flavonoids
  anthraquinones, 9
  anti-inflammatory activity, 231
  anti-spasmodic activity, 231
  arbutin, 139
  aurones, 9

phenolic metabolites (*cont.*)
  batatasin I, 13
  benzenoids, 14
  benzofurans, 9
  benzoquinones, 9
  Berberidaceae, 242
  bergapten, 141, 231
  biphenols, 13
  caffeic acid, 60, 259, 262, 264
  calcium antagonists, 231
  cancer, 242, 243
  catechol, 19, 78
  chalcones, 9
  chlorogenic acid, 13, 60, 259, 263
  chromones, 9
  cinnamic acid, 44, 61, 74
  cinnamyl alcohols, 43
  columbianadin, 141
  columbianetin, 141
  columbianetin acetate, 141
  coniferyl alcohol, 270
  4-coumaric acid, 44, 61
  p-coumaric acid (*syn.* 4-coumaric acid), 44, 61
  coumarins, 9, 13, 20, 139-141, 231, 240
  m-cresol, 9
  p-cresol, 14, 15
  curcumin, 259
  cyanidin 3,5-diglucoside, 15
  delphinidin 3,5-diglucoside, 15
  depsides, 22
  depsidones, 22
  dermatological diseases, 230
  Digitalis purpurea (foxglove), 151
  dihydroconiferyl alcohol, 13
  dihydroflavonols, 9
  dihydrophenanthrenes, 9
  Etoposide®, 242, 243
  eugenol, 14
  expectorant, 231
  ferulate, 259, 262, 363
  ferulate-5-hydroxylase, 363

phenolic metabolites (*cont.*)
- furanocoumarins, 11, 12, 16, 65, 66, 230, 252, 254
- gallic acid, 13
- Gentianaceae, 9
- Guttiferae, 9
- herniarin, 240
- hydroxybenzoic acids, 10
- hydroxycinnamates, 259, 262
- *p*-hydroxycinnamic acid, *see* 4-coumaric acid
- hydroxycinnamic acids, 259, 263
- imperatorin, 231
- isocoumarins, 268
- isoeugenol, 14
- isopimpinellin, 231
- Lastet®, 243
- Leguminosae, 9
- leukaemias, 243
- lichens, 9
- lignans, 9, 242, 243, 268, 270
- lignin, 355, 363, 364
- luteone, 18, 19
- lymphomas, 243
- matairesinol, 270
- Meladinine®, 230
- Memphis Chemical Company, 230
- methyl benzoate, 14
- microbial degradation, 270
- monolignols, 43, 44, 48
- monophenols, 13
- naphthoquinones, 9
- osthol, 141
- Oxsoralen®, 230
- pelargonidin 3,5-diglucoside, 15
- 5-pentadecylresorcinol, 19
- peroxidases, 10
- phenolases, 10
- phenolic acids, 9, 13
- phenolic ketones, 9
- phenols, 9, 159

phenolic metabolites (*cont.*)
- phenylindan, 259
- phenylpropanoids (*see also* hydroxycinnamic acids, flavonoids), 9, 10, 13, 44, 49, 52, 53, 60, 61, 63, 65, 66, 77, 268, 270, 318, 355, 356, 360, 363
- podophyllin, 242
- podophyllotoxins, 242, 292
- *Podophyllum peltatum*, 242, 243
- polyvinylpolypyrrolidone (PVPP), 151
- psoralens, 141, 230, 231, 254
- PUVA (psoralens-UV-A), 231
- quercetin, 10
- quinones, 10
- quinonoids, 9
- salicin, 17
- salicylate, 10, 11, 75, 78, 79, 299
- sarcoma, 243
- scopoletin, 231
- secoisolariciresinol, 270
- sinapate, 363
- sinapic acid, 363
- sinapoyl choline, 363
- sinapoyl malate, 363
- sinapoylcholine, 363
- sinapoylmalate, 363
- stilbenoids, 9, 13
- tannins, 9, 10, 36, 118, 220, 221, 224, 355
- teniposide, 243
- α-tocopherol, 264
- 3,4,5-trihydroxybenzoic acid, 12
- umbelliferone, 13, 141, 240
- *Uvae ursi*, 139
- vanillic acid, 231
- Vepeside®, 243
- vitamin E, 261-263
- warts, 242
- xanthones, 9

phenolic metabolites (*cont.*)
xanthotoxin, 12, 141, 231
phenolic substances, 56
phenols, 9, 159
phenylpropanoids, *see phenolic metabolites*
phenylacetaldehyde, 315
phenylalanine, 8, 9, 32, 44, 52, 358, 360, 363, 364
phenylalanine ammonia lyase, *see* PAL
phenylindan, 259
pheromones, 15
phloroisobutyrophenone synthase, 46
phloroisovalerophenone synthase, 46
Phlox drummondii, 15
Phomopsis perniciosa, 19
phorbol esters, 264, 265
phosphatases, 74
phosphoenolpyruvate, 358
6-phosphogluconate dehydrogenase, 61
phosphorylation protein, 68, 70, 74
photosensitisation, 255
photomorphogenesis, 76
phytases, 340, 341
phytate, 340
phytoalexins, 20, 46, 52, 54, 55, 61, 65, 66, 77, 255
4-ipomeanol, 255
furanocoumarins, 65, 66
furanoterpenoids, 255
glyceollin, 41, 46
glyceollin I, 65
ipomeamarone, 255
isoflavonoids, 20
medicarpin, 54, 65
phaseollin, 65
pisatin, 65
pterocarpans, 61, 65
phytochemical literature, 94
phytochrome, 71, 75, 76
phytoecdysones, 21
Cyperus iria, 21
ecdysone, 21
phytoecdysones (*cont.*)
20-hydroxyecdysone, 20, 21
Polypodium vulgare, 20
sedge, 21
Taxus baccata (yew), 20
yew, 20
phytoestrogens, 268-271
abortifacient, 21
apple, 21
formononetin, 21
isoflavone, 21
lambing, effects on, 21
miroestrol, 21
pomegranate, 21
sheep, effects on, 21
phytohaemagglutinin, 72
Phytophthora infestans, 52
phytosterols, 5, 224
piperidine alkaloids, 7, 22, 29, 30, 34
piperine, 104
pisatin, 65
plant breeding, 365
plant cell culture, 283-286, 287, 288, 296, 298
plant collection, 218
plant growth substances, 12, 37, 39, 75
abscisic acid, 12, 13, 39, 75
auxins, 12, 39, 51, 71, 75, 76
brassinosteroids, 12, 13, 39, 75, 76
cytokinins, 13, 75, 76
ethylene, 12, 13, 75
gibberellic acid, 12, 13
gibberellins, 12, 13, 39, 75, 76
indole-3-acetic acid, 12, 13
polyamines, 12, 13
zeatin, 12
plant oestrogens, *see* phytoeostrogens
plant screening, 218
Plantago media, 125
Plasmodium, 238
Plasmodium falciparum, 231
Plasmodium vivax, 346
Plasmodium yoelii, 346

plastids, 49
platelet-aggregating factors, 240
pleiotropic effects (of genetic manipulation), 59
Plocamium, 310, 311
podolactone E, 13
podophyllin, 242
podophyllotoxins, 242, 292
Podophyllum peltatum, 242, 243, 292
pollination, 14, 15
pollination, compounds important in,
- anthocyanins, 14
- carotenoids, 14, 15
- carvone oxide, 15
- *p*-cresol, 15
- cyanidin 3,5-diglucoside, 15
- delphinidin 3,5-diglucoside, 15
- 2,4-dithiapentane, 15
- dimethyl disulphide, 15
- ethyl acetate, 15
- flavonoids, 14
- geraniol, 15
- indole, 15
- isoeugenol, 15
- methyl anthranilate, 15
- methyl sulphides, 14
- patuletin, 15
- pelargonidin 3,5-diglucoside, 15

poly A-binding proteins (PABPs), 72
polyamide, 120, 121
polyamines, 12, 13, 60, 268
polyethylene glycol, 36
polygodial, 132, 133
Polygonum hydropiper, 132
polyhydroxy alkaloids, 222, 223, 226
polyketides, 46, 365
Polymerase Chain Reaction® (PCR®), 40, 62
polymerisation, 174
polyphenols, 121
Polypodium vulgare, 20
polypropene, 283
polysaccharides, 220, 221
polyurethane, 284
polyvinylpolypyrrolidone (PVPP), 36, 151
pomegranate, 21
poplar, 55
post-harvest processing, 241
post-transcriptional regulation, 70
potassium hydroxide, 163
potato breeding, 367
potato virus X (PVX), 347
potatoes, 53, 254, 257, 332, 334, 335, 253, 360, 362
potential difference (PD) , 257
PR-1b, 333
PR-S, 341
precipitation, in extraction procedures, 151
- alkaloids, 151
- benzylisoquinoline, 151
- *Digitalis purpurea*, 151
- flavanone glycosides, 152
- ginkgolide A, 151
- ginkgolide B, 151
- ginkgolide C, 151
- ginkgolides, 152
- glycosides, 152
- hesperidin, 152
- luxandrine, 151
- Mayer's reagent, 151
- mercuric iodide, 151
- Pb(II) ions , 151
- polyvinylpolypyrrolidone (PVPP), 151
- *Pseudoxandra sclerocarpa*, 151
- quaternary alkaloids, 151
- Valser's reagent, 151

prenyltransferases, 36
preparative capillary gas chromatography, 148
preparative HPLC, 128
preparative LC, 148

preparative TLC, 124, 137, 138, 174
primaquine, 231
primary metabolism, relationship to secondary metabolism, 52
progoitrin, 267
proline, 29
promoters, 53, 63, 66, 68-70, 72
promoter methylation, 70
prostaglandins, 77
prostatic cancer, 270, 271
proteases, 303, 304
proteasomes, 74
protein degradation, 73, 74
protein kinases, 74, 76
protein phosphatases, 76
protein phosphorylation, 361
protein turnover, 74
proteinase inhibitors, 77, 256, 340
proteins, 328, 347
proto-oncogenes, 264
protoplast fractionation, 48
protoplasts, 63, 72, 339, 343
Provir®, 221
Pseudomonas, 78, 319
Pseudomonas putida, 319
Pseudoxandra sclerocarpa, 151
psilocybin, 140
psoralens, 140, 141, 230, 231, 254
psoralen synthase, 318
psychotropic effects, 235
pterocarpans, 51, 54, 61, 65
Pueraria mirifica, 21
Pulicaria gnaphalodes, 150
pungency, 299
purification, *see* extraction, enzyme purification
purines, 7
putrescine, 29, 37, 41, 46, 55, 295, 364
putrescine *N*-methyltransferase (PMT), 37, 41, 55, 61, 363
PUVA (psoralens-UV-A), 231
PVX (potato virus X), 347
Pyricularia oryzae, 18, 19
pyridine alkaloids, (*see also* nicotine) 57
pyrimidines, 7
pyrrolidine alkaloids, (*see also* nicotine) 7, 60
pyrrolizidine alkaloids, 7, 22, 56, 252
Pyrus communis, 19

Q

qinghaosu, 144, 231
quassinoids, 4, 5
quaternary alkaloids, 151
quercetin, 10, 240, 259, 261-264
Quercus falcata, 11
Quercus robur, 18
Quinh Hao, 231
quinic acid, 32, 60, 259
quinine, 1, 29, 115, 156, 238
quinoline alkaloids, 7, 29
quinolizidine alkaloids, 7, 22
quinones, 10, 159
quinonoids, 9, 22

R

R factor, 67
R proteins, 67
rabies virus, 335
racemates, 302
radical scavenging, 261, 263
radical-coupling reactions, 291
radicals, 263
radio-HPLC, 37
radio-TLC, 37
radiometric assays, 37
radish, 314
raffinose, 256
raffinose oligosaccharides (RFOs), 256
rainbow trout growth hormone, 333
Ranunculaceae, 227, 238
rape, 331, 348

rape pollen, 188
raumacline, 289
Rauvolfia, 34, 289, 298
Rauvolfia serpentina, 289, 298
receptor-based assays, 225
red pepper, 259
redox potentials, 262, 263
redox reactions, 300
refractive index detection, 155, 156
regioselectivity, 280, 296, 304, 312, 315
regulatability, 59
regulatory approval, 348
regulatory capacity, 59, 61
regulatory considerations, 348
regulatory genes, 62, 66
relayed COSY, 197, 198
reporter-genes, 53, 63, 68
Research Triangle Institute, 225
respiratory conditions, 231
respiratory diseases, 230
resveratrol synthase, 46
reticulates, 285
(*S*)-reticuline, 49, 291
reversed-phase (RP) chromatography, 115, 121, 126-132, 139
RFOs (raffinose oligosaccharides), 256
rheumatism, 227, 229
rheumatoid arthritis, 258
Rhizobium, 22, 56, 75
Rhone-Poulenc, 244
RI detectors, 158
Ribes nigrum, 19
ribulose bis-phosphate carboxylase/ oxygenase (Rubisco), 73
rice blast, 18
ripening, 40, 57
ripening-associated genes, 40
risk assessment, 366
RNA plant viruses, 329, 346
RNAse protection, 53, 63
RNAses, 72
ROESY, 203
Roman chamomile, 118
root cultures, 30, 55, 56, 288
Rosa, 129, 130
Rosaceae, 6, 19, 20
Rosmarinus officinalis, 105
rotation planar chromatography (RPC), 136
rotation locular counter current chromatography (RLCC), 119
rotenone, 16, 18
Royal Botanic Gardens, Kew, 222, 226
rubber, 55, 331
rubber tree, 331
rubber-particle protein, 318
Rubiaceae, 238
Rubisco, 73
Rudbeckia hirta, 14, 15
Ruminococcus flavefaciens, 339, 340
Ruptiliocarpon caracolito, 188
Ruta graveolens, 38
Rutaceae, 254
rutin, 60, 198

S

Saccharomyces, 360
Saccharomyces cerevisiae, 360
safety concerns, 366
safflower, 236
saffron, 146
safranal, 146
sakuranetin, 19
salicin, 17, 299
salicyl alcohol, 299
salicylamide, 241
salicylate, 10, 11, 75, 78, 79
salicylate hydroxylase, 78
salicylate radicals, 79
salicylate: UDPglucose glucosyltransferase, 79
salicylate-binding protein, 79
Salix, 11, 17
salutaridinol-7-*O*-acetate, 44

sample clean-up, 131
Sandoz, 242
sanguinarine, 54
sapogenin, 19
saponins, 4, 18, 118, 190, 192, 224, 229, 253, 257
SAR (systemic acquired resistance), 78, 79
sarcoma, 224, 243
SAUR, 72
saxifragolin B, 191
SC (secretory component, of secretory IgA), 336
scFv antibodies, 338
sclareol, 19
scopolamine, 29, 31, 56, 165, 233, 234, 296, 364
scopoletin, 20, 231
Scopolia, 233
scoulerine, 49, 291
screening, 224, 225
Scrophulariaceae, 238
SDS-PAGE, 37
secoisolariciresinol, 270
secologanin, 34, 35, 41, 42, 51
secoterpenoids, 188
secretory component (SC, of secretory IgA), 336
secretory IgA (SIgA), 336, 337
sedge, 21
seed storage proteins, 331, 332
seeds, 361-363, 365
selective ion monitoring, 161
β-selinene, 134
Senecio, 17, 22
Senecio jacobaea, 17
senecionine, 16, 17
senescence, 77
sense-inactivation, 70
separation mechanisms, 120
separatory funnels, 114, 117, 131-133
Sephadex® LH-20, 118, 120, 123, 124
Sephasorb®, 120
sequencing, 190
serotonin, 362, 363
serpentine, 165
β-sesquiphellandrene, 148, 149
sesquiterpenes, 3, 5, 7, 14, 49, 120, 126, 150, 159, 161, 174, 162, 231, 233, 240, 306
    alcohols, 107
    hydrocarbons, 129, 134, 135, 144, 148, 149
    lactones, 5, 13, 99
SFC (Supercritical Fluid Chromatography), 142-144, 146, 148, 153, 167
SFC detector, 158
SFC-MS, 160
SFC-NMR, 162
sheep, 21
shikimate pathway, 10, 32, 46, 52, 53, 61, 355, 358-360, 363
shoot cultures, 288
sialic acid, 349
Siberian ginseng, 119
SIgA (secretory IgA), 336, 337
signal molecules, 75, 78
signal peptides, 333, 338-341
signal sequences, 333
signal transduction pathways, 69, 74-76, 79, 81, 264-271
signalling, 268
signalling pathways, 75-77
silencing, 329
silica, 120, 121, 123-127, 129, 132-134, 136, 137, 150, 162, 172, 175
silkworms, 20
silver(I), 123, 129
silylation, 147
sinapic acid, 363
sinapoyl-CoA, 44
sinapoylcholine, 363
sinapoylmalate, 363

1-sinapoyl: L-malate sinapoyltransferase, 51
single-chain Fv (scFv), 338
sinigrin, 16, 17, 19, 267, 268
site-directed mutagenesis, 314, 315, 320
β-sitosterol, 144, 145
slugs, 18
small auxin upRNA (SAUR), 72
sniffing ports, 169
snowshoe hare, 16, 17
Solanaceae, 30, 233, 253
α-solanine, 253
Solanum (see also potatoes), 18, 367
Solanum demissum, 18
solid phase extraction (SPE), 114, 131, 133
solid phase microextraction, 112
solvent extraction, 99
solvent microextraction, 113
solvent partition, 114
    alkaloids, 114
    cichoriin, 116
    flavonoids, 114
    frachinoside, 116
    glycosides, 114, 116
    partition coefficient K, 114
    partitions, 175
    quinine, 115
    separatory funnels, 114, 131, 132, 133
    silica, 175
    strychnine, 115
solvents, 96, 97, 99, 113 *et seq.*
sorghum, 305, 363
Sorghum bicolor, 305
Southern hybridisation, 41, 62
Soxhlet extraction, 101, 102
soyasapogenols, 257
soybean (*Glycine max*), 46, 65, 72, 256, 257, 269-271, 314, 315, 317
soybean lipoxygenase, 315, 317
SPE (solid-phase extraction), 114,
SPE (solid phase extraction) (*cont.*), 131-133
SPE sample clean-up, 131
spermine, 12
Spherisorb®, 129
spin-spin coupling measurements (J), 193
spinach (*Spinacia oleracea*), 259
spirocaracolitones, 188, 189, 190
Spirodela oligorrhiza , 303
Spodoptera frugiperda, 319
spray reagents, 163, 165
squamocin, 190, 191
stachyose, 256
starch, 338
steam distillation, 106-108, 171, 172
steam distillation extraction (SDE), 108, 109
stereochemistry, 42, 48, 192, 319
stereoselectivity, 280, 296, 301, 302, 305, 310, 312, 314, 315
steroid saponins, 5
steroids (*see also* isoprenoids), 5, 8, 20, 32, 192, 253, 257, 268, 271
stilbene, 46
stilbene synthase, 47
stilbenoids, 9, 13
strawberries, 112, 259
Streptococcus mutans, 337
Streptomyces, 46, 365
stress, 356
strictosidine, 35, 41, 42, 51
strictosidine β-D-glucosidases, 35, 42
strictosidine synthase, 34, 35, 37, 41, 42, 51
Striga, 22
strychnine, 1, 2, 96, 97, 115, 187-190
Strychnos, 3, 96, 97, 99
styrene, 313, 315, 316
subcellular compartmentation, 48
subcellular localisation, 48, 79
suberin, 48, 291
substrates, 32, 35, 37, 41

subtractive hybridisation, 61
sulphides, 315
sulphoraphane, 267
sulphoxidation, 314
sulphoxides, 315
sulphuric acid , 163
supercritical fluid chromatography (SFC), 142-144, 146, 148, 153, 167
supercritical fluid extraction (SFE), 103-105
superfamilies, 41
suppressing agents, 268
surfactants, 307
suspension cultures, 61
Swainsonia canescens, 223
swainsonine, 223
sweet Annie, 231
sweet potato, 255, 259
sweet wormwood, 231
symbiotic associations, 21
  brevitoxin A, 22
  depsides, 22
  depsidones, 22
  lichen, 22
  piperidine alkaloids, 22
  pyrrolizidine alkaloids, 22
  quinolizidine alkaloids, 22
  quinonoids, 22
  Rhizobium, 22
systemic acquired resistance (SAR), 78
systemin, 77

T

T-DNA, 336
Tabernaemontana, 42, 164
Tabernaemontana divaricata, 42
tabernaemontanine, 192
tabersonine, 51
Tagetes patula, 41
Tanacetum parthenium, 3, 18, 19, 92
tannins, 9, 10, 18, 36, 118, 220, 221, 224 355
TAS (tetrabutylammonium sulphate), 122
Taxaceae, 243
taxanes, 171, 244
taxine alkaloids, 172
taxines, 171, 173
Taxol®, 32, 219, 225, 243, 244, 281
Taxotere®, 244
Taxus, 20, 171, 243, 244, 281
Taxus baccata, 20, 171
Taxus brevifolia, 243, 281
TBSV (tobacco bushy stunt virus), 346
TCD (thermal conductivity detector), 166
TDC (tryptophan decarboxylase), 360, 362, 363
tendril-coiling, 77
teniposide, 243
teratogenic effects, 254
terpene cyclases, 49
terpene trilactones, 158
terpenes, *see* terpenoids, isoprenoids
terpenetrilactones, 136
terpenoid biosynthesis, 46, 51
terpenoid esters, 302
terpenoid indole alkaloid, 35
terpenoid pathway, 46, 51
terpenoids, 3-5, 14, 32, 49, 51, 56, 129, 146, 156, 163, 168, 177, 240, 268, 287, 296, 302, 318, 356
testicular teratoma, 237
tetrabutylammonium sulphate (TAS), 122
12-*O*-tetradecanoyl-phorbol-13-acetate (TPA), 264
tetrahydrocannabinols, 235, 236
tetraketide, 29
tetranortriterpenes, 175
tetraterpenoids, 3
tGH-II (trout growth hormone-II), 333, 334
thebaine, 44
thermal conductivity detector (TCD), 166

thin layer chromatography (TLC), 133, 162
thioanisole, 314
9-thiomethylanthracene, 92
thiophenes, 41
three-directional TLC, 135
thrombin, 348
Tinospora cordifolia, 44
TLC, 125, 133-136, 153, 162, 163, 165
TMV, 342, 343, 346, 349
tobacco (*Nicotiana tabacum*), 53, 60, 67-69, 268, 330, 332-341, 343, 344, 346, 347, 356, 361-363
tobacco etch virus (TEV), 334
tobacco mosaic virus (TMV), 342
tobacco PR-S protein, 338
tocopherols, 236, 259, 264
TOCSY, 198
tolu balsam, 129
tomatine, 19, 57, 254
tomato (*Lycopersicon*), 40, 52, 56, 57, 69, 74, 76, 254, 259, 335, 336, 344
tomato bushy stunt virus (TBSV), 346
tonoplast, 51
toothpick plant, 230
topoisomerase I, 225, 234
topoisomerase II, 225
topotecan, 234
total correlation spectroscopy (TOCSY), 198
toxicants, 252, 255, 272
toxins (*see also antifeedants*), 75, 252, 255
  azadirachtin, 17
  brevitoxin A, 22
  canavanine, 11
  cardiac glycosides, 16
  cyanogenic glycosides, 16
  demissine, 18
  fluoracetate, 16
toxins (*cont.*)
  fluoracetic acid, 17
  furanocoumarins, 11, 12
  lactupicrin, 18
  linamarin, 18
  miserotoxin, 16, 17
  mycotoxins, 20
  β-ocimene, 18
  oenanthetoxin, 17
  ouabain, 17
  papyriferic acid, 17
  rotenone, 18
  salicin, 17
  senecionine, 16, 17
  sinigrin, 16, 17
  xanthotoxin, 12, 16, 18
TPA (12-*O*-tetradecanoyl-phorbol-13-acetate), 264
traditional healers, 226
transesterification, 172
transcription factors, 60, 62, 63, 67-69
transcriptional control, 62, 66-68, 70
transduction mechanisms, 74
transformation, 329
transformed plants, 331
transformed root cultures, 32, 61
transition metals, 259
translation, 72
translational enhancer, 332, 334
transport, 55, 56, 61
transposon tagging, 39, 62
triacylglycerols, 158
2,4,6-tribromophenol, 314
Trichoderma reesei, 340
trichomes, 36, 55, 69
trichothecins, 22
Trifolium subterraneum, 21
triglycerides, 230
trihomojasmonate, 77
3,4,5-trihydroxybenzoic acid, 12
4,2',4'-trihydroxychalcone, 46

4,8,12-trimethyl-1,3E,7E,11-tridecatetraene, 111
tripdiolide, 225, 294
Tripterygium wilfordii, 294
triptolide, 294
triterpenes, 3, 4, 192, 257
triterpenoid saponins, 5
trolox, 263
tropane alkaloid biosynthesis, 296
tropane alkaloids, 7, 29, 30-32, 37, 55-57, 167, 233, 234, 296
tropic acid, 57
tropine, 28, 30
tropinone, 29
tryptamine, 34, 35, 41, 42, 51, 360, 362, 363
tryptophan, 32, 38, 59, 288, 358, 360, 362
tryptophan decarboxylase, 39, 51, 360
tubulin, 225, 244
tulips, 19
Tulipa, 19
tulipalin A , 19
1-tuliposide A, 19
6-tuliposide A, 19
tumour cells, 268
tumour necrosis factor-a, 264
tumour suppressor genes, 265
tumours, 244, 268
turmeric, 259
turnip (*Brassica napus*), 267, 314
two-component regulators, 76
two-dimensional GC, 147
two-dimensional NMR, 194, 195, 206, 300
two-dimensional TLC, 134, 135
tyramine, 362
tyrosine, 32, 358, 362
tyrosine decarboxylase, 54
tyrosine kinase, 264, 271

U

ubiquitin, 74
ulcer-protective activity, 241
Umbelliferae, 254
umbelliferone, 13, 141, 240
uv detection, 144
uv light, 53, 54, 56, 61, 63, 254
uv protectants, 75
uv spectroscopy, 153, 154
Uvae ursi, 139

V

vaccines, 334, 344, 345, 349
vacuoles, 51
vacuum liquid chromatography (VLC), 124
vallesiachotamines, 42
Valser's reagent, 151
vanadium, 311-313
Vanilla, 300
vanillic acid, 231
vanillin, 300, 317
vanillin-β-D-glucoside, 300
vanillin-sulphuric acid, 163
vanillyl alcohol , 300
vanillylamine, 299
Vepeside®, 243
verbascose, 256
Vicia faba, 216, 338
vinblastine, 51, 219, 225, 236, 237, 293, 294
Vinca rosea (syn. Catharanthus roseus), 11, 236
vincamine, 119
vincine, 119
vincoside, 42
vincristine, 51, 219, 225, 236, 237
vindoline, 51, 293, 294
viral infections, 221

viral vectors, 327, 329, 330
Virend®, 221
virus, 343, 346
virus-based vectors, 329, 341, 349
  and:
  α-trichosanthin, 342
  ACEI, 343, 344, 346
  ACEI-modified coat protein, 344
  angiotensin-I-converting enzyme inhibitor (ACEI), 343
  cauliflower mosaic virus (CaMV), 341
  coat proteins, 342-344, 346, 347
  cowpea mosaic virus (CPMV), 342, 344, 345
  enkephalin, 343
  epitope-presentation systems, 344, 346
  epitopes, 345, 346
  foot and mouth disease virus (FMDV), 344, 347
  gene II, of CAMV, 341
  green fluorescent protein (GFP), 347
  human immunodeficiency virus (HIV), 345, 346
  human interferon αD (IFN αD), 341
  human rhinovirus 14 (HRV-14), 345
  influenza virus, 346
  interferon, 341
  interferon αD, 341
  leu-enkephalin, 343
  metallothionein, 341
  mink enteritis virus (MEV), 345
  MTII (metallothionein II), 341
  Nicotiana benthamiana, 343, 347
  peptides, 343, 348
  Plasmodium vivax, 346
  Plasmodium yoelii, 346
  potato virus X (PVX), 347
virus-based vectors (*cont.*) *and*:
  protoplasts, 343
  RNA plant viruses, 346
  tobacco, 344, 347
  tobacco mosaic virus (TMV), 342, 346
  tomato bushy stunt virus (TBSV), 346
  tomatoes, 344
  vaccines, 344, 345
virus-like particles (VLPs), 334
vitamin A, 162
vitamin C, 259, 262, 264
vitamin E, 236, 259, 261-263
VLC (vacuum liquid chromatography), 124-126, 175
VLPs (virus-like particles), 334
voacamidine, 164
voacamine, 164

W

W couplings, 196
warts, 242
water-pepper, 132
watercress, 266, 268
western hare, 17
wheat bran, 259
wheat germ, 259
wild parsnip, 11, 16
woad, 278
wounding, 53, 77

X

X-H COSY, 199
X-ray crystallography, 188, 189
  brassinolide, 188, 189
  heavy atoms, 189, 190
  neutron diffraction, 190
  Ruptiliocarpon caracolito, 188

X-ray crystallography (*cont.*)
  secoterpenoids, 188
  spirocaracolitones, 188, 189, 190
  strychnine, 188-190
XAD resins, 123, 287
xanthine oxidase, 263
xanthinin, 13
xanthones, 9
xanthotoxin, 12, 16, 18, 140, 141, 231
xenobiotics, 277
xylanases, 340
xylans, 340
XYLD-C, 339
xyloglucans, 340
xylose, 340
xynD gene, 339, 340
xynZ, 340

Y

yew (*Taxus*), 20, 243

Z

Zanthoxylum budrunga, 125
zeatin, 12
α-zingiberene, 148, 149
zooecdysones, 21
Zygogynum, 15